T5-AMY-949

SCIENTISTS IN ORGANIZATIONS

REVISED EDITION

SCIENTISTS IN ORGANIZATIONS

Productive Climates for Research and Development

REVISED EDITION

DONALD C. PELZ

Director, Center for Research on
Utilization of Scientific Knowledge
Professor of Psychology

FRANK M. ANDREWS

Program Director, Survey Research Center
Associate Professor of Psychology

Institute for Social Research
The University of Michigan
Ann Arbor, Michigan

ISR Code No. 2487–rev. ed.

Part One of this volume was originally published as *Scientists in Organizations: Productive Climates for Research and Development.*

Library of Congress Catalog Card No. 76-620038
ISBN 0-87944-208-5

Published by the Institute for Social Research
The University of Michigan, Ann Arbor, Michigan 48106

First Published 1976
Second Printing 1978

PREFACE TO THE REVISED EDITION

The original 1966 edition of *Scientists in Organizations* was allowed to go out of print by its publishers. The volume, however, continues to be in demand as a treatment of how factors of individual motivation, group structure, and organizational process bear upon the performance of scientists and engineers in R & D organizations. The Institute for Social Research, therefore, has decided to reissue the volume in an expanded and updated form.

The original text–reporting findings from 1300 scientists and engineers in 11 governmental, industrial, and academic laboratories–remains unchanged. To it we have added five articles which appeared between 1967 and 1975 and which elaborate or extend some of the themes. The first article appears as an introduction to the revised edition.

Creative Tensions–An Overview. As first published, the book lacked a summary. A few broad features as sketched in the original introductory chapter (p. 7) suggested that scientists and engineers performed well under conditions that were not fully comfortable, that contained contradictory pressures. The new introductory chapter broadens that theme.

When findings in the various chapters were scrutinized, it appeared that the optimum climate was not necessarily some compromise between extremes but rather a combination of factors that seemed antithetical. On the one hand, technical staff were effective when faced with demands from their environment that required disruption of established patterns–*challenge.* On the other hand, technical staff were also effective when protected from their environment by assurances of stability and continuity–*security.* It appeared that high-performing scientists and engineers experienced simultaneously (or perhaps successively–our measurements were too coarse to discriminate time sequences) sources of stability or security on the one hand and of disruption or challenge on the other. This conjunction of opposites we have called a "creative tension"; perhaps "creative contradiction" would be more apt. The introductory article summarizes eight such tensions or contradictions which emerged from findings throughout the book.

The remaining four articles are reprinted in Part Two of the revised edition.

Problem Solvers vs. Decision Makers. In the summaries at the end of each chapter of the original volume we grappled with practical implications, often resorting to conversation with an imaginary reader. The data as such are correlational—how facts about technical people are related to their performance during the previous five years. But practical interpretation requires inferences of cause and effect, the imaginary reader properly questions, for example, whether contacts with colleagues enhance performance or whether high performance increases a scientist's contacts (p. 52).

In a frankly speculative view of the dynamic process relating individual characteristics and the institutional environment to technical achievement, the first article of Part Two envisions a circular, iterative process. The individual brings essential personal ingredients from his endowment or training: competence, curiosity (an internal source of challenge), and self-confidence (an internal source of security). When these are enhanced by involvement and resources, the result is creative achievement or solving of a problem. If the achievement is publicly recognized by the institution, the individual is exposed to new and unsolved problems (an external source of challenge) and gains prestige (an external source of security); these in turn enhance self-confidence, curiosity, competence, resources, and involvement, and the stage is set for a further round of problem-solving achievement.

This view sees technical performance as the result of neither individual resourcefulness nor environmental facilitation, but of repetitive reinforcement between them.

Creative Process. In Chapter 9 we examined how the individual's creative ability as measured by the Remote Associates Test related to technical performance, and concluded that the effect depended heavily on the nature of the situation. The second article in Part Two reports a follow-up study on 115 medical sociologists which confirmed that creative ability was virtually unrelated to judges' ratings of either innovativeness or productiveness of project output, and that in the presence or absence of organizational support and individual motivation the correlation between creative ability and innovativeness of output varied from positive to negative. The results suggest that organizations seeking innovation may face a "security dilemma"; creative activities may erode professional security, yet without such security the individual is unlikely to utilize his creative potential.

Time Pressure. In Chapter 12 of the original book we reported that in loosely coordinated settings it was necessary that the individual remain strongly motivated or stimulated if he were to continue achieving. One form of stimulation may be the presence of deadlines or other forms of time pressure. The role of this motivation in technical performance is addressed in the third article of Part Two. In contrast to the synchronous data of the original study the data here were longitudinal—involving two measurements of motivation and performance on 100 NASA scientists over a five-year interval. Time pressure was found associated with above-average performance in the subsequent period, but the converse was not found. We conclude that a sense of time pressure—within the bounds felt appropriate by the persons involved--can enhance technical achievement.

Supervisory Practices. In Chapter 2 of the original volume we described the role of several decision-making echelons in deciding technical assignments, but did not deal directly with leadership practices in technical teams. This gap is filled by the final article of Part Two which reports on a study of 21 small teams in a NASA research center, where ten composite measures of supervisory behavior were related to group scores on innovation. Surprisingly, we found that human relations skill mattered little; among other conditions, innovation occurred under supervisors who knew technical details of their subordinates' work, could critically evaluate that work, and could influence work goals.

We hope that the revised edition—with the five new pieces which clarify or elaborate the earlier conclusions—will continue to supply a conceptual and practical tool to promote productive climates in research and development organizations.

D.C. Pelz
F.M. Andrews

May 1976

PREFACE

This book is addressed to scientists and engineers, to administrators of research and development, and to all others who are concerned about the effects of organizations upon the work of their members. This book is one of the first major studies to examine the relationship between a scientist's performance and the organization of his laboratory. Unlike many previous expositions about the best environment for technical people, the findings resulted from extensive analysis of factual data from a wide range of research personnel.

Our concern with the topic of stimulating laboratory environments was aroused in the fall of 1951 when two members of the executive staff of the National Institutes of Health stopped in for a visit. Would the Survey Research Center be interested in studying attitudes and environmental factors related to the performance of NIH scientists? The Center was, and Pelz took on the job of project director.

Work progressed over the next four years, and a number of intriguing results began to emerge.[1] But as these were discussed with other investigators studying different kinds of R & D laboratories, discrepancies appeared. It became clear that a broader study was needed before one could be sure what constitutes a stimulating environment for research personnel. We set out to design a study in which standardized instruments would be administered to scientists and engineers in several types of laboratories.

The years 1956 to 1957 were spent in devising methods and raising

[1]The technical report containing these results is out of print (D. C. Pelz, R. C. Davis, G. D. Mellinger, and H. Baumgartel, *Interpersonal Factors in Research,* Part I, 1954; Part II, 1957 (mimeo); Institute for Social Research, University of Michigan, Ann Arbor. Results, however, are summarized in several places: D. C. Pelz, "Some Social Factors Related to Performance in a Research Organization," *Administrative Science Quarterly,* 1956, vol. 1, pp. 310–325, reprinted in B. Barber and W. Hirsch (eds.), *The Sociology of Science,* The Free Press of Glencoe (Macmillan), New York, 1962, pp. 356–369. See also D. C. Pelz, "Relationships between Measures of Scientific Performance and Other Variables," in C. W. Taylor and F. Barron (eds.), *Scientific Creativity: Its Recognition and Development,* John Wiley and Sons, New York, 1963, pp. 302–310. A short but full summary of results by 1956 from the NIH study and elsewhere appears in D. C. Pelz, "Motivation of the Engineering and Research Specialist," *American Management Association, General Management Series, No. 186,* 1957, pp. 25–46 (reprint available as Publication #1213 from the Survey Research Center, University of Michigan).

money to conduct such a study. Under a grant from the Foundation for Research on Human Behavior, Pelz and his associates, Wallace P. Wells and Stewart West, interviewed 150 scientists in two industrial laboratories, a university defense-oriented institute, and several academic departments. These interviews explored the kinds of motivations which pulled or pushed these people, and the ways their performance was affected by colleagues and supervisors.

In the fall of 1958 the Carnegie Corporation of New York gave a sizable grant to launch this present study. Our various questionnaires were administered to 144 scientist-professors in seven departments of a large midwestern university late in 1958. In 1959 similar data were collected from 526 scientists and engineers located at five industrial laboratories. Early in 1960, 641 research personnel in five government laboratories were added to the study. The next five years were spent in analyzing these data and testing interpretations of the emerging results on a variety of technical audiences.

Organization of the Book

Chapter 1 sets the context for the study. It also provides certain basic information about the varieties of research personnel who participated and briefly describes our methods. The chapter concludes with a short series of statements which summarize some of the most general findings.

The remaining twelve chapters are reasonably self-contained descriptions of research results. The reader is encouraged to pick and choose among them as he pleases. The order in which these chapters occur, however, is not completely arbitrary.

Chapters 2 through 9 are similar in that each examines various characteristics of scientists and/or their laboratories and shows how these characteristics related to their performance. In the chapters in this group, parallel analyses were carried out for five separate groups: Ph.D's in development labs, Ph.D's in research labs, and three other groups which have been labeled "engineers," "assistant scientists," and "non-Ph.D scientists." These chapters are all concerned with scientists' performance *relative* to other scientists with similar training and experience.

Chapters 10 and 11 are related in that both are concerned with the type of laboratory environment which was optimum for scientists at different *ages*. These chapters report results for the same five groups of scientists used previously.

In Chapter 12 we look again at some of the factors which showed pronounced relationships with performance in previous chapters. Instead of examining the five groups of scientists separately, we asked whether they worked in labs where the work was tightly coordinated or where coordination was rather loose. This amounted to cutting the same data in a different way and provided further insights about the effects of different laboratory environments on scientists' motivations and performance.

The last chapter, 13, turns from individual scientists and engineers to the performance of the groups or teams of which they were members. Performance of the typical group was found to decline as its members stayed together longer, and the chapter identifies a number of factors which seemed to help groups resist this decline.

The reader who is curious about methods will find details in the appendices.

July 1966

Donald C. Pelz
Frank M. Andrews

ACKNOWLEDGMENTS

For their help in launching and sustaining this project, we are greatly indebted to the following:

Officials of the eleven participating organizations in industrial, government, and university settings who gave not only their permission but their wholehearted backing.

The 1300 members of their staffs who conscientiously filled out questionnaires and evaluated performance.

The foundations and agencies without whose generous financial support the project would have been impossible: the Foundation for Research on Human Behavior (Ann Arbor); the Carnegie Corporation of New York, the National Science Foundation, and the U. S. Army (Materiel Command and later Army Research Office, Durham)—these three providing the major support; U. S. Public Health Service, the National Aeronautics and Space Administration, and two of the participating companies.

Our collaborators at various stages of exploration, design, and preliminary analysis: Wallace P. Wells, S. Stewart West, Allen M. Krebs, and George F. Farris.

Albert F. Siepert, then executive officer at the National Institutes of Health, whose enthusiastic faith sustained our earlier study there, on which succeeding steps were based; Eldon Sweezy, whose keen interest and steady encouragement opened government doors during the present study; and Rensis Likert, director of the Institute for Social Research and student of management process, who was always confident that we were doing something important even when the prospect seemed dubious to us.

The wizards of data processing whose machines performed miracles: John Sonquist, Mrs. Kathleen Goode, Keith Mather, and David Schupp of the Institute's data-processing section, and Dr. Bernard A. Galler of the University's computing center, for devising an indispensible program.

Our tireless assistants in research who kept administrative details under control: Mrs. Maria Chiarenza, Mrs. Betty Sears, and Mrs. Virginia Lawrence; and our faithful secretaries Mrs. Mary Scott and Miss Mary Hope who, with untiring skill, saw to the meticulous production of several thousand pages of letters, tables, preliminary reports, and the final manuscript.

Finally, we are grateful to two university physicists who, at a formative stage in the study, courteously declined to participate, saying that although they realized much good might come of a study like this, it implied the premise that rules can be found for dealing with scientists as a class, whereas they profoundly believed that each man must be dealt with essentially and entirely as himself, a unique individual.

This view has both chastened and spurred us. As an ethical principle in an age of organization, it deserves our deepest respect. But as a scientific principle, we must—from the perspective of investigators seeking regularities in behavior—just as respectfully dissent. We have not, to be sure, sought generalities applying to *all* research personnel. Rather we have sought regularities applying within limited categories: scientists with the doctoral degree and those without; those oriented toward knowledge and those toward application; low in status or high; younger or older; in structures coordinated closely or not at all. If consistencies appear across these categories, so much the better; if not, the complexities of the scientific enterprise must be recognized.

In addition to the hundreds of persons who participated, then, we acknowledge two who did not, and thereby sharpened some philosophical and scientific issues in this search for an understanding of stimulating atmospheres in research organizations.

Donald C. Pelz
Frank M. Andrews

CONTENTS

CREATIVE TENSIONS—AN OVERVIEW

Technical Achievement of Scientists and Engineers Was High under Conditions that Seemed Antithetical.

Donald C. Pelz

What kinds of climate in research and development organizations are conducive to technical accomplishment? What is the optimum degree of freedom versus coordination? of pure research versus practical development? of isolation versus communication? of specialization versus diversification?

To find some answers, my colleagues and I studied 1300 scientists and engineers in 11 research and development laboratories. Since the answers in different kinds of settings might vary, we included five industrial laboratories, five government laboratories, and seven departments in a major university. Their objectives ranged from basic research to product development.

Among the findings appeared a number of apparent inconsistencies. The optimum climate was not necessarily some compromise between extremes. Rather, achievement often flourished in the presence of factors that seemed antithetical.

Some examples are given below and summarized in Table 1.[1] As we pondered these findings, it seemed possible to fit many of them under two broad headings. On the one hand, technical men were effective when faced with some demand from the environment—when their associates held divergent viewpoints or the laboratory climate required disruption of established patterns. These might be called conditions of challenge.

On the other hand, technical men also performed well when they had some protection from environmental demands. Factors such as freedom, influence, or specialization offer the scientist stability and continuity in his work—conditions of security.

It seemed reasonable to say that the scientists and engineers of our study were more effective when they experienced a "creative tension" between sources of stability or security on the one hand and sources of disruption or challenge on the other. The term was suggested by T. S. Kuhn in a paper entitled "The essential tension: tradition and innovation in scientific research."[2]

This article originally appeared as "Creative Tensions in the Research and Development Climate" in *Science*, Vol. 157, No. 3785, July 1967, pp. 160-165 and is reprinted here with permission of the publisher. Copyright 1967 by American Association for the Advancement of Science.

TABLE 1 *Eight Creative Tensions*

Security	Challenge
Tension 1	
	Effective scientists and engineers in both research and development laboratories did not limit their activities either to pure science or to application but spent some time on several kinds of R & D activities, ranging from basic research to technical services
Tension 2	
Effective scientists were intellectually independent or self-reliant; they pursued their own ideas and valued freedom . . .	. . . But they did not avoid other people; they and their colleagues interacted vigorously
Tension 3	
a) In the first decade of work, young scientists and engineers did well if they spent a few years on one main project . . .	. . . But young non-Ph.D.'s also achieved if they had several skills, and young Ph.D.'s did better when they avoided narrow specialization
b) Among mature scientists, high performers had greater self-confidence and an interest in probing deeply . . .	. . . At the same time, effective older scientists wanted to pioneer in broad new areas
Tension 4	
a) In loosest departments with minimum coordination, the most autonomous individuals, with maximum security and minimum challenge, were ineffective . . .	. . . More effective were those persons who experienced stimulation from a variety of external or internal sources
b) In departments having moderate coordination, it seems likely that individual autonomy permitted a search for the best solution . . .	. . . to important problems faced by the organization
Tension 5	
Both Ph.D.'s and engineers contributed most when they strongly influenced key decision-makers . . .	. . . but also when persons in several other positions had a voice in selecting their goals
Tension 6	
High performers named colleagues with whom they shared similar sources of stimulation (personal support) . . .	. . . but they differed from colleagues in technical style and strategy (dither or intellectual conflict)
Tension 7	
R & D teams were of greatest use to their organization at that "group age" when interest in narrow specialization had increased to a medium level . . .	. . . but intererest in broad pioneering had not yet disappeared
Tension 8	
In older groups which retained vitality the members preferred each other as collaborators . . .	. . . yet their technical strategies differed and they remained intellectually combative

Necessity is said to be the mother of invention, but our data suggest that invention (technical achievement) has more than one parent. Necessity might better be called the father—since necessity is one form of challenge, a masculine component. The role of mother is, rather, some source of security. When both are present, the creative tension between them can generate scientific achievement.

Methods

The findings were not obtained by polling scientists concerning what climate they preferred. Rather, we obtained measures of each man's scientific performance, including his scientific or technical contribution to his field of knowledge in the past 5 years, as judged by panels of his colleagues; his overall usefulness to the organization, through either research or administration, also as judged by his colleagues; the number of professional papers he had published in the past 5 years (or, in the case of an engineer, the number of his patents or patent applications); and the number of his unpublished reports in the same period.

The performance measures were modified in several ways. Since distributions of papers, patents, and reports were skewed, a logarithmic transformation was applied to normalize them. Systematic variations with level of education, length of working experience, time in the organization, and type of institution were removed by adding constants so as to equalize the means. Each scientist, that is, was scored relative to others with similar background.

Characteristics of the climate were obtained on a carefully tested questionnaire. The two sets of data (on performance and on climate) were analyzed to find those conditions under which scientists actually performed at a higher or lower level.

Since optimum conditions might differ in different settings, all analyses were replicated within five subcategories: Ph.D.'s in research-oriented laboratories; Ph.D.'s in development-oriented laboratories; non-Ph.D.'s in research-oriented and in development-oriented laboratories (for convenience the latter have been called "engineers"); and non-Ph.D.'s in laboratories where 40 percent or more of the staff members held a doctoral degree (because of the limited influence and promotional opportunity of these non-Ph.D.'s we have called them "assistant scientists").

Science versus Application

For the first illustration, consider a tension not between factors of security and challenge but rather between science-oriented and product-oriented activity. The respondent estimated the proportion of his technical time (that is, time spent on research or development, as opposed to administration or teaching) that he allocated to each of the following five "R & D functions":

Research (discovery of new knowledge, either basic or applied):
- General knowledge relevant to a broad class of problems ___%
- Specific knowledge for solving particular problems ___%

Development and invention (translating knowledge into useful form)
- Improving existing products or processes ___%
- Inventing new products or processes ___%

Technical services (either analysis by standardized techniques or consultation and trouble-shooting) ___%

Some interesting trends appeared. For instance, Ph.D.'s in both research-oriented and development-oriented laboratories were judged most effective, on the basis of several criteria, when they devoted only half their technical time to research as such (first two categories above) and the rest to activities described as development or technical services. Similarly, Ph.D.'s in development-oriented laboratories were most effective when they spent only one-quarter or one-third of their time on activities labeled "development."

Another way to summarize the same data is illustrated in Fig. 1, where technical contribution is plotted against the number of R & D functions to which the individual devoted at least a little time (6 percent or more). Similar curves (not shown) were obtained for other measures of achievement usefulness, publications, patents, and unpublished reports. Even in laboratories devoted to pure research the best performers carried on four functions; they did not concentrate on research alone, but spent some time on development or service fuctions. Performance dropped if Ph.D.'s or assistant scientists tried to perform all five functions, although engineers flourished under this condition.

Effective scientists, in short, did not limit their efforts either to the world or pure science or to the world of application but were active in both (see Table 1, tension 1).

Is this involvement with both worlds a genuine tension? I am inclined to think so. As time invested in one increases, investment in the other must decrease. Demands for solution of practical problems can interfere with long-range research.

Why, then, should such a tension be creative? Several writers have proposed that a creative act occurs when a set of elements not previously associated is assembled in a new and useful combination. Diversity in technical activities may broaden the range of elements from which the scientist or engineer can draw in synthesizing new combinations.

Other findings reinforced the importance of diversity. Individuals performed better when they had two or three "areas of specialization" within their scientific discipline, rather than one. The Ph.D.'s did their best work not when they devoted full time to technical activities but when they spent about one-quarter of their time in either teaching or administration.

In the framework of challenge versus security, diversity in the task may also be viewed as a source of disruption and hence a condition of challenge. For data on specialization versus diversity, see Table 1, tension 3.

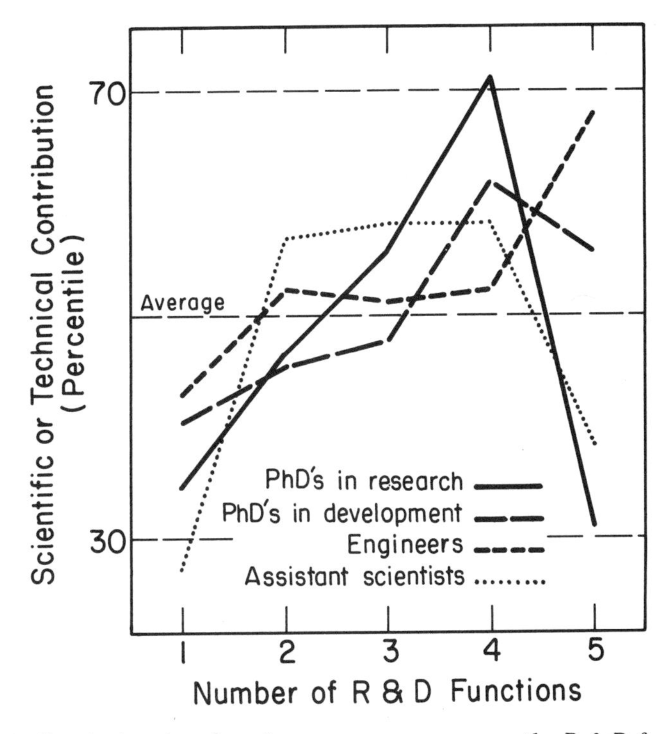

Figure 1. *Graph showing that the more numerous were the R & D functions, up to four, performed by Ph.D.'s and assistant scientists in development-oriented and research-oriented laboratories, the higher was their scientific or technical contribution as judged by colleagues; engineers did best when they had five R & D functions.*

Independence versus Interaction

Scientists place high priority on freedom. To measure this need, an index of "motivation from own ideas" was constructed, from self-reported (i) stimulus by one's previous work, (ii) stimulus by one's own curiosity, and (iii) desire for freedom to follow one's own ideas. This score—the index might also be labeled intellectual independence—was analyzed in relation to the four performance measures within each category of scientific personnel. A series of positive correlations appeared. Among the 36 correlation coefficients, 25 were positve (r = +.10 or larger) and none were negative; this was one of the most stable trends in the analysis, and was consistent with other research. As stated by Anne Roe, [3] "almost all studies of scientists agree that the need for autonomy, for independence of action, is something that seems to be particularly strong in this group."

In what seemed an inconsistency, however, effective scientists did not avoid other people; they and their colleagues interacted vigorously. High performers conferred with their most important colleagues several times a week or daily; they regularly conferred with several colleagues in their own section and often with ten or more elsewhere in the organization.

In our speculative framework, independence or self-reliance is a source of security. Interaction with colleagues is a source of challenge, for they may criticize and prod. The high contributor experienced a creative tension between independence and interaction (Table 1, tension 2).

The skeptic may ask, Are the two conditions antithetical? In terms of their occurrence in our data, not necessarily. Yet in common experience it is often difficult to maintain one's independence under social pressure. As Ralph Waldo Emerson put it over a century ago in his essay "Self-Reliance": "It is easy in the world to live after the world's opinion; it is easy in solitude to live after our own; but the great man is he who in the midst of the crowd keeps with perfect sweetness the independence of solitude." The aphorism fits our effective scientists today. In the midst of the crowd they retained—with enough sweetness to be creative—the independence of solitude.

Age, Specialization, Diversity

In one analytical study we considered the question, Under what conditons can younger or older scientists, respectively, do their best work? Andrews and I had speculated that younger scientists already face challenge because their work is new; mainly they need security. Older scientists, we thought, possess security and mainly need challenge. To test these ideas we correlated several measures of climate against performance within successive age brackets.

The findings were far from simple. The overall conslusion, however, was that, among younger and older scientists alike, *both* security and challenge were required for achievement.

In the youngest age categoies (up to age 34), positive correlations appeared between technical performance and length of time the scientist or engineer had spent in his main project. Devoting 2 or 3 years to one undertaking is a source of security. It enables the young man to build contributions in which he can take pride. But, at the same time, young non-Ph.D.'s were effective when they had several areas of specialization, and young Ph.D.'s did better when they were *not* preoccupied with "digging deeply in a narrow area." A diversified task provides challenge (Table 1, tension 3*a*).

After age 40, a somewhat different set of measures accompanied high performance. Older individuals achieved only when self-confident—when motivated from their own ideas and willing to take risks. After age 50, achievement was also linked with an interest in probing deeply. These factors both suggest security. On the other hand, achievement after 50 was also linked strongly with interest in mapping broad features of new areas (Table 1, tension 3*b*). Thus, among older scientists, positive correlations appeared between performance and *both* penetrating study and wide-ranging study. The tension in this case was genuine; self-ratings of the two interests were found to be negatively correlated.

One wonders whether, in the creative tensions discussed thus far, the opposing conditions occur simultaneously or successively. Does the effective scientist pursue one narrow specialization at the same time he is exploring several new frontiers, or does he alternate between these postures? Does he retreat one month to his own ideas and engage in dialogue the next, or does he do both at the same time?

Our data contain no means of distinguishing. My hunch is that many creative scientists are flexible; they are able to alternate between contrasting roles.

The Individual and the Organization

We saw previously the importance of desire for independence. But to desire independence does not mean that one *is* independent. We therefore measured the individual's freedom to choose his own research or development tasks by asking who exerted weight in deciding what his technical goals or assignments were to be. The more weight exerted by the technical man himself, relative to that exerted by his chief, his colleagues, or higher executives or clients, the greater his perceived autonomy. The measure appeared valid: it was highest for Ph.D.'s in research, and lowest for "assistant scientists."

Now the more autonomy an individual has (the more weight in selecting his own assignments), the greater should be the stability and continuity of his work—the greater his security. And we found that, as autonomy increased, so did performance—up to a point. We were puzzled, however, to observe that when Ph.D.'s in both research-oriented and development-oriented laboratories had more than half the weight in choosing their goals their performance dropped, whereas in the case of non-Ph.D.'s, as their autonomy increased their performance continued to rise. Why?

In one search for answers we examined an organizational variable: the tightness or looseness of coordination within the department, measured by nonsupervisory scientists' ratings of the coordination within their section and supervisors' ratings of coordination between sections. (Individual autonomy and departmental looseness are of course interrelated, but within a given department the freedom of individuals can vary.) A loose organization does not make demands on its members; it provides high security with little challenge.

We found first that, in the most loosely coordinated departments, highly autonomous individuals actually experienced *less* stimulation, from either external or internal sources. They withdrew from contact with colleagues; they specialized in narrow areas; they even became less interested in their work. In these settings, maximum autonomy was accompanied by minimum challenge.

Yet in the most loosely coordinated settings, we also found, it was essential that the person be challenged if he were to achieve. It was here that the strongest correlations appeared between performance and various stimulating factors: diversity in the work, communication with colleagues, competition between groups, involvement in the job.

In these loosely coordinated settings, the most autonomous individuals were able to isolate themselves from challenge. A nondemanding organization permitted them to withdraw into an ivory tower of maximum security and minimum challenge. There they atrophied (Table1, tension *4a*).

What about the more demanding organizations–those of moderately tight coordination? Why was autonomy an asset here and not a handicap? We found that autonomous persons here had more diversity in their work, not less. One can speculate that in these departments the technical man had to face problems important to the organization; personal freedom enabled him to find the best solutions. Again a creative tension: the organization itself presented challenges; autonomy provided security for solving them (Table 1, tension 4*b*).

Influence Given and Received

The question used to measure autonomy also indicated the weight exerted by other persons in the choice of an individual's assignments. The "decision-making sources" were grouped into four categories: the individual, his immediate supervisor, his colleagues or subordinates, and higher executives or clients. We scored for each scientist how many of the four sources were said to have had at least some weight (10 percent or more) in selecting his technical goals.

Now, to discuss one's projects with persons in several positions is to run the risk of criticism and disruption. The more sources there are involved in decision, the greater is the likelihood of challenge.

For the scientist to allow other people some weight in his assignments does not, however, mean that he is powerless. He can *influence* the decision-shapers, and influence provides security.

We divided respondents into those who felt they exerted strong influence over key decision-makers and those who felt they exerted little. Responses on this item appeared valid; the highest influence was reported by Ph.D.'s in research laboratories, and the lowest by assistant scientists.

The results were clear: both Ph.D.'s and engineers performed well when all four sources had some voice in shaping their goals but when, at the same time, the individual could influence the main decision-makers. From this arose creative tension 5 (Table 1): influence received from several others (challenge) combined with influence exerted on others (security).

The reader may ask, To what extent are the receiving and giving of influence antithetical? In conventional views of bureaucracy, each is seen as restricting the other; the size of the "influence pie" is considered a constant, so that if superiors have more, subordinates have less. Likert[4] argues, however, in a fashion compatible with our results, that the total amount of influence is not fixed. When everyone exerts more—when total control rises—performance is likely to improve.

But why should participation enhance the scientist's performance? Mainly, I suspect, because it helps him to avoid the narrow or trivial, to select tasks of *significance*, either to the organization or to science. Diverse contacts may also turn up unrecognized problems, or suggest new approaches to old ones. Finally, the interest of others in the scientist's work will enhance his own involvement in it.

"Dither"

Another way in which a man's colleagues can provide challenge is through questioning his ideas. An apt label was borrowed by Warren Weaver[5] from British colleagues who built into antiaircraft computing devices a "small eccentric or vibrating member which kept the whole mechanism in a constant state of minor but rapid vibration. This they called the 'dither.'. . . We need a certain amount of dither in our mental mechanisms. We need to have our ideas jostled about a bit so that we do not become intellectually sluggish."

A scientist's colleagues may jostle his ideas if they and he approach a problem differently. To test this hypothesis, we measured similarity or dissimilarity between the scientist and his colleagues in several ways. One method was subjective—the respondent's perception of how his own technical strategy resembled that of his co-workers. Other measures were objective, in the sense that we examined the approaches reported by the respondent and by each of his colleagues and numerically scored the similarity among them.

How much dither or disagreement is healthy? In our data the answer depended on the kind of dither. One objective measure concerned the source of motivation—whether one's superior, the technical literature, or some other source. Scientists who responded to the same sources were somewhat more effective—perhaps because they had similar interests.

On three other measures we found the opposite to be true. Scientists and engineers did somewhat better when they saw themselves as different from colleagues in technical strategy, and when as scored objectively, they differed from colleagues in style of approach (when, for example, the individual stressed the abstract, his colleagues the concrete) or differed in career orientation.

How to reconcile this paradox? In some preliminary data obtained by Evan[6] for industrial R & D groups, the teams he found most effective reported personal harmony or liking among members, but intellectual conflict.

Colleagues who report the same sources of motivation as the scientist's own probably provide personal harmony and support—a form of security. When they argue about technical strategy or approach, they provide dither or challenge (Table 1, tension 6).

Group Age

Another portion of our analysis concerned the age of groups—the average tenure of membership in a given section or team. A reasonable hunch is that, as a group gets "older," security is likely to rise and challenge is likely to diminish. If this is so, what conditions are needed to maintain vitality as the group ages?

To study this question, Wallace P. Wells identified 83 sections or teams in industrial or government laboratories (ranging in number of members from 2 to 25, with a median of 6). He averaged the measures for scientific contribution and usefulness of members in each group and adjusted the averages to rule out the effects of individual age, percentage of Ph.D.'s, and type of setting.

When he plotted the adjusted measures against group age, Wells found that group performance generally declined as group age increased, although usefulness was highest for groups with an average tenure of 4 to 5 years.

Why the decline after 5 years? In a search for clues, Wells examined several measures of the group's climate in relation to its age. Two of these measures are plotted in Fig. 2. The average preference for "deep probing of narrow areas" (a source of security) rose steadily as group age increased, while the interest in "broad mapping of new areas" (a source of challenge) dropped. Note in Fig. 2 that usefulness was highest shortly beyond the point where the two curves cross, where both interests were present in some degree (Table 1, tension 7). The finding is similar to that for tension *3b* and may partly overlap it, since older groups tend to contain older individuals.

Not all older sections declined in vitality; some continued to be both useful and technically creative. Why? Wells examined other measures of group climate. One he called "cohesiveness"; a group scored high on this measure if its members listed other members of the team as their main colleagues. If group members prefer one another as collaborators, they are undoubtedly secure.

Wells found that in older groups (average group age, 4 years or more), cohesiveness was correlated strongly with usefulness and technical contribution. That is, if an older team continued to be cohesive, it stayed effective. Also, those older groups whose members communicated freely with one another performed better than younger ones did.

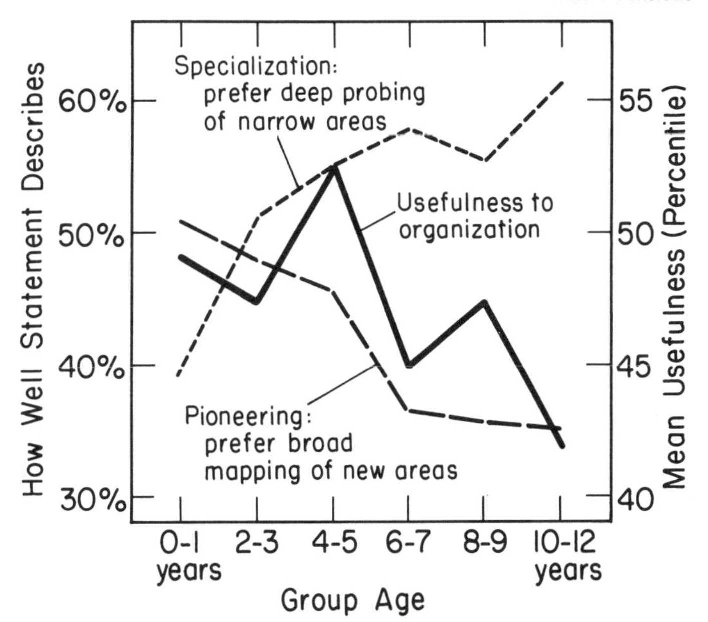

Figure 2. *Graph showing that R & D teams were most useful at that group age when the members wanted both to specialize and to pioneer.*

Yet the climate in effective older groups could hardly be called relaxed. On the measure of felt similarity to colleagues in technical strategies, Wells found that, in older groups, the more dissimilar the approach was, the higher was the performance.

One other measure proved surprising. Scientists rated the "hesitance to share ideas" within their section (for convenience we have called it "secretiveness"). Usually such hesitance was absent or mild. When some of this feeling was present in new groups, it was a handicap; it hindered their work. But this feeling *enhanced* the performance of older groups.

On reflection, this contrast makes sense. A new, insecure group must suspend criticism while it searches for new ideas. An old, secure group, on the other hand, will profit from criticism. If it stays effective it is not a club where one can lower his intellectual guard. On the contrary, there is competition in ideas; members sharpen their wits and marshal their evidence before speaking. Such a climate indicates challenge rather than insecurity.

Creative tension 8 (Table 1)–intellectual combativeness among colleagues who value each other–resembles tension 6. To prefer one's section members as collaborators is a sign of personal support, while the atmosphere of combativeness indicates intellectual conflict.

Practical Implications

Before considering practical implications I should raise the question, What is cause and what is effect? Does a combination of security and challenge help to generate achievement? Or do scientists who achieve experience more security and sense of challenge?

My own speculation is that a feedback loop exists. Usually a high performer has not only ability but also personality traits of curiosity and confidence. He is attracted to diverse problems and to contact with colleagues (a source of challenge) and at the same time insists on freedom and a voice in decision (conditions of security.) He thus exposes himself to conditions which in turn stimulate him to achieve. If this is the case, might lower achievers surround themselves with a similar climate and so enhance their own performance? Can R & D managers help to create such environments? I believe they can, and offer the following suggestions.

Conditions of Security

An important quality (see Table 1, tension 2) is self-reliance and pursuit of one's own ideas. But in development-oriented laboratory the manager cannot give each man a free hand: how then can he build an individual's pride in his own work? One way perhaps is to insure that once or twice a year each man produces a product which bears his own name—even if this requires that a jointly prepared document be broken into parts. It was disturbing to find in our sample that two out of five non-Ph.D.'s in research had not published a single paper in 5 years; among engineers the figure was four out of five. Half the engineers had not a single patent to their credit in the past 5 years, and one out of five had not authored even an unpublished report. How can a scientist feel confident of his own ideas if he has no output in which to take a fatherly pride?

Consider how the method of rewarding performance may affect self-reliance. Typically a single chief assigns tasks, judges results, evaluates performance, and recommends promotions. What better way to stamp out independent thought? To build self-reliance there must be multiple channels for recognizing achievement. Make sure that each subordinate has a chance once or twice a year to explain his work to colleagues *outside* his group. In review sessions with executives or clients, include the engineer who is doing the work and let him do some of the talking.

Another security factor is autonomy—substantial weight exerted by the individual in choice of assignment (see Table 1, tension 4). Such weight does not mean, however, that the individual should be completely on his own. From a further analysis (not reported above) it appeared that a technical worker in a development-oriented laboratory performed best when he and his supervisor *jointly* determined assignments. For Ph.D.'s in research laboratories, an effective condition was joint determination by the scientist and his colleagues. Assignment by the supervisor alone was the worst condition in all settings.

Security can be provided by the opportunity to influence others who decide one's assignments (Table 1, tension 5). Organizational structure plays a part here. Such influence is probably weaker in a many-leveled impersonal organization where each level has a veto. The individual's voice counts more in an organization of flat structure with fewer levels, where there is a chance for face-to-face contact with the people who shape his assignments.

Security increases with the length of time an individual spends on a given project (tension 3), particularly in the case of the younger man. Give him a year or two to dig into his main project, instead of shifting him every 3 months. He must have time to build a solid contribution.

One's colleagues can also be a source of security. In forming teams, managers can put together individuals who have similar sources of motivation—who are interested in the same kinds of problems (tension 6).

As R & D teams get older they can remain productive if they stay cohesive (tension 8). The supervisor can encourage cohesion by giving credit to the group rather than to himself. He can build mutual respect by publicizing the contribution of each member. He can strengthen teamwork through promoting competition with other groups in the solution of technical problems.

Conditions of Challenge

Scientists and engineers performed well not only when they had continuity and stability but also when they were challenged by demands from their environment. Frequent contact with one's colleagues (tensions 2 and 5) can be an important source of challenge. Such contacts can stimulate the individual in many ways. They can point to significant problems, suggest new approaches, or correct errors in a present approach.

How can the R & D manager encourage fruitful interaction? Often simply by knowing who in the organization or the field is doing what; he can steer the scientist to others who can give or use help. He can invite the individual to talk to a seminar, set up study teams and evaluation groups, pose problems which require consultation for their solution.

To encourage friendly disagreement, the R & D manager can invite members of an older group to look for flaws in each other's presentations (tension 8). When forming a new project committee he can include individuals who like each other but who use different strategies (tension 6). Periodic regrouping of teams—always with the consent of the persons involved—may help in maintaining a vital atmosphere.

Specialization lends security but diminishes challenge; some degree of diversity is required (tensions 1, 3, and 4). The manager should beware of letting some individuals focus exclusively on research, others exclusively on development. He should encourage his staff to tackle some jobs in both areas.

A younger scientist needs more than one area of specialization (tension 3*a*). In addition to a main continuing assignment, give him each year a second, shorter assignment which demands that he learn a new skill. Keep the older man's interest in broad areas strong by tempting him with problems on the pioneering edges of his field (tension 3*b*). Set up refresher courses; arrange sabbatical exchanges with a university.

Teams as well as individuals can become too specialized and lose interest in pioneering (tension 7). The R & D manager should not assume that one group has become *the* expert group in a specific area. As problems in this area arise, occasionally he will give one of them to a different team. He will challenge the expert group now and then with a task outside its specialty.

In the short run, such a policy may not be the most efficient way to manage a laboratory. It may cost more and take more time. But in the long run it will make for breadth and flexibility, and these will continue to open doors for creative advances.

SUMMARY

As Andrews and I examined the conditions under which scientists and engineers did effective work, we observed a number of apparent paradoxes. Achivement was high under conditions that seemed inconsistent, including on the one hand sources of stability or confidence (what I have called "security") and on the other hand sources of disruption or intellectual conflict (that is, "challenge"). It appears that, if both are present, the creative tension between them can promote technical achievement.

Refences and Notes

1. Data concerning the various tensions of Table 1 appear in the following chapters: tension 1, chap. 4; tension 2, chaps. 3 and 6; tension 3, chap. 11; tension 4, chap. 12; tension 5, chap. 2; tension 6, chap. 8; tensions 7 and 8, chap. 13; the performance measures are described in appendices A-C.
2. T. S. Kuhn, in *Scientific Creativity: Its Recognition and Development*, C. W. Taylor and F. Barron, Eds. (Wiley, New York, 1963), pp. 341-54.
3. A. Roe, *ibid.*, p. 135.
4. R. Likert, *New Patterns of Management* (McGraw-Hill, New York, 1961), especially chap. 4.
5. W. Weaver, *Science*, 130, 301 (1959).
6. W. Evan, *Ind. Management Rev.* 7, 37 (1965).

7. Collaborating in this research were W. P. Wells, S. S. West, A. M. Krebs, and G. R. Farris. The work has been supported by grants from the Carnegie Corporation of New York, the National Science Foundation, the U.S. Army Research Office (Durham), the Foundation for Research on Human Behavior, the U.S. Public Health Service, The National Aeronautics and Space Administration, and industrial laboratories. This article is based on a lecture presented in Miami in August 1966 before a meeting of the American Sociological Association.

PART ONE

Given the intentional heterogeneity of the laboratories studied, we wondered how the research personnel should be divided into more homogeneous subgroups. There were numerous bases on which these R & D people might be divided. Which was best? University versus industry versus government? Basic versus applied? Ph.D versus non-Ph.D? Chemist versus physicist? Old versus young? It would have been impractical to use all. After considerable exploration,[2] we settled on three dominant factors.

1. Orientation of the Scientist's Department toward Research or Development

A "department" was a subdivision of the laboratory usually containing 20 to 60 members and several sections, with two or three levels of supervision (except for academic departments). "Research-oriented" departments were those in which staff members agreed that executives esteemed and rewarded *scientific publication.* In "development-oriented" departments, the staff agreed that what really counted was development of *new or better products or processes.* These values were recognized, even though the activities of individuals might vary. The staff of the pharmaceutical laboratory, for example, published heavily, but agreed that the big rewards went to the man who developed a profitable drug. Individual motivations, as well as the type or quantity of output (papers, patents, reports), differed substantially among departments which varied on this dimension.

2. Possession of the Ph.D

Ph.D's differed noticeably from non-Ph.D's in their motivations and the quality and quantity of their output. (Possession of a master's degree, on the other hand, made little difference.) We were convinced that doctoral scientists were distinct unto themselves—whether by selection or by training—and should be analyzed separately.

3. Domination of the Department by Ph.D's

For nondoctoral scientists, it made a difference whether the department was run by Ph.D's (in some departments, 40% or more of the staff held the doctorate), or not so dominated (in other departments, fewer than 10% were Ph.D's).

Nondoctoral scientists in Ph.D-dominated laboratories tended to be permanently subordinate. They felt they had less autonomy and influence

[2] A log of these investigations is included in the following working papers, available from the Survey Research Center, University of Michigan: Analysis Memo #7, "Dimensions of Organizational Atmosphere" (Publication #1825); Analysis Memo #9, "Organizational Atmosphere as Related to Types of Motives and Levels of Output" (Publication #1826); Analysis Memo #10, "How Motives Relate to Three Kinds of Output in Various Types of Laboratories" (Publication #1826).

than other nondoctorals, they saw their professional opportunities as more limited, and they seldom held jobs above a middle status. Where Ph.D's were scarce, on the other hand, nondoctorals could rise to the top of the status ladder.

From these three factors, five "primary analysis groups" were defined as follows; these will be used throughout the book.[3]

A. *Ph.D's in Development-Oriented Laboratories.* Half of these were located in industry, half in government.

B. *Ph.D's in Research-Oriented Laboratories.* Two-thirds were in the university (all of our academic scientists were in this category) and one-third in government.

C. *Non-Ph.D's in Development-Oriented Labs not Dominated by Ph.D's.* Because the majority of these people had been trained in engineering specialities, it was convenient to call them *"engineers."* About three-quarters were in industrial locations, one-quarter in government.

D. *Non-Ph.D's in Ph.D-Dominated Laboratories* (either research- or development-oriented). They were part of the professional staff, but because of their subordinate status, we have called them *"assistant scientists."* Half were in government, half in industry.

E. *Nondoctoral Scientists in Research-Oriented Labs not Dominated by Ph.D's.* All of these were in government settings.

Sources of Data

The data described in this book may be classified into two broad types: (a) information about the scientist and the conditions which prevailed in his laboratory, and (b) information about his performance.

Since each of the succeeding chapters considers its own particular set of laboratory conditions or characteristics of research personnel, time will not be taken at this point to describe the various measures in detail. In most instances, they were derived from a carefully tested questionnaire. It had been shown to give results which were highly consistent within themselves and reasonably reliable over time.[4] The questionnaire was administered under conditions of complete confidentiality and inquired

[3]Differences in motivations and attitudes of these five groups are described in D. C. Pelz and F. M. Andrews, "Organizational Atmosphere, Motivation, and Research Contribution," *American Behavioral Scientist,* December 1962, vol. 6, pp. 43–47.

[4]The stability of questionnaire responses is often questioned. Are answers likely to fluctuate from day to day, depending on how the individual is feeling? We investigated this. After data from our main study had been collected, 418 other industrial scientists completed the questionnaire. Two months later, a random sample of 52 of these scientists repeated many of the same items. One important result was the very high stability of the mean scores for this group. On 89 items consisting of five- or seven-point scales, the correlation between the two sets of group means was .97. Further details appear in Appendix F; also in Preliminary Report #9, "Reliability of Selected Questionnaire Items," available as Publication #1991R from the Survey Research Center, University of Michigan.

only about subjects which most scientists had no fears of discussing. For these reasons, we think the data truly reflect the perceptions of the respondents.

Performance. Since information collected about each scientist's performance would be crucial to the success of the study, great efforts were made to ensure validity. Two different but complementary approaches were used.

Some of the performance data were based on *judgments* of each man's work. These judgments were made by people in the same laboratory who knew the man's work and who felt qualified to compare it with the work of others in that laboratory. Judges were senior people from both the nonsupervisory and supervisory levels within each laboratory.

Each judge provided two different rankings of the people he felt qualified to compare: first according to their *contribution to general technical or scientific knowledge in the field* (within the past five years), and secondly according to their *over-all usefulness in helping the organization carry out its responsibilities* (also within the past five years).

Nearly all people were judged by several different judges. Although each judge worked individually without knowledge of rankings made by other judges, there was substantial agreement among the judges.[5]

Although the judgments formed one important source of information about performance, they had certain limitations. Since judges were instructed to compare each person with others in the same laboratory, the averages for the different laboratories (when judgments had been converted into percentiles) necessarily were the same. There was no way of telling whether the performance of the top man in one laboratory was above or below the performance of the top man in another laboratory. Also, it was possible for the evaluations to be influenced by subjective factors, such as the judges' liking for the individual.

It therefore seemed wise also to measure performance by more objective (but not necessarily better) criteria: the numbers of various scientific products which the individual had produced within the past five years. Each participant indicated the number of *papers* he had published in professional journals within the past five years, the number of his *patents or patent applications,* and the number of his unpublished technical *reports* or manuscripts within the same period. A later check showed that respondents' claims were reasonably accurate.[6]

[5]Appendix A presents details of the procedures used, the method of combining judgments from several different judges, and the extent of their agreement.

[6]It turned out that only the engineers produced a significant number of patents and that engineers rarely claimed published papers. Reports, however, were a relevant form of output for all five analysis groups.

Appendices B and C present details about the validity and reliability of these measures and describe the steps used in preparing them for analysis.

After deleting certain types of output which were irrelevant to some groups, we had four separate measures of each participant's performance: judged contribution, judged usefulness, output of reports, and output of papers or patents. Not surprisingly, people who scored high on one tended to score high on the others.[7] Nevertheless, since the agreements were far from perfect (the measures were intended to measure different aspects of performance and therefore were not expected to show perfect agreement), all analyses were carried out and reported in quadruplicate, one for each performance measure. When results based on the judgments were corroborated by results based on the objective output of scientific products, our confidence in the findings was heightened.

Adjustment of Performance Measures. Before beginning our analyses, it seemed important to remove the effects of certain factors which accounted for some of the differences in performance but which were extraneous to our main concerns.

For example, government Ph.D's in research published 50% more than did university Ph.D's; and "assistant scientists" in government published twice as much as those in industry. Was this the result of a better climate in government? We suspected it stemmed rather from the obligation of government labs to let the public know where its money was going. Similarly the relatively low publication rate among scientists in industrial labs could be attributed to "company security."

We decided to eliminate differences in the output measures which were attributable to the type of setting (we retained, however, differences among laboratories within the same type of setting). Accordingly, we added constants to equalize the *average* output from the following categories: (a) the five "primary analysis groups" described earlier; and also (b) university, government, and industry.

However, certain other effects, extraneous to our primary concerns, were still present. We wanted to remove these also.

For example, we expected to find performance related to experience. If younger scientists were enthusiastic, whereas older ones were calm, and if younger ones had achieved less because of their youth, we might find enthusiasm seeming to inhibit achievement! Clearly one should compare young scientists with their contemporaries before making inferences about preferred forms of motivations and working relationships.

Examining the data on *time since degree,* we did indeed find that performance started low for people recently past their degree, rose rather steadily to 15 or 20 years beyond the degree, and then started to drop again. Length of *time in this lab* also showed a general upward trend, independent of time since degree. Effects due to these factors were also

[7] Appendix D presents the interrelationships between the various performance measures.

removed by adding constants to equate the average performance of groups of scientists who differed in experience.[8]

The effect of the various adjustments was a set of performance measures which showed whether a scientist or engineer was performing high or low *relative to his peers,* that is, relative to other professionals with the same level of formal education, the same amounts of experience, and located in similar settings.

Results: Some Over-all Impressions

To conclude this chapter we present a series of statements, abstracted from succeeding chapters, which outline a few of the broad features which characterized the environments of the most productive scientists and engineers.

- Effective scientists were self-directed by their own ideas, and valued freedom. But at the same time they allowed several other people a voice in shaping their directions; they interacted vigorously with colleagues.
- Effective scientists did not limit their activities either to the world of "application" or to the world of "pure science" but maintained an interest in both; their work was diversified.
- Effective scientists were not fully in agreement with their organization in terms of their interests; what they personally enjoyed did not necessarily help them advance in the structure.
- Effective scientists tended to be motivated by the same kinds of things as their colleagues. At the same time, however, they differed from their colleagues in the styles and strategies with which they approached their work.
- In effective older groups, the members interacted vigorously and preferred each other as collaborators, yet they held each other at an emotional distance and felt free to disagree on technical strategies.

Thus in numerous ways, the scientists and engineers whom we studied did effective work under conditions that were not completely comfortable, but contained "creative tensions" among forces pulling in different directions.

[8]Appendix C describes the procedures by which these adjustments were carried out, and a rather complex computer program which was used. This appendix also describes two other factors for which adjustments were made. Since adjusting for the latter actually made little difference, they have not been detailed here.

In Chapters 10 and 11 we were particularly interested in effects associated with age, and therefore used performance scores which had not been adjusted for length of experience. Chapter 13 is concerned with the performance of groups, and a somewhat different process of performance adjustment was applied there; the details are in Appendix C.

2

FREEDOM[1]

Is Coordination Compatible with Freedom? Best Performance Occurred When Both Were Present.

Research laboratories and their directors face a major dilemma. How can inner motivation be maintained in a large R & D organization? Scientists say they want freedom, and desire for self-direction is essential to high performance. But the laboratory must accomplish the objectives for which it is financed. And this means coordination of technical staff toward specific goals. How are these two needs—of the individual and the organization—to be reconciled?

We therefore examined our data with these questions: How much actual freedom (in contrast to desire for freedom) goes with high performance? Can freedom and coordination co-exist, or must they impede each other? And how do the answers vary in different kinds of labs or for different levels of scientific personnel?

As a guide to the reader, we sketch here the main threads of the chapter. When decisions are being made about an individual's technical assignments, other people or groups can enter such decisions in a bewildering variety of patterns. Two ways of simplifying these patterns were adopted. When we looked simply at the *number* of "decision-making sources"—the man himself, his immediate chief, his colleagues, and higher executives or clients—we found that in general the more of these who were involved, the better he performed. This was especially true when the man himself could influence key decision-makers. A second technique was to examine what *combination* of sources exerted major weight. High performance accompanied weight exerted by the scientist himself jointly with his chief (in development labs), or jointly with his colleagues (Ph.D's

[1]The analysis described here was largely supported by grants from the U. S. Army Research Office (Durham). The basic data were first reported at a conference of research administrators in Estes Park, Colorado, September 1963: D. C. Pelz, "Autonomy versus Coordination in Scientific Laboratories," in *17th National Conference on the Administration of Research*, University of Denver, 1963, pp. 97–105. An expanded version by Pelz appeared as "Freedom in Research," *International Science and Technology*, February 1964, pp. 54–66.

in research labs); performance was low for all groups when the chief alone decided.

A puzzling result was the fact that high autonomy benefited only non-Ph.D's. When we came back to this question later, we found that autonomous scientists and engineers did well if their interests were broad and diversified, but not if specialized. The chapter closes with some practical speculations.

Measures of Autonomy

A basic tool for measuring autonomy in the individual's own situation was Question 29 shown in the following box. We asked the respondent to tell us who had a hand in deciding his scientific goals and objectives. If he assigned himself a large portion of the weight in deciding his technical goals, we classified him as a man having high autonomy.

Question 29. Consider the choice of goals or objectives of the various technical activities for which you are responsible (either your own work, or work which you supervise or coordinate). Who has weight in deciding on these goals and objectives? Estimate the relative percent of weight exerted by each of the following, to nearest 5–10%.

	Percent of weight in deciding goals
Myself	____%
Subordinates	____%
Colleagues—other persons without supervisory authority over me	____%
My immediate chief	____%
Higher-level technical supervisors in this organization	____%
Nontechnical executives	____%
Clients or sponsors	____%
Other: ______________________________	____%
Total (should add to 100%)	(____%)

The reader might wonder how accurate these reports were. Did scientists or engineers actually have as much (or as little) autonomy as they said? We compared reports from persons in different situations. Ph.D's in research reported the most autonomy (largest weight for themselves, on the average); next in order were Ph.D's in development, followed by "engineers" (non-Ph.D's in development); lowest were "assistant scientists"—nondoctorals in Ph.D-dominated labs. Also, the higher the career level (apprentice, junior, senior, and supervisor) in each group, the greater the autonomy reported.

Thus the answers corresponded closely to what we would expect from people in these situations. On the average, the measures were realistic.

(The ways we measured performance, and the factors whose effects were removed from the performance measures, were described in Chapter 1 and will not be repeated here.)

Preliminary Results

As happened many times throughout this study, we made several attempts before finding a general pattern that made sense. Some of our preliminary results were puzzling. We had expected that people with high autonomy would have higher-than-average performance. Correspondingly, we had expected that a scientist who allowed his colleagues to have a hand in deciding his goals would perform less well than a man who set his goals himself. Control exerted by higher echelons over goals, and certainly control by nontechnical executives or clients, ought to inhibit performance.

But preliminary results did not bear out these expectations. Some weight exerted by several other groups even appeared to help slightly. In a very tentative way, it seemed that scientists performed better when influence on their important decisions was *shared* with several persons at various levels.[2]

But shared with *whom?* Might some combinations—self plus colleagues, or self plus chief, for instance—work better than others? Given the seven decision-making sources listed in Question 29, hundreds of different patterns of influence were possible. How could we abstract the basic kinds of patterns which made a difference for performance?

Number of Decision-Making Sources

One approach was to ask whether the sheer number of different sources affecting technical goals made a difference in performance. Since "subordinates" had relatively little weight, they were combined with "colleagues" as a single source. Weight exerted by the last three groups—higher-level technical supervisors, nontechnical executives, and clients or sponsors—varied in different kinds of labs, so these also were combined and considered as one composite source. The scientist himself, and his immediate chief, constituted two other sources, making four possibilities in all.

For each participant, we went back to his questionnaire and recorded

[2]These preliminary results are recorded in a working paper by D. C. Pelz, "Time and Influence Factors in Laboratory Management, as Related to Performance," Analysis Memo #18, September 1962, available as Publication #1993 from the Survey Research Center, University of Michigan.

how many of these four decision-making sources exerted at least a slight weight (10% or more) on his decisions about technical goals.[3] Charts 1-A through 1-E show how the number of sources related to several measures of performance.

The following charts (1-A and 1-B) show how scientific performance of scientists and engineers varied when different numbers of "decision-making sources" were involved in setting their technical goals. A decision-making source might be the man himself, his immediate chief, his colleagues plus subordinates taken together, or higher executive levels. Men whose goals were influenced by only one source having 10% weight or more—himself alone, for instance, or his chief alone—have been grouped under "one decision-making source."

Plotted vertically is mean performance on four separate measures, all referring to the previous five-year period. As described in Chapter 1, "scientific contribution" and "over-all usefulness" were based on judgments by senior scientists. The contribution and usefulness scores are expressed as percentiles within each laboratory; the mean is necessarily 50. The mean output of published papers (or of patents and patent applications for "engineers," group C) and unpublished reports is based on logarithmic transformations. All scores have been adjusted so as to remove effects due to length of experience, and have been superimposed so that the 70th and 30th percentiles for all scores coincide.

To give the reader some idea of the magnitude of these relationships, the "correlation ratio," or eta, has been computed. If the means were to lie in a straight line, eta would equal *r*, the ordinary coefficient of correlation.

An asterisk beside the eta indicates that it is "statistically significant"—a variation this large is not likely to happen by chance more than five times out of a hundred. (To compute means and etas, the data have been weighted so as to compensate for different sampling rates in different parts of our population. Significance tests, however, are based on unweighted or actual number of cases.)

Chart 1-A shows results for Ph.D scientists in development-oriented laboratories (labs where scientists agree that executives valued development of better products, rather than scientific publication; half of these were in industry, half in government). As more sources were involved at

[3] Previous reports of these data used the term "echelons" rather than "sources." That term is misleading, we now feel. It suggests several hierarchical layers exerting veto power, whereas by "sources" we simply mean distinguishable parts of an organization (either individuals or groups) from which influence emanates.

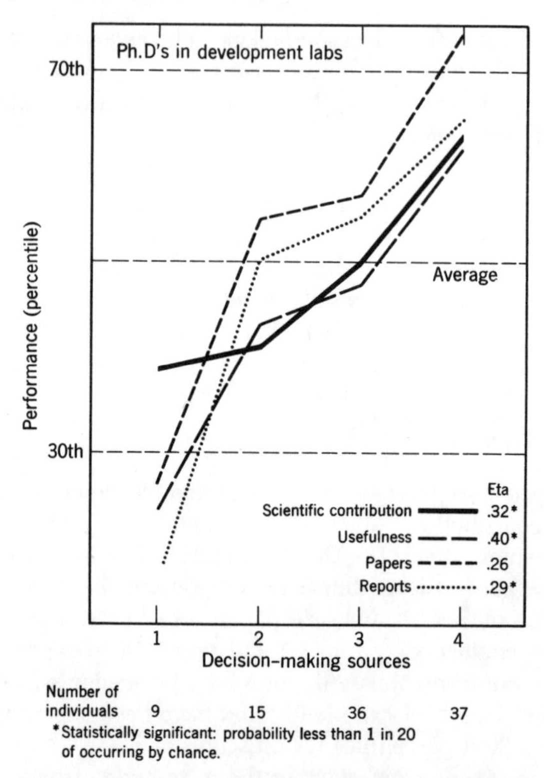

Chart 1-A. *For these Ph.D's in labs where executives placed higher value on development of better products than addition to scientific knowledge, it appeared that as more decision-making sources were involved, performance rose on all four criteria. The 30th and 70th percentiles for papers meant, respectively, about 4 and 12 per man, for a five-year period; for reports, 6 and 16.*

least slightly in deciding the scientist's assignments, colleagues ranked the man more highly both on scientific contribution and usefulness to the organization; he also wrote more papers and reports.

Chart 1-B gives the picture for Ph.Ds in research-oriented labs (that is, labs where executives stressed scientific publication; two-thirds were in university departments, one-third in government). Here we see some similarity to Chart 1-A, plus a puzzling difference. In terms of producing papers and reports, the trend was about the same as for the development scientists: people whose goals were affected by four sources published more papers and wrote more reports than those who were influenced by fewer groups. But the curves for colleague evaluations of contribution and usefulness showed that scientists in the one-source category also performed well. We can only speculate on this discrepancy. Perhaps the high-performance, single-source scientists talked more than they wrote; perhaps they described their work to colleagues rather than taking time to write it up.

The next three charts concern nondoctoral personnel. Chart 1-C shows those in development-oriented laboratories where Ph.D's were in the minority. We have called them "engineers" since over half had been trained in engineering specialties. Except for one deviation, the same trend existed as we saw for development Ph.D's: the more sources involved in setting goals, the higher the performance. Whether the high patent output for the one-source group is meaningful or is a random departure, we can't be sure. Although the single source was not necessarily "self," we shall see in a later chart that autonomous engineers were indeed high on patents.

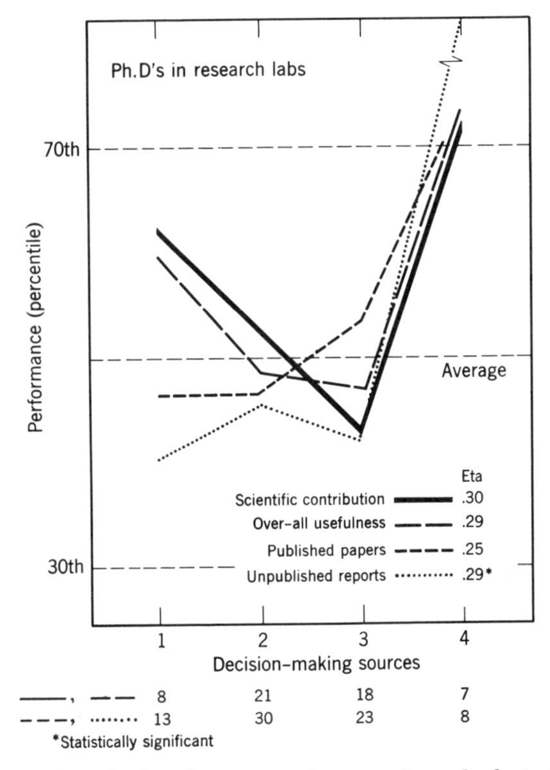

Chart 1-B. *Among these Ph.D's in departments whose executives valued scientific publication, performance on all measures was maximum with involvement of four sources, but a few made useful contributions where there was only one source (which for this group was the man himself). The 30th and 70th percentiles of paper publication meant about 5 and 13 per man over five years; for reports, 5 and 15.*

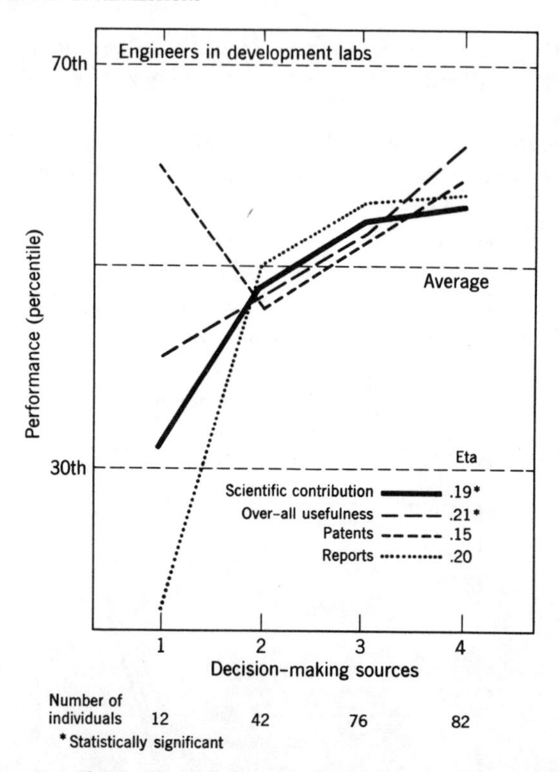

Chart 1-C. *These non-Ph.D's also worked in development-oriented labs, but in departments where few staff members were doctorals. Most were engineers. Instead of papers published, we plotted patents; 30th and 70th percentiles meant about 0 and 3 per man over five years; for reports, the values were 4 and 11. Except for patents in the one-source category, all measures rose as more sources were involved.*

Chart 1-D shows nondoctoral scientists who worked in Ph.D-dominated labs where 40% or more of the staff held the doctoral degree. (No distinction was made here between research and development orientation.) Among these "assistant scientists," as we have called them, the same trend appeared as before: with two exceptions, the more sources involved in setting goals, the higher was the performance. (Again exceptions arose for the one-source group. As we found for the engineers, we shall see later that autonomous "assistant scientists" were indeed productive.)

The final chart (1-E) concerns the relatively small group of non-Ph.D's in research-oriented labs in government where few of the staff held doctorates. Only one person reported a single decision-making source, so comparison could not be made for this category. Among the remainder, an interesting contrast appeared. Performance was generally higher when *three* rather than four sources were involved (although usefulness was still highest at four). Again we can only speculate. Perhaps these men occupy what the sociologists call "marginal" roles. Their aspirations and values are those of the scientist. But they lack the Ph.D required for admittance to the world of science. A marginal group is likely to be somewhat insecure. To argue their goals with three other groups may create more tension than stimulation.

Some Possible Objections

Let us pause to examine some objections to the main finding thus far—the more sources involved in deciding the scientist's technical goals, the better his performance. One might argue that scientists whose work is scrutinized by several groups are likely to be senior people; they are more productive *not* because they are involved with more people, but because they have greater experience.

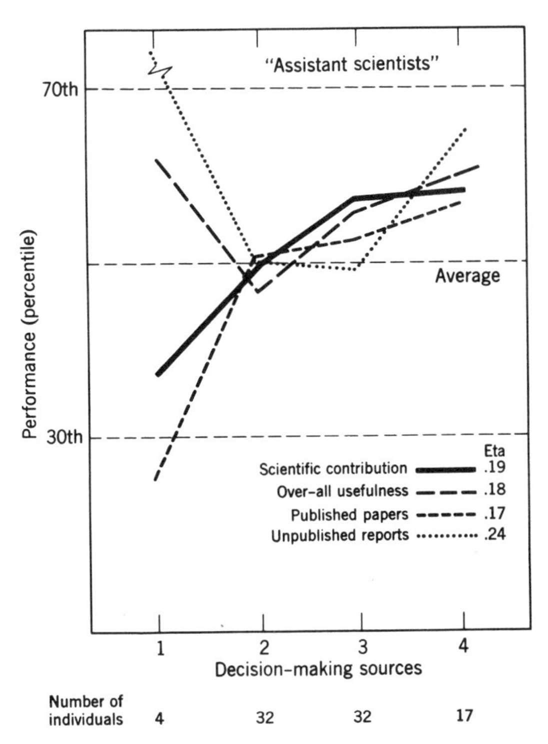

Chart 1-D. *These "assistant scientists" worked in departments dominated by Ph.D's; their promotional opportunities and general status were correspondingly limited. Even so, they tended to be more productive when more sources were involved in their decisions, with two exceptions in the small, single-source category. Here the 30th and 70th percentiles for papers published over five years were roughly 1 and 4 per man; for reports, 1 and 8.*

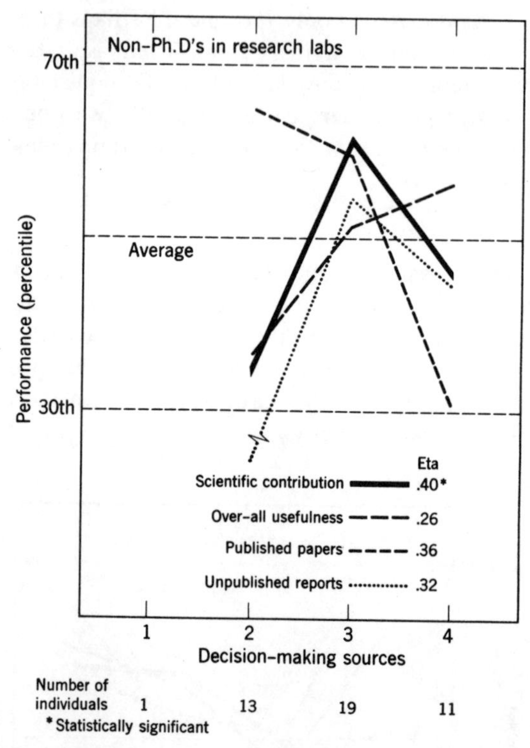

Chart 1-E. *For this small group of nondoctoral scientists in research labs not dominated by Ph.D's, performance tended to peak when three rather than four sources were involved. Perhaps people in "marginal" roles like this were inhibited by too much outside scrutiny. The 30th and 70th percentiles for paper publication over five years were 1 and 4 per man; for reports, 3 and 8.*

We tried to eliminate this objection by adjusting the data (as we mentioned in Chapter 1) so as to rule out achievement due mainly to length of experience.

Even so (the reader might object), it is not simply length of experience that counts, but ability. Better people are promoted to higher levels, and the higher the level, the more the individual's work becomes known throughout the organization.

Again we tried to answer this one by looking separately at people occupying different career levels. In general, the same results appeared among people having the same level of responsibility, especially among nonsupervisors. The "levels" factor did not explain away the trend.

But one might still argue that regardless of level, the fruitful worker gets attention both from colleagues and higher-ups. Their involvement in his work is a result of his high performance, not a cause of it. This is a simple and gratifying explanation; many scientists like it. And with our data we cannot disprove it.

But let us argue the opposite case: that involvement of one's colleagues, research executives, or client representatives can actually *stimulate* high performance. An experience by one of the authors seems relevant. For several years, Pelz has taught an introductory course in survey research methods. Believing in the value of autonomy for education, he had always asked each student to design a hypothetical survey on a topic of his choice.

Recently he tried a different approach: a single project on why local voters first rejected, and then accepted, requests from the Board of Education for higher taxes to support a larger school system.

It struck him later that the class procedures involved four decision-making sources: a citizens committee serving as an eager client; the individual student, with considerable autonomy to contribute his own ideas; his classmates, who interacted vigorously on committees and in the class; and the instructor, in the role of immediate supervisor.

What was the effect? The quality of the class committees' work was first-rate. A colleague, who knew one of the students, reported that their out-of-class involvement was "fantastic." Pelz was convinced that the multisource approach provided far more excitement and challenge than had been the case with individual projects in previous years.

Influence

Let us come back to the original dilemma. Is the scientist's freedom eroded as more people influence his technical assignments and decisions about his goals?

This brings up the question of *influence.* To what extent can a scientist influence key decision-makers? And can he have high influence even though he lacks full autonomy or freedom to go his own way?

To the last question, the answer was "yes." High autonomy tended to coincide with high influence, but the two were not identical. Many scientists reported that their goals were heavily affected by several other groups (and correspondingly had little pure autonomy), but nevertheless felt able to control the actions of key decision-makers.

Question 31, used to measure influence, is shown in the box. Again the reader may wonder how accurate these reports were. We compared them with other facts. Ph.D's in research reported the most influence; development Ph.D's and engineers stood next; assistant scientists were lowest. Also, the higher the individual's career level, the more influence he said he had. These facts all supported the view that participants, on the average, were answering realistically.

Question 31. Please write the name of the one person or group (other than yourself) who has the most weight in choice of your work goals. . . . To what extent do you feel you can influence this person or group in his recommendations or decisions concerning your technical goals? CHECK ONE.

____ I can exert almost no influence	____ Considerable
____ A little influence	____ Great influence
____ Moderate	____ Irrelevant, since no one else affects my choice of goals

To study the connection of influence with performance, we divided participants into two groups: those with a high feeling of influence ("considerable" or more) over important decision-makers, and those who felt lesser influence ("moderate" to "almost no influence"). Within the "high

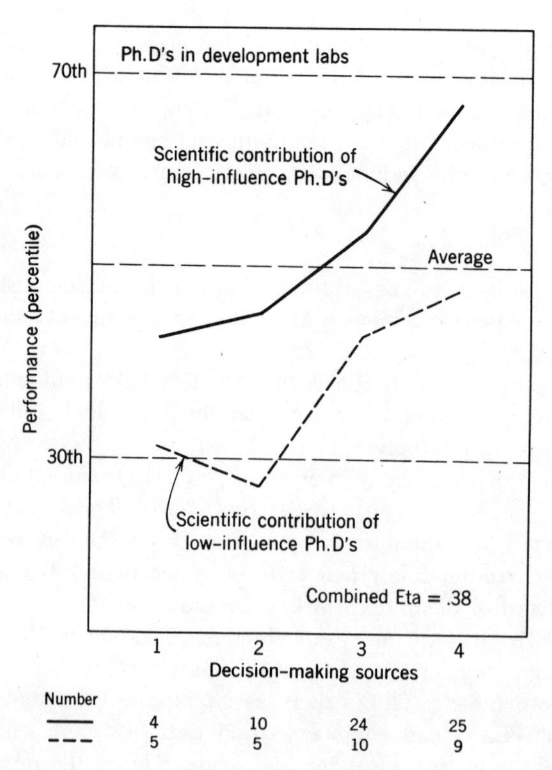

Chart 2-A. *Scientists rated their influence on key decision-makers. Here we see that maximum contribution occurred when these Ph.D's in development had both high influence and the involvement of several others in setting their goals.*

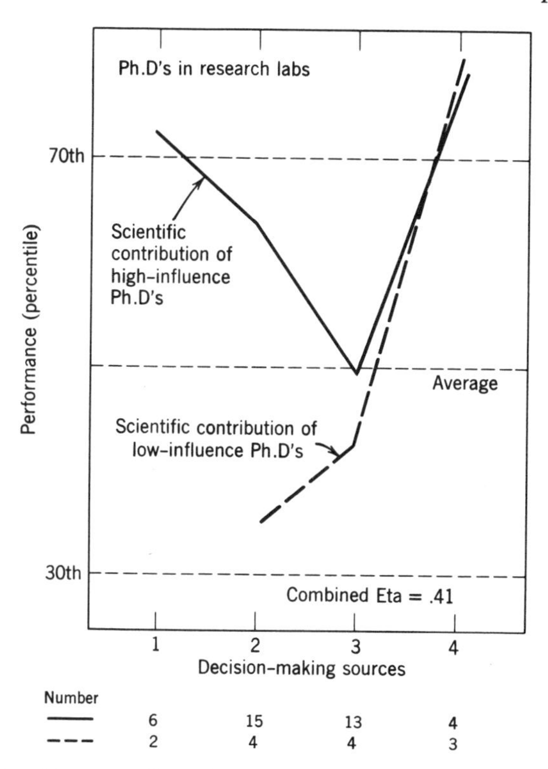

Chart 2-B. *A few high-status Ph.D's in research, possessing high influence, were effective when no one else helped decide their goals, but those with lower status or influence worked best when several others were involved.*

influence" and "low influence" subgroups, respectively, we re-examined the relationship between number of decision-making sources and scientific contribution. Charts 2-A through 2-D show the results.

In Chart 2-A, for example, we see scientific contribution plotted for those with high influence and those with less influence among Ph.D scientists in development laboratories. Perhaps not surprisingly, we found the high-influence people rated higher on contribution. However, note that with both groups, the same upward slope appeared as in the earlier charts; the more sources involved, the better. This was particularly true for high-influence scientists. With a substantial voice in their goals, they appeared to benefit if their planning was shared with several other groups.

Maximum performance occurred when the scientist had *both* high influence and the involvement of several others. On the other hand, if he lacked influence, multisource involvement was helpful, but less so.

Chart 2-B gives data for Ph.D's in research laboratories. The high-influence scientists showed the same trend as in Chart 1-B. It may be that those men who command respect, that is, high influence, can work effectively when fully self-determining. On the other hand, it seemed that low-influence scientists in research should not work in isolation, but should be in touch with at least two other sources besides themselves.

In Chart 2-C we see data for engineers in development laboratories. Again, those with high influence showed a rising trend in contribution as more sources were involved. For engineers with less influence, three sources rather than four seemed desirable. Why is this so? It may be, as in the case of the nondoctoral scientists (see Chart 1-E), that the low-influence engineer is in a "marginal" or insecure position. To discuss his work with a wide audience may threaten as well as stimulate. Thus, three sources—perhaps himself, his chief, and a third party, such as a respected colleague—seem optimal.

This speculation gains support from the data in Chart 2-D for "assistant scientists" in Ph.D-dominated laboratories. When we broke these down

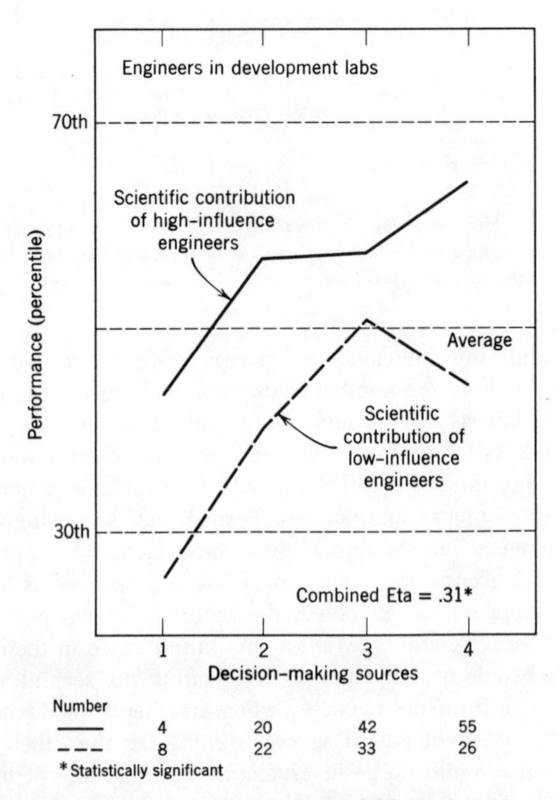

Chart 2-C. *High-influence engineers, like the Ph.D's, did best when four sources helped to shape their decisions. Those with low influence seemed to perform best with three sources.*

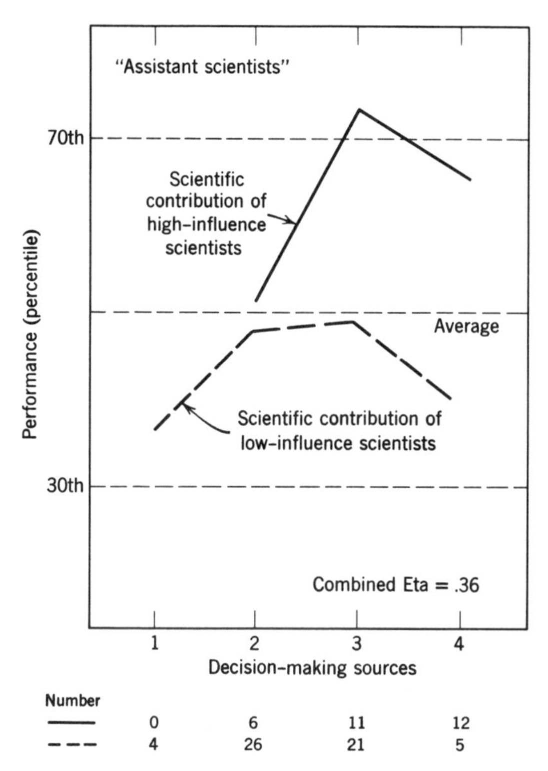

Chart 2-D. *When "assistant scientists" in Ph.D-dominated labs were divided into those with high and low influence, both groups did best when three rather than four sources shaped their decisions. In such marginal roles, too much stimulation may inhibit.*

by influence level, the curvilinear pattern appeared at both levels. (It was obscured in the previous Chart 1-D by the distribution of cases in the various subgroups; see numbers at bottom of chart.) Again we gain the impression that for these subordinate-status scientists, discussion of their work with two other sources besides themselves was stimulating, but more than that began to inhibit. (Because of the few instances in the remaining group of nondoctoral scientists in research labs, we shall not plot the data, but can simply say that the same curvilinear pattern appeared at both influence levels; highest performance was shown by those with high influence and three sources of involvement, not four.)

Parallel Evidence

Pelz noted a parallel between these findings and procedures he had observed in a petrochemical laboratory where he interviewed a number of technical men in 1955. It seemed to him that the company had been successful in building strong motivation. It had done this by giving a large measure of individual responsibility to its technical people. One of them called the process "controlled freedom." A general problem area was sketched out for the technical man; he was shown what mountain to climb, and then it was up to him to get to the top. But he was not ignored; he met periodically with other people to whom his work was important, such as R & D directors, customers, or manufacturers.

These meetings kept the research man on his toes, funneled useful information to him, and in particular gave his upper-level executives a chance to appreciate what he was doing. The feeling that others are interested in your work, we feel, is an excellent way of sustaining your own interest.

Our findings are consistent also with results obtained by Gerald Gordon in a study of 223 projects in medical sociology.[4] Reports from each project were evaluated by panels of leaders in medical sociology on four criteria, one of which was innovation—contribution of "new theory or findings" not explicit in previous work, or "new methods of research."

On the basis of a questionnaire from each project director, it was ascertained whether or not the project director had an "administrative superior" who bore some responsibility for the research and, if so, the extent of discussion with this superior about the research, and the latter's influence on its funds and design.

Twice as many projects were judged highly innovative if (a) the project director had a superior with whom he discussed the research but who did not determine the procedures, compared to situations in which there was either (b) no administrative superior, or in which (c) a dominant superior substantially determined procedures.

Gordon and an associate, Selwyn W. Becker, suggest that innovation is more likely where "consequences are visible" (represented in his data by a superior who keeps in touch), but where at the same time the researcher has freedom (is not dominated by the superior).

In our data, involvement of several decision-making sources should increase visibility of consequences; and the individual's influence guards against domination by one superior. Under this combination, we found performance to be highest.

Major Sources of Weight in Decisions

Thus far we have described results with one approach—one abstraction from the complex patterns obtained on Question 29. We have simply looked at the *number* of persons or collections of persons having some weight in determining a man's goals. We have not specified who was

[4]G. Gordon and Sue Marquis, "The Effect of Differing Administrative Authority on Scientific Innovation," Working Paper No. 4 from project on Organizational Setting and Scientific Accomplishment, Graduate School of Business, University of Chicago, 1963, 13 pp.

involved. When the chart said "one decision-making source," for instance, did it mean the man himself? His boss? A top executive?

To answer such questions, we used another scoring system which is illustrated in the following charts. Previously we asked who had at least "slight" weight, meaning 10% or more. Now we set a stiffer standard and asked who had "major" weight, meaning 30% or more. If the participant said that he alone was the source—had at least 30% weight in setting his goals, and no other source had this much—we classified him in the "mainly self" category in the next series of charts. If he said that he and his colleagues both had at least 30% weight, we classed him in the "self and colleagues" category, and so on.

The "mainly self" category was essentially a condition of *autonomy*. How did scientists perform under autonomy, compared with other patterns of weight in deciding assignments?

For Ph.D's in development labs (Chart 3-A), we see that "mainly

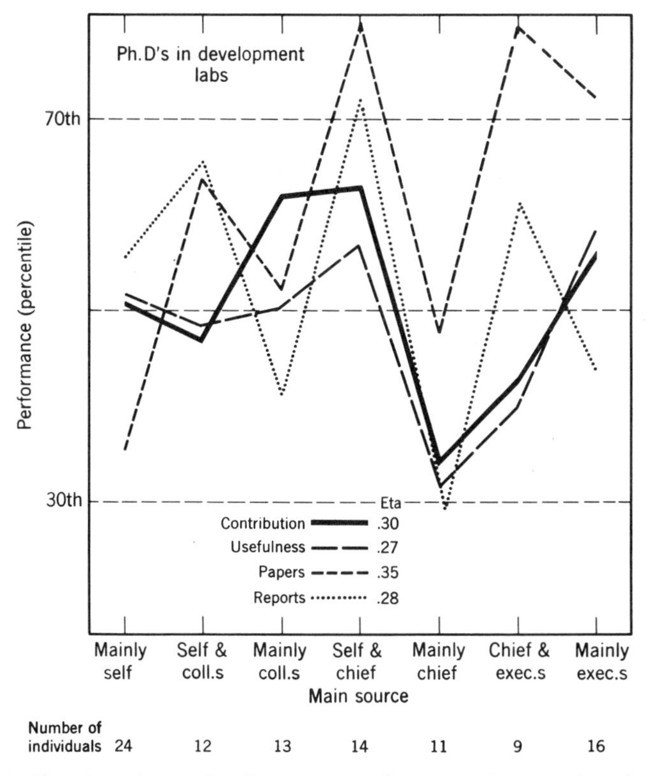

Chart 3-A. *This chart shows what happens to performance when specific individuals or groups exert major weight on decisions. Among Ph.D's in development labs, performance was best when the scientist's technical goals were decided jointly by himself and his chief; performance was lowest when the chief alone decided.*

self" was not in fact very fruitful by any of the four performance measures. The most favorable condition was the combination of "self and chief," where the scientist and his immediate supervisor had a large mutual voice in setting his goals. Just as clearly, the condition of "mainly chief" was unfavorable by all four criteria.

For the other combinations, the picture was mixed. When "self and colleagues" set the goals, many papers and reports were written, but these were of mediocre scientific value or usefulness. A similar pattern appeared when "chief and executives" set the goals without the scientist's participation. But let us beware of reading too much meaning here, for the discrepancies may result from a variety of extraneous factors.

Corresponding data for Ph.D's in research laboratories are shown in Chart 3-B. Only three of the ten possible combinations of "major weight" occurred here with any frequency, and therefore this chart has fewer

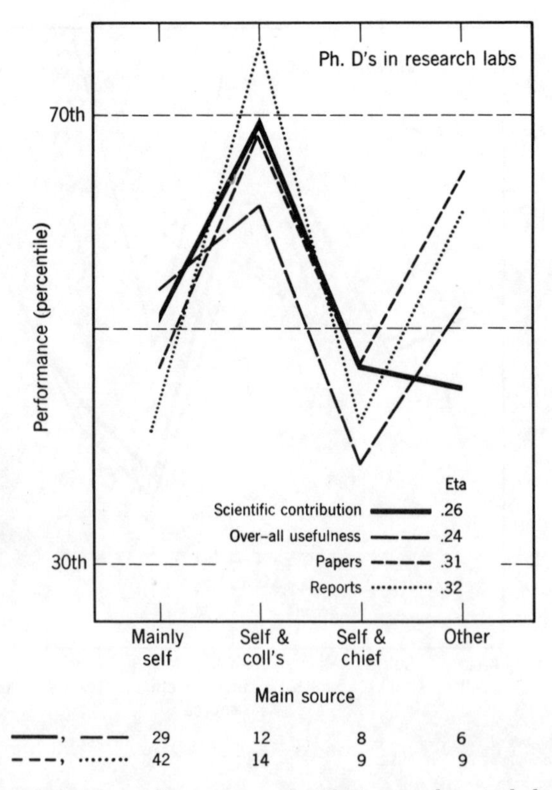

Chart 3-B. *Typical decision-making patterns for Ph.D researchers include the scientists themselves in some way, hence the brevity of the chart. Unlike the development scientists, these men performed best when both "self and colleagues" established technical goals.*

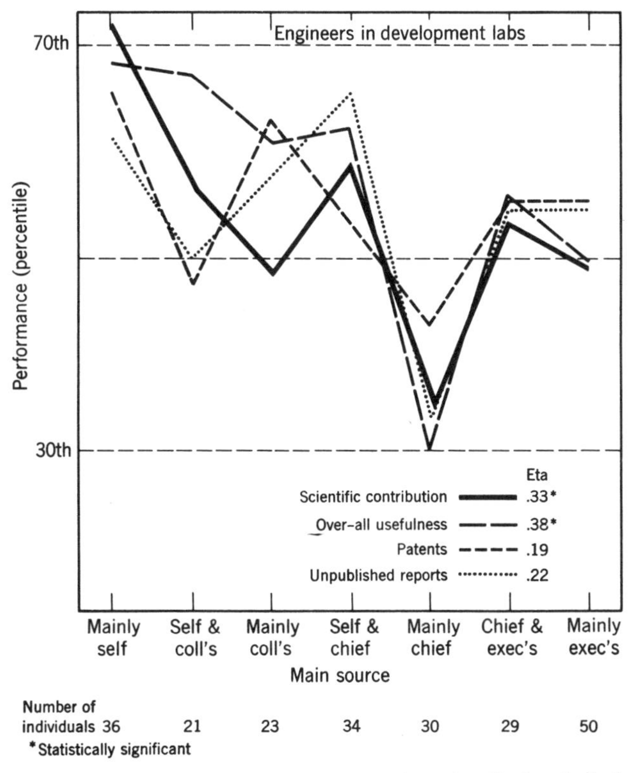

Chart 3-C. *Like the development Ph.D's, engineers performed well when both chief and self set goals; did poorly when mainly the chief decided what the goals were to be. But unlike Ph.D's, autonomous engineers (those who set goals themselves) did even better.*

categories. (This is what one might expect in research labs, where technical decisions typically are made by the scientist himself, or jointly with his colleagues or immediate chief.) The most fruitful pattern was that in which "self and colleagues" set the goals. In contrast with the previous chart, goal-setting by both "self and chief" was unproductive by any criterion.

What puzzled us in this chart was the fact that autonomy, enjoyed by half the sample, yielded only average output. How do we reconcile this with data shown in Chart 1-B where scientists with one decision-making source performed well? (We checked to make sure that the single source was indeed the man himself.) Note that the latter scientists were only a dozen out of the 40 in the "mainly self" category. Our conclusion: some research scientists can be creative when completely self-determining; but in our sample they formed a small minority.

Chart 3-C shows the data for engineers in development laboratories.

There were some similarities with the Ph.D scientists, along with some differences. Like the Ph.D's, the engineers' contribution and report-writing were high when "self and chief" set goals; performance was low by all standards when the goals were set by "mainly chief." But note a major difference from the Ph.D's: autonomous engineers ("mainly self") made highly significant contributions.

Among "assistant scientists" in Ph.D-dominated departments (Chart 3-D), we found the same trend as for engineers: autonomous men contributed highly and wrote numerous reports. But when their decisions were made by the chief alone, their contribution dropped. Performance was worst of all when the goal-setting was done by executives. (For the few nondoctoral scientists in research labs—data not shown—the trends for contribution and usefulness were similar: high when autonomous, dropping when chief alone or higher executives decided.)

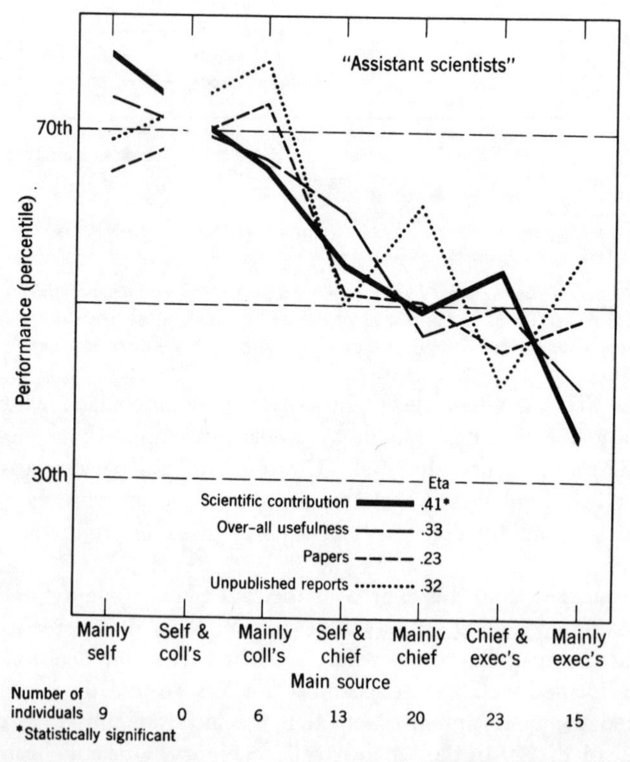

Chart 3-D. *"Assistant scientists" (in Ph.D-dominated labs) did well both in quantity and quality when colleagues determined their goals, and contributed best when autonomous (like the engineers). Goal-setting by the chief alone, and especially executives alone, were unfavorable conditions.*

The reader could well argue that low performance is not an effect of goal-setting by the chief, but rather its cause. If the chief has a mediocre subordinate, what else can he do but assign his tasks?

Possibly so. But by this same behavior, the chief may be stifling the chance of future growth. We have found that inner motivation is essential for achievement—a desire for self-direction, for thinking independently. (For more evidence on this point, see Chapters 6 and 10.) Goal-setting by the chief alone, or higher echelons alone, is likely to weaken this essential ingredient.

Now here is a real puzzler: Why did autonomy work for nondoctorals but not for doctorals? One speculation is that the autonomy of a nondoctoral is likely to be "controlled freedom," in the sense discussed earlier. There are visible mountains in the areas in which he is assigned. Autonomy releases his energy to climb them. On the other hand, except for a gifted but rare minority, the typical Ph.D who is isolated from colleagues and chief may spend his autonomy looking for the mountains, or climbing irrelevant ones.

Autonomy

In the previous charts, an individual was considered "autonomous" if he himself had at least 30% of the weight in setting his own technical goals, and no other person or group had this much. What would appear, we wondered, if we examined the full range of the autonomy scale? The result is given in Chart 4 which plots the mean scientific contribution of scientists exerting various degrees of weight in setting their goals within each of the five primary analysis groups.

Over the first half of the autonomy scale, with own weight ranging from 0 to 49%, contribution rose with increasing autonomy in a similar way within every group. But above the 50% point on autonomy, the scores for Ph.D's slanted downward, whereas those for non-Ph.D's continued to rise, at least up to 80% autonomy. (Few Ph.D's had more than this; a small group of completely autonomous engineers also dropped in contribution.) The curves for other performance measures (not shown) presented a similar shape.

These curves were reassuring. At least the groups resembled each other at lower autonomy levels. But why did the Ph.D curves drop with higher autonomy? Chapter 12 will explore a number of leads. Here we shall raise a related question. Were there some conditions under which the performance of autonomous Ph.D's would *not* drop? We examined several measures which later chapters found related to performance.

One promising result appeared with an item on interest in *breadth.* The questionnaire presented 19 statements and asked the respondent to

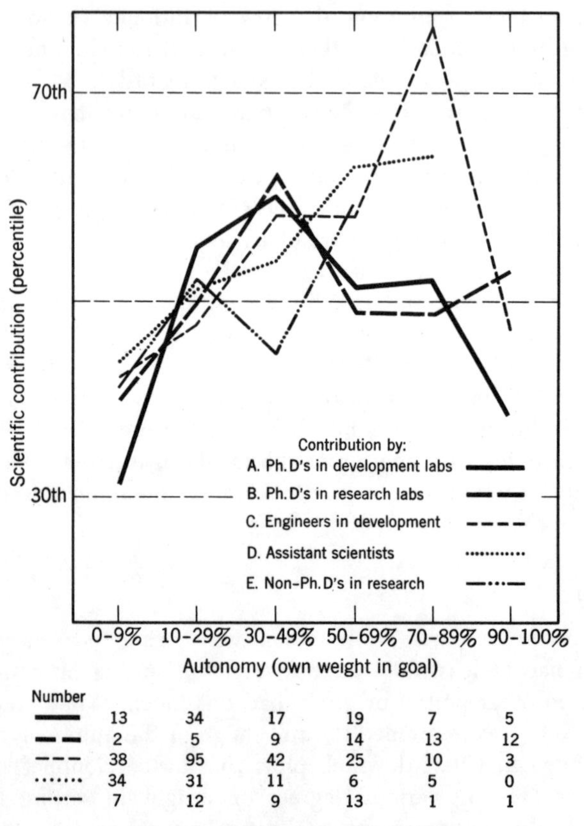

Number	0-9%	10-29%	30-49%	50-69%	70-89%	90-100%
A (━━━)	13	34	17	19	7	5
B (━ ━)	2	5	9	14	13	12
C (– – –)	38	95	42	25	10	3
D (······)	34	31	11	6	5	0
E (—···—)	7	12	9	13	1	1

Chart 4. *Autonomy by individuals in setting their goals is plotted against their scientific contribution, in the five primary groups. Performance of most groups improved in parallel fashion as autonomy increased from low to moderate; but above 50% autonomy, performance of non-doctorals continued to improve while that of Ph.D's dropped.*

rate how closely each statement described the approach he preferred to use in his work. Item 19E was: "I prefer to map out *broad features of important new areas,* leaving detailed study to others." This item will play an important part in Chapter 13, and its over-all relationship to performance will be reported in Chapter 6 on motivations. (Question 19 is given in a box in Chapter 6, page 92.)

Would scientists who maintained a broad interest in new areas be able to utilize autonomy to better purpose than scientists who specialized in narrow areas? We subdivided each analysis group into those with relatively strong or weak interest in breadth, and again examined autonomy in relation to scientific contribution. Results for the two groups of Ph.D's are shown in Charts 5-A and 5-B.

Among Ph.D's in development labs (Chart 5-A), scientists with a relatively weak interest in breadth (but instead an interest in depth or detail) performed better with only moderate autonomy in setting their goals. If these individuals had considerable leeway, their performance suffered. In contrast, scientists with a strong interest in mapping broad new areas did best when they had rather high autonomy.

A similar pattern appeared for Ph.D's in research labs (Chart 5-B—because of the few cases, those with less than 50% of own weight have been combined into a single category). Scientists with moderate autonomy performed somewhat better if their interest in breadth was weak (or, conversely, they were interested in detail), but at high levels of autonomy the reverse occurred: the "broad" individuals performed quite well indeed, and the "detailed" ones poorly.

Autonomous engineers (data not shown) also performed better if they had a strong interest in broad exploration. But a sharper pattern, shown in Chart 6, appeared with another measure based on the number of different kinds of R & D functions on which the individual spent at least

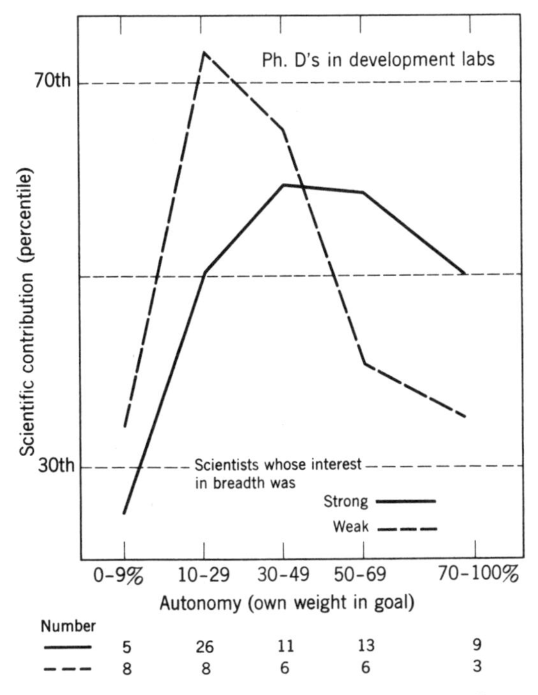

Chart 5-A. *Scientists were subdivided according to strong or weak interest in exploring broad new areas. In development labs, autonomous Ph.D's performed well if they retained an interest in breadth.*

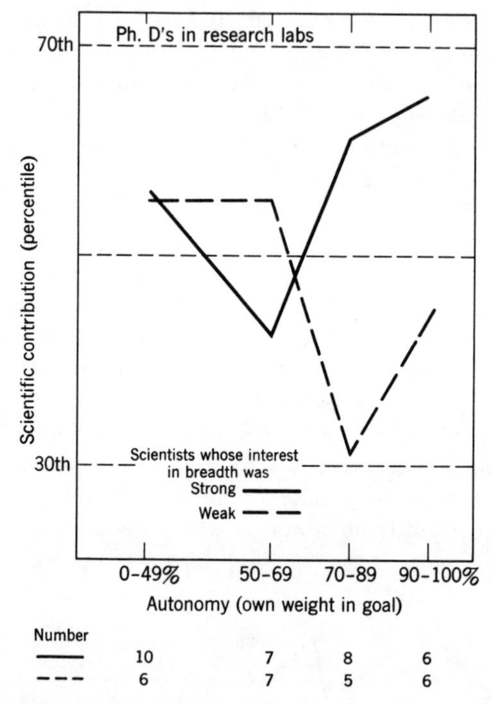

Chart 5-B. *In research laboratories, similarly, autonomous Ph.D's contributed highly if they kept a strong interest in broad exploration, but contribution dropped if their interest was in depth rather than breadth.*

some time. (This measure is discussed in detail in Chapter 4.) R & D functions include basic research, applied research, invention of new products or processes, improvement of existing ones, and technical services. The more different activities of this kind performed by the individual, the more diversity may be said to exist in his work.

If the engineer maintained high diversity (five R & D functions), the more autonomy he had, the better his contribution. With moderate diversity (three or four functions), increasing autonomy brought some improvement in performance, but not so much. With limited diversity (concentration on one or two R & D functions), high autonomy did not help. Thus engineers showed much the same pattern as did Ph.D's: with breadth of interests, autonomy was put to good use; with narrow or specialized interests, considerable autonomy did not help (although it did not hinder the specialized engineer as much as it did the specialized Ph.D).

Charts 5-A and 5-B do *not* say that autonomy necessarily brings with it a loss in breadth. (By examining the number of cases, the reader can

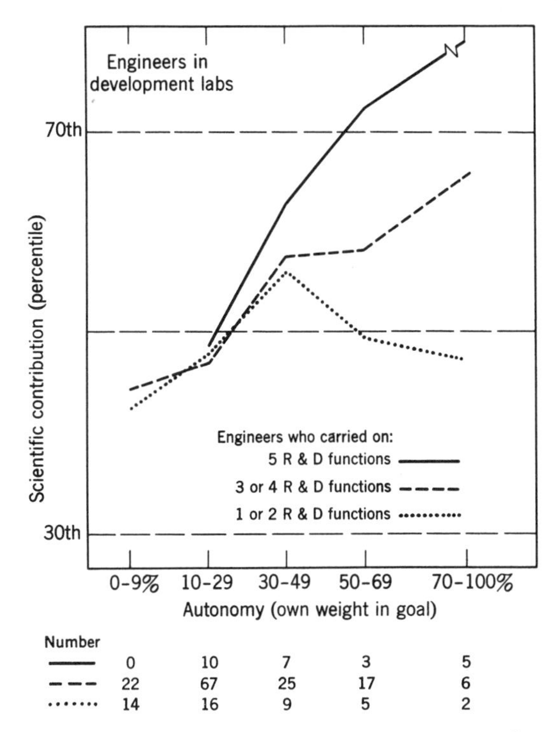

Chart 6. *Engineers in development labs who carried on a variety of R & D functions contributed more and more as their autonomy increased. The fewer their functions, however (the narrower their scope), the less they benefited from high autonomy.*

see that there was slight if any connection between the two variables for Ph.D's. For engineers, on the other hand, increasing autonomy was accompanied by some increase in diversity; or, conversely, those with low autonomy had less diversity.) But regardless of the connection between autonomy and breadth, this much can be said: Freedom helped those who did not withdraw into a narrow specialization, but kept open a lively interest in a *variety of research problems.*

SUMMARY AND IMPLICATIONS

In order to tie together the main threads of analysis and to speculate on practical implications, we shall use throughout the book the format of a dialogue with the reader. The questions put in his mouth are sometimes contrived in order to help the explication, but often they stem from actual questions raised by technical men and research supervisors in audiences to whom we have presented the results.

Would you summarize your results in a nutshell?

Much of the writing about research organizations assumes that freedom and coordination are incompatible. Our data suggest that some combination of both is not only feasible, but helpful for the scientist himself, that is, when he involves several other people in shaping his assignments, but keeps substantial influence over the decision process.

> I'm groping for a general framework on which to hang these points. Do they fit in with, or do they contradict, existing theories of managing organizations?

We did not design the research to test a systematic organizational theory. As it turned out, though, the foregoing results tied in well with some of Rensis Likert's ideas, formulated mainly from data on nonscientific organizations.[5] In a national service organization, for example, Likert found that employees in high-producing departments felt they exerted more influence on decisions affecting them, but also that other echelons exerted influence too. In short there was more *total* influence than was the case with low-producing employees.[6] An older view of organizations implied a fixed quantity of influence on decisions; if subordinates had more, superiors would have less. Likert's concept of management by overlapping groups is consistent with our results.

> I'm still skeptical as to whether involvement of several other sources in your data isn't the result of high performance rather than a cause. Why should it stimulate a person to have others telling him what to do?

Perhaps the idea of a man "interacting" with others is more appropriate than having others "tell him what to do." The man who was "told" by his chief alone, with little voice himself, was not effective.

As to reasons, we can offer several speculations. Through these contacts the man can learn the major *goals* that are important to the organization—what mountains need to be climbed. "Everybody must know what the over-all goal is," points out an executive in the Bell Labs, "so that within each man's area he can look for those solutions which are most relevant to the goal."[7] These contacts also provide what Gordon calls "visibility of consequences"; the relevance of the scientist's efforts to the solution of important problems is directly apparent.

Then there is the stimulation of having other people interested in what the scientist is doing. They can provide a testing ground, a chal-

[5] See *New Patterns of Management*, McGraw-Hill, New York, 1961, especially Chapter 4.

[6] Arnold S. Tannenbaum has found similar results in labor unions, Leagues of Women Voters, and business firms. See his "Control in Organizations: Individual Adjustment and Organizational Performance," *Administrative Science Quarterly*, 1962, vol. 7, pp. 236–257.

[7] J. A. Morton, "From Research to Technology," *International Science and Technology*, May 1964, pp. 82–92.

lenge to sharpen his ideas, providing they do not hold a veto power. And when he has done a good job, other people will know about it and give recognition and appreciation. (In Chapter 7 we'll talk more about the connection between recognition and performance.)

Finally, contact with a number of people in different roles will expose him to a diversity of viewpoints, and we shall see in Chapter 4 that diversity can be a stimulus to achievement.

> Say that I'm a research manager. How do I put these ideas into practice? Take the matter of multiple involvement in deciding assignments. Does this imply a lot of meetings?

Not a lot of them, but certainly periodic meetings with critical people at critical points. The procedures of "controlled freedom" that were used by the petrochemical laboratory provide an illustration. The article just cited by Morton provides good illustrations on procedures used in the Bell Telephone Labs.

> What about influence? How can you give individuals more influence and still preserve coordination?

Organizational structure plays some part here. One way to *reduce* influence is to establish a tall organization with several levels of review, each with a veto. The individual can exert more influence in a flat structure with fewer levels. There has to be a chance for *face-to-face interaction* between the scientist and other significant people in his R & D system. There needn't be frequent meetings, but see to it that two or three times a year, the scientist participates in small conferences in which he and key decision-makers have a chance to review his work and his future directions. Colleagues should be represented in these sessions as well as supervisors.

> One point in your data still bothers me. You show that performance is low where the chief alone sets the scientist's assignments. What is cause here and what is effect? Doesn't this situation simply mean that the chief has a subordinate who isn't very good, and has to make most of the decisions himself?

Possibly. But some chiefs may overdo it. If you expect the subordinate to grow, watch out. Continued direction by the chief will stunt initiative and independence, and these are qualities basic to scientific achievement. If direction comes from one man only, the supervisor, who also evaluates the subordinate's performance and determines his pay, this can be deadly.

> Suppose some of my section chiefs are like that—able men who allow little leeway to their subordinates? What can I, as a research director, do?

That really calls for administrative creativity! If possible, see that the section members are given a chance to serve on panels outside their immediate group. Let them attend discussions involving outside colleagues, executives, or sponsors. After a few years of service, see that section members are rotated to other locations. By broadening the channels of communication, you can help to keep open the potential for growth.

3

COMMUNICATION[1]

Effective Scientists Both Sought and Received More Contact with Colleagues

One view of a laboratory is that it is a facility which provides services and equipment so that its scientists can conduct R & D activities. A somewhat different view is that it is a system of interacting scientists (and other components) in which the inhabitants stimulate each other to produce high-quality R & D.

Under the assumption that a laboratory could be such a stimulating environment, we set out to answer some basic questions about contacts between people. Did it matter how often a scientist contacted his colleagues? If so, what was the optimum amount of contact? Did it matter how the contacts originated? For whom were contacts most useful? Answers to these questions have important implications for the organization of existing laboratories and for the establishment of new labs or research teams.

In this chapter we report an examination of the interaction among members of a laboratory. The hypothesis was that by interacting with one another, scientists can contribute to each other's effectiveness. The first part of the chapter describes findings which suggested that contacts with colleagues contributed to a man's performance. Later some explorations about optimum forms of colleague contact and groups for whom contacts were most relevant are described.

One problem in answering questions about the amount of a man's communication with his colleagues was determining how to measure "communication." Which co-workers should be considered colleagues? Should communication be restricted to face-to-face conversations? Does attendance at a meeting constitute communication?

Since there was no obvious best way to measure communication, we made some arbitrary decisions and asked scientists to respond to a variety of questionnaire items. "Communication," we decided, would be defined

[1]This analysis was mainly supported by a grant from the National Aeronautics and Space Administration.

as broadly as possible and therefore included contacts which occurred via memos and meetings, as well as direct conversations.

"Colleagues" were defined as other professionals with whom a man worked within the lab. His supervisors were specifically excluded (separate questions were asked about them), as were subprofessional assistants. Subordinates who were themselves professionals, however, could be claimed as colleagues. Some questions were restricted to a man's most important colleagues—he could name up to five. Other questions asked about the entire set of people with whom he exchanged useful information. Still other questions probed his general preferences for working alone or with others and the over-all level of coordination in his lab.

In all, eight possible measures of communication with colleagues were examined. Reasonably consistent relationships with performance appeared for four of them.

Frequency of Contact with Colleagues

The average *frequency* with which a man claimed to contact his most important colleagues was related to his performance. The source of the data is shown in the box below; the results appear in Charts 1-A to 1-C.

Question 41. As a general rule, how frequently do you communicate with each of your . . . colleagues on work-related matters? (Whether by conversation, memos, seminars, etc.)

[Having named his five most significant colleagues, the respondent rated the frequency of communication with each using a four-point scale ranging from "few times a year or less" to "daily." From these data, an average frequency of communication with colleagues was computed for each respondent.]

Results were clear for three of the five major groups of scientists described in Chapter 1. Among Ph.D's in development labs, Ph.D's in research labs, and engineers, those who had relatively frequent contact with colleagues tended to perform at higher levels than those with less frequent contact. (All performance measures used in this chapter were adjusted to remove effects related to differences in length of experience—see Chapter 1 for explanation.)

Among Ph.D's in development labs, the average frequency of contact was quite high; over half the respondents contacted colleagues several times a week or more. Chart 1-A shows that judgments of technical contribution, judgments of usefulness, and output of reports all were highest for development Ph.D's who had daily contact with colleagues. Output of papers, however, was highest when the scientist contacted his colleagues about weekly.

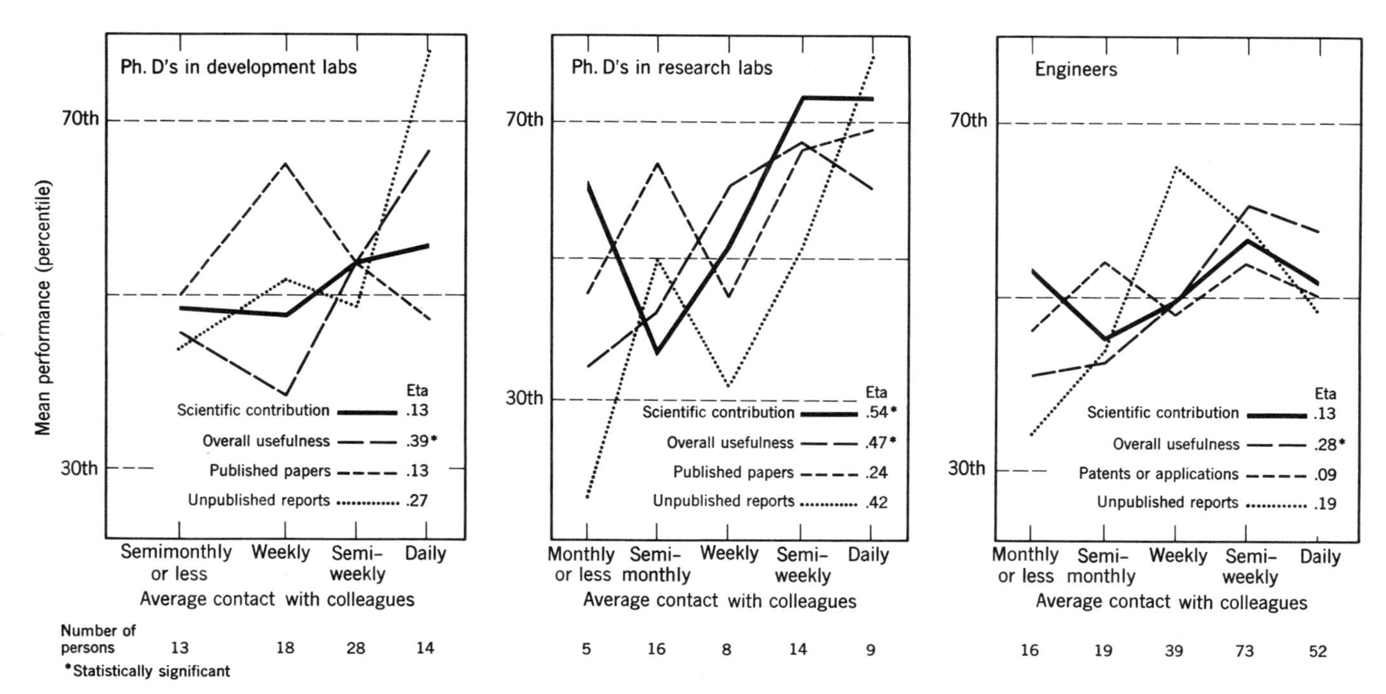

Charts 1-A, 1-B, and 1-C. *Among Ph.D's in development or research labs and engineers, those who contacted their colleagues with above-average frequency tended to show high scientific performance. For Ph.D's in development labs, the optimum frequency was daily for three of the four performance measures. For Ph.D's in research, performance was better with semiweekly or daily contact than with lower frequencies. For engineers, the optimum frequency was semiweekly for most measures of performance.*

Among Ph.D's in research labs, the average frequency of contact was lower; the median frequency was about weekly. The best performance, however, was shown by those who contacted colleagues semiweekly or daily (see Chart 1-B).

Engineers were like Ph.D's in development in that the average frequency of contact was fairly high; several times a week was typical. Although many engineers claimed average frequencies as high as daily, Chart 1-C shows that this may have been too frequent for this group. Semiweekly contact seemed the optimum amount.

Results for the other two major groups, non-Ph.D scientists and assistant scientists, were also examined. The relationships, however, proved to be much less consistent. Some of the curves rose as frequency of contact increased, but about an equal number jumped around in a jagged fashion or declined. Since this was also the case when the performance of these groups was related to other measures of contact (described later), it was difficult to interpret the results. It may simply have been that contacts with colleagues were less important for these groups. Whatever the cause, the lack of clear relationships suggested that these groups be omitted from this first (and also subsequent) series of charts.

The general finding for Ph.D's and engineers, however, seemed clear: scientists who saw their most important colleagues rather frequently (several times a week or daily) tended to perform at higher levels than those who had less contact with their colleagues.

This finding suggested a variety of interesting questions and possible explanations. Before examining them, however, the comparable relationships for other measures of contact will be described.

Time Spent Contacting Colleagues

The second way we attempted to measure the amount of contact with colleagues was to ask how much *time* was spent on this activity. The questionnaire item appears in the following box. This measure proved

> *Question 42.* In the course of a normal week, about how much time all together do you spend talking or communicating with each of these persons on work-related matters? (Whether on or off the job.)
>
> [The respondent rated the hours per week spent contacting each important colleague on a six-point scale ranging from "less than one hour per week" to "more than 20 hours."]

to be highly related to the first one on frequency of contact.[2] On the whole, scientists who spent a great deal of time seeing colleagues tended

[2]Correlations between the two measures ranged from .4 to .7 across the three groups of scientists.

to see them frequently (that is, long, infrequent meetings were unusual), and those who spent little time tended to see their colleagues only rarely. Since the measures were highly related, it was not surprising to find that the relationships with performance were also similar. Charts are not shown, but the findings can be quickly summarized: For both Ph.D groups and the engineers, the more time a man spent contacting his colleagues (up to a point), the higher his performance. The optimum time for Ph.D's seemed to be somewhere between six and ten hours a week per colleague (very few indicated amounts greater than these); the optimum for engineers was slightly higher, somewhere between eight and fifteen hours a week per colleague. In all three groups, the scientists who performed at the highest levels spent considerably more time communicating with their colleagues than was typical for their group.[3]

Number of Colleagues

Another indication of the amount of colleague contact was the number of people with whom the scientist exchanged information. Data from two questionnaire items were examined. One asked about people in the scientist's own group; the other asked about those he contacted elsewhere in the organization. The wording is shown in the following box.

> *Question 28.* About how many people in the following situations do you work with closely—in the sense of exchanging detailed information from time to time that is of benefit to you or to them? (Exclude subprofessional assistants or clerical personnel.)
>
> [The respondent checked seven-point scales ranging from "None" to "20 or more" to indicate the number of people "In my immediate groups (sections, projects, teams, etc.)" and "In other technical groups within this organization."]

How did these measures of colleague contact relate to each other and to those described previously (frequency and time)? There was a moderate tendency for scientists to see many colleagues outside their own group (but within their organization) if they also worked closely with many colleagues within their own group.[4] Whether this reflected consistency in their behavior (perhaps a professional "sociableness") or exigencies of their work was not clear. Whatever the cause, the positive relationship

[3] The median number of hours spent contacting each important colleague per week was about two for Ph.D's and three for engineers. Of course, some contacts may have involved several colleagues simultaneously.

[4] Correlations between these two items ranged from .3 to .5 across the three groups of scientists.

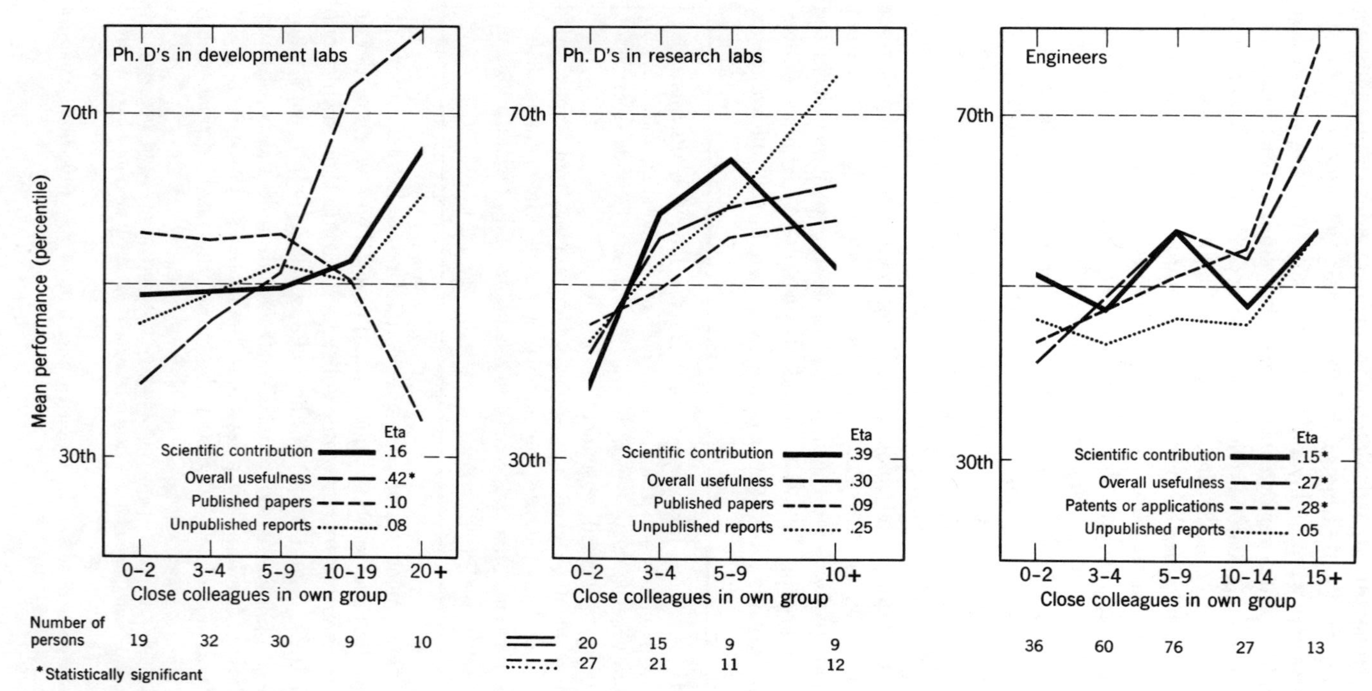

Charts 2-A, 2-B, and 2-C *show that among three groups of scientists, high performance tended to be associated with having many colleagues in one's own group. Groups as large as 20 or more provided effective environments for Ph.D's in development labs. Although groups this large could not be separately examined for Ph.D's in research labs and engineers, those scientists claiming the largest number of close colleagues tended to show the highest performance.*

between the two items was only moderate, and results will be shown for each separately.

It was interesting to discover that the total number of colleagues a man worked with was only mildly related to the time he spent contacting his most important colleagues and the frequency with which he contacted them.[5] Thus it appeared that data about the number of colleagues provided a rather different approach to measuring the amount of contact a man had with colleagues. In spite of this different way of measuring amount of colleague contact, however, the previous findings again appeared and are now described.

Number of Colleagues in Own Group. The results of relating this item to performance are shown in Charts 2-A to 2-C. With but two exceptions in the 12 curves, the larger a person's immediate group of co-workers, the higher his performance. For Ph.D's in development labs, groups as large as 20 or more people seemed to provide the most effective environments. Although there were insufficient instances to examine groups this large for Ph.D's in research labs and engineers, the largest groups we could examine (ten or more, and 15 or more, respectively) again tended to be the most effective ones. Thus for three rather different types of scientists, there seemed to be a consistent trend for those who exchanged information with many people to perform at higher levels. (Although data from assistant scientists and non-Ph.D scientists did not generally contradict these findings, the data did not consistently support them either.)

Number of Colleagues outside Own Group but within Organization. In Charts 3-A to 3-C, relationships are shown between scientists' performance and the number of people they exchanged information with outside their own group (but within their organization). As in the previous series of charts, the general finding was that high performance was shown by the scientists who had high amounts of colleague contact. Conversely, scientists who exchanged information with very few people outside their own groups tended to have low performance.

Other Measures of Colleague Contact

In addition to the four measures of colleague contact just described, the relationships between performance and four others were also examined. These included two indications of the scientist's preference for working alone or with others, an indication of whether the work in his lab was

[5]Correlations between the two items inquiring about number of colleagues and the item concerning time spent contacting important colleagues ranged from .3 to $-.1$ across the three groups (median $r = .1$). Correlations between the two items inquiring about number of colleagues and the item concerning frequency of contacting important colleagues ranged from .3 to .0 across the three groups (median $r = .2$).

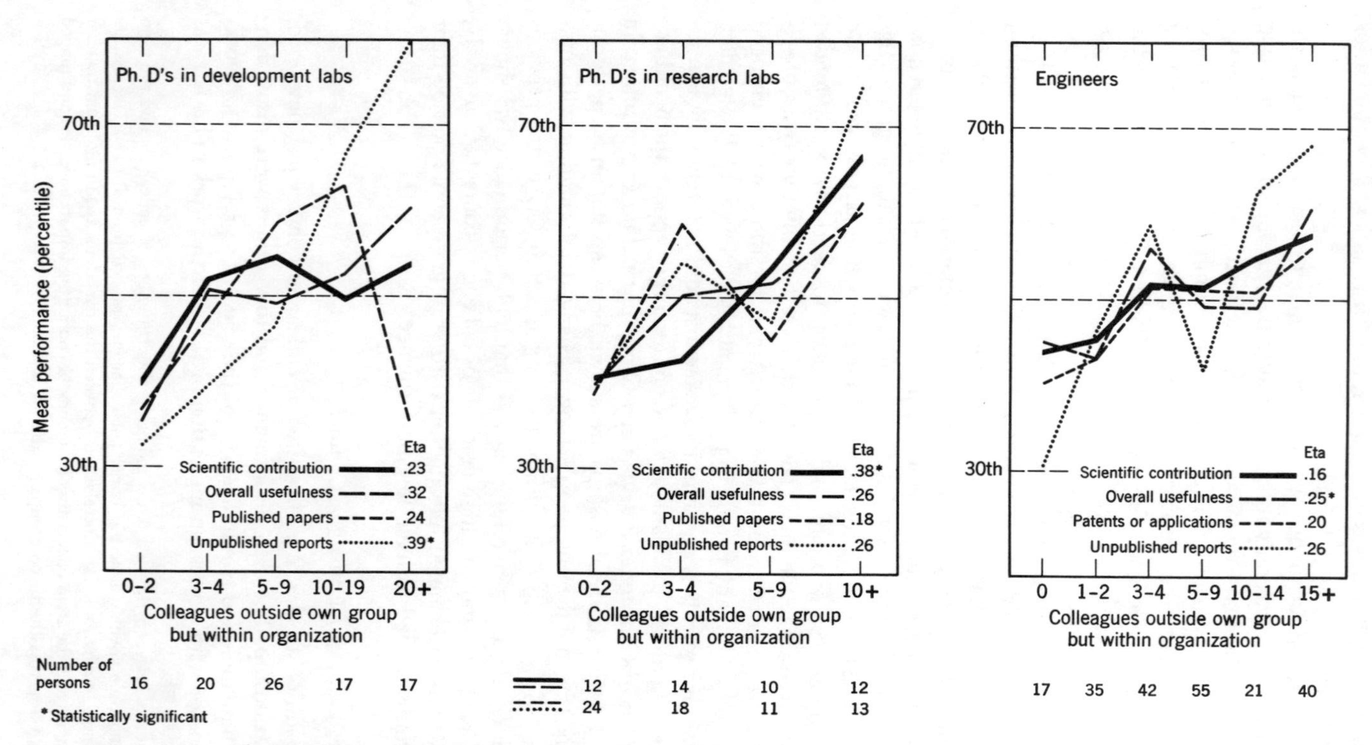

Charts 3-A, 3-B, and 3-C *show the relationship between several measures of performance and the number of colleagues a man exchanged information with who were located outside his own group but within his organization. Among three groups of scientists, the highest performance was shown by those having contacts with many colleagues.*

organized on an autonomous or coordinated basis, and information about who originated colleague contacts. None of these, considered alone, showed consistent relationships with performance. This was not surprising; at best, these could have been only indirect measures of colleague contact. Apparently, scientists could perform at high (or low) levels regardless of their general preferences for working with others or the degree of team orientation within their labs.

Was the Contact-Performance Relationship an Artifact?

Up to this point, it has been shown that four indications of a man's contact with his colleagues were all positively related to a variety of different measures of scientific performance. But perhaps these relationships were in some way artifactual. Could they be explained by some third factor which was related to both performance and contact?

Effect of Experience. It was conceivable, for example, that high performance appeared to go with large amounts of colleague contact simply because the people who had been in the lab longer had had more time to build up both their range of acquaintances and their performance. Although this undoubtedly occurred, it did not wholly account for the relationships we found.

The reason the performance scores were adjusted to remove effects associated with length of experience was precisely so that this kind of possibility could be rejected. The fact that the relationships emerged even after the performance scores had been adjusted indicated that it must have been something more than just differences in experience which accounted for them.

Effect of Supervisory Status. Even if relationships were not due simply to differences in experience, they might, perhaps, have been caused by differences between supervisors and nonsupervisors. Not surprisingly, supervisors tended to have higher performance than nonsupervisors (most organizations selected their research supervisors partly on the basis of past performance). It was also true that supervisors tended to have markedly more contacts with colleagues than nonsupervisors. (Note that colleagues had been defined to include professional-level subordinates.) Had these differences accounted for the appearance of the relationships?

To check this, the relationships among the various measures of colleague contact and the various measures of performance were examined for the nonsupervisors and supervisors separately. As shown in Table 1, the positive relationships between contact with colleagues and performance again appeared when the nonsupervisors were examined alone (see Part *a*). This relationship also appeared for the Ph.D supervisors when they were examined alone, but did not appear for the supervisors who were engineers (Part *b*).

TABLE 1 *Shown below are correlations among four measures of contact with colleagues and four measures of scientific performance (adjusted for length of experience). Data are shown separately for nonsupervisors and supervisors (Parts a and b). The predominance of +'s in Parts a and b indicates that performance was positively associated with contact for both groups.*

(A + or – sign appears when the correlation was at least ±.15; double signs appear when the correlation was statistically significant at the .05 level. P = papers for Ph.D's and patents for engineers; R = unpublished reports; C = scientific contribution; U = usefulness. N's vary somewhat owing to missing data.)

	Frequency of contacting colleagues				Time spent contacting colleagues				Number of close colleagues in own group				Number of colleagues in other groups in org'n.				Scorecard	
a. Relationships for nonsupervisors	*P*	*R*	*C*	*U*	*P*	*R*	*C*	*U*	*P*	*R*	*C*	*U*	*P*	*R*	*C*	*U*	+'s	–'s
Ph.D's in dev. labs ($N = 34$)	0	0	0	0	+	++	0	0	0	0	0	+	0	++	0	+	7	0
Ph.D's in res. labs ($N = 30$)	0	+	+	+	+	+	+	+	+	+	+	+	+	0	+	+	14	0
Engineers ($N = 110$)	0	0	0	+	0	+	0	0	+	0	0	0	0	+	+	+	6	0
																Totals	27	0
b. Relationships for supervisors																		
Ph.D's in dev. labs (N = 35)	0	0	0	+	0	++	+	+	0	0	0	+	0	+	0	0	7	0
Ph.D's in res. labs ($N = 14$)	+	++	++	++	0	++	+	+	0	+	0	0	0	++	++	+	17	0
Engineers ($N = 82$)	0	0	0	0	+	0	–	– –	+	0	–	0	0	0	0	0	2	4
																Totals	26	4

Thus Table 1 provided increased assurance that the positive relationships between contacts with colleagues and performance were more than just an artifact.

Were Colleague Contacts a Result of High Performance?

We were exploring to determine whether contacts with colleagues might have stimulated scientific performance. But it was possible that the relationship worked the other way: perhaps men who performed at high levels were then sought out by others and thus achieved their high interaction as a result of their high performance. Unfortunately, there was no way to test the time order of the events in these data. Information was available, however, about the chief origin of the contacts which occurred.

If one looked only at scientists who themselves initiated the contacts (thus omitting those people who were primarily being sought out by others) would the relationship hold up? Briefly, the answer was "yes." The data are shown in Part *a* of Table 2.

The table is based on an item in the questionnaire (reproduced in the following box) which suggested different ways in which contacts could originate: the man himself might originate them, the man's colleagues might originate them, or the laboratory might originate them (by scheduling meetings, seminars, or other gatherings). A final category permitted scientists to indicate an unplanned contact—one that arose spontaneously when the man and his colleague chanced to come together. As the question was asked, respondents could indicate the proportion of their contacts with each colleague which originated in each way.

Question 43. How does the communication with each person usually originate? Estimate the percent occurring in the following ways, to nearest 5–10%.

[For each of five most important colleagues, the respondent entered percents for the categories that follow.]

I visit or contact him.
He visits or contacts me.
We both attend a meeting or seminar.
Conversation arises spontaneously when we see each other.
Other ways: ________________

Of course, some scientists indicated mixtures. For example, about a third of the contacts might have been originated by the scientist, about a third by his colleagues, and the other third split between spontaneous conversations and formal meetings. Since it was hard to handle the data from such scientists, they were omitted from the analysis. Table 2 in-

TABLE 2 *These data show that contacts with colleagues related to scientific performance regardless of how the contacts originated (note predominance of +'s in Parts a to d). Contacts were more related to performance, however, when they were initiated by the respondent or his colleagues (Parts a and b) than when initiated by the organization or unplanned (Parts c and d).*

(A + or − sign appears when the correlation was at least ±.25; double signs appear when the correlation was statistically significant at the .05 level. P = papers for Ph.D's and patents for engineers; R = unpublished reports; C = scientific contribution; U = usefulness. All performance measures have been adjusted for length of experience. N's vary somewhat owing to missing data.)

Principal origin of communication	Frequency of contacting colleagues				Time spent contacting colleagues				Number of close colleagues in own group				Number of colleagues in other groups in org'n.				Scorecard		
	P	*R*	*C*	*U*	*P*	*R*	*C*	*U*	*P*	*R*	*C*	*U*	*P*	*R*	*C*	*U*	+'s	−'s	Net
a. Respondent contacts colleagues																			
Ph.D's in dev. labs (*N* = 32)	0	0	++	++	0	0	0	++	−	+	0	++	0	++	0	+	12	1	
Ph.D's in res. labs (*N* = 6)	+	+	−	0	+	+	−	−	0	+	0	+	+	0	+	+	9	3	
Engineers (*N* = 74)	0	0	0	++	0	++	0	0	++	++	0	++	++	++	++	++	18	0	
																Totals	39	4	+35
b. Colleagues contact respondent																			
Ph.D's in dev. labs (*N* = 8)	++	+	+	++	+	+	+	+	−	0	0	0	0	+	+	+	13	1	
Ph.D's in res. labs (*N* = 7)	0	+	++	+	0	+	+	+	−	+	+	++	0	+	+	+	14	1	
Engineers (*N* = 32)	0	0	0	0	0	0	0	+	0	0	0	0	0	−	−	0	1	2	
																Totals	28	4	+24
c. Both attend seminar, etc.																			
Ph.D's in dev. labs (*N* = 7)	0	+	−	0	+	++	+	−	0	0	0	0	0	+	−	−	6	4	
Ph.D's in res. labs (*N* = 4)	0	+	+	+	0	0	0	0	0	0	+	−	0	0	++	0	6	1	
Engineers (*N* = 13)	−	0	0	0	−	0	0	0	+	0	0	0	++	+	+	++	7	2	
																Totals	19	7	+12
d. Unplanned conversations																			
Ph.D's in dev. labs (*N* = 10)	−	0	−	−	0	+	−	−	0	0	0	0	0	+	0	0	2	5	
Ph.D's in res. labs (*N* = 14)	0	+	+	++	0	+	+	+	+	0	+	++	+	−	+	0	13	1	
Engineers (*N* = 40)	0	0	+	++	0	0	0	0	0	−−	0	0	0	0	0	0	3	2	
																Totals	18	8	+10

cludes only those scientists for whom one type of origin clearly stood out above the others.

In Table 2, Part *a*, one may note the strong predominance of positive relationships among all three major groups of scientists. This indicated that large amounts of colleague contact tended to go with high performance even when one looked only at scientists who themselves were the primary initiators of the contacts. Under these conditions it was difficult to believe that the contacts were primarily the *result* of previous high performance. Thus the hypothesis that contacts with colleagues stimulated performance seemed to be supported.

Optimum Forms of Contact

In addition to increasing confidence in the hypothesis that colleague contacts enhanced performance, Table 2 is of interest for another reason. It provides some clues as to the kinds of contacts that were most helpful to scientists.

One of the striking findings contained in the table is that colleague contacts were positively related to performance regardless of how the contacts originated. From the laboratory's point of view, this was a hopeful finding; even organization-initiated contacts seemed to enhance performance.

Note, however, that the net number of scorecard pluses (which, in this instance, provided a rough indication of the strength of the underlying relationships) differed according to the primary origin of contact. They were highest when the respondent contacted his colleagues (Part *a*); next highest when colleagues contacted the respondent (Part *b*). Although the trend was still positive, it was weaker when contacts were originated by the laboratory, or unplanned. This seemed to make good sense; if the contacts were purposefully originated by the people directly concerned—the man himself or his colleagues—they tended to have higher payoff than if they were unplanned or originated by some third party.

Numerous other questions concerning the optimum forms of colleague contact came to mind. If a man saw many colleagues within his own group, did it matter how many he saw outside? If he saw his most important colleagues frequently, did it matter how long he spent communicating with them? By taking the various measures of colleague contact in pairs and examining their *combined* effects, these questions could be answered.

It was discovered that three of the four contact measures seemed to have stimulating properties which could "accumulate." These three were: frequency of contact with important colleagues, number of colleagues in own group, and number of colleagues in other local groups. Charts 1-A through 3-C showed that performance was positively associated with each of these three. In addition, we now discovered that performance tended

TABLE 3 *This is based on correlations among three measures of contact with colleagues and four measures of scientific performance (adjusted for length of experience). Part a shows that among scientists who spent little time contacting colleagues, those who had many colleagues and/or contacted them frequently tended to have higher performance. Among scientists who spent much time on communication (Part b), however, number of colleagues or frequency of contacting them was not consistently related to performance. Part c shows that colleague contacts were especially important to those who spent little time on such contacts.*

(A + or − sign appears when the correlation or difference was at least ±.15; double signs appear when the correlation or difference was statistically significant at the .05 level. P = papers for Ph.D's and patents for engineers; R = unpublished reports; C = scientific contributions; U = usefulness. N's vary somewhat owing to missing data.)

	Frequency of contacting colleagues				Number of close colleagues in own group				Number of colleagues in other groups in org'n.				Scorecard	
	P	*R*	*C*	*U*	*P*	*R*	*C*	*U*	*P*	*R*	*C*	*U*	+'s	−'s
a. Relationships for those averaging *0–2 hours* per week contacting each colleague														
Ph.D's in dev. labs ($N = 38$)	0	0	0	+	−−	0	0	+	0	++	+	++	7	2
Ph.D's in res. labs ($N = 29$)	0	0	+	++	+	++	++	++	++	+	++	+	16	0
Engineers ($N = 97$)	+	0	0	++	++	0	0	++	++	++	++	++	15	0
Totals													38	2
b. Relationships for those averaging *3 or more hours* per week contacting each colleague														
Ph.D's in dev. labs ($N = 31$)	−	+	0	+	−	0	+	+	0	+	−	0	5	3
Ph.D's in res. labs ($N = 15$)	0	0	+	−	−	0	−	+	0	0	0	+	3	3
Engineers ($N = 101$)	−−	−−	0	0	++	0	−	++	0	+	0	+	6	5
Totals													14	11
c. Difference: *r* for those spending little time minus *r* for those spending much time														
Ph.D's in dev. labs	+	−	0	0	−	0	−	0	0	+	+	+	4	3
Ph.D's in res. labs	0	0	0	++	+	+	+	+	+	+	+	0	9	0
Engineers	++	++	+	++	0	0	0	0	+	0	0	0	8	0
Totals													21	3

to be higher if a person had high scores on two of these measures than if he scored high on just one (data not shown). Thus *frequent* contact with *many* colleagues was preferable to frequent contact with just a few (and the lowest performance of all came from those who saw few colleagues only rarely). Similarly, having many colleagues in one's own group *and* many colleagues in other local groups was preferable to having many colleagues in one's own group only or in other local groups only. (Again, the lowest performance was obtained from those who had few colleagues in both their own and other groups.)

The measure which referred to the amount of time spent contacting colleagues, however, showed a different pattern of relationships. When the time measure was combined with other measures of contact, it was discovered that these other measures were especially important for scientists who spent relatively *little* amounts of time on communication. The data are shown in Table 3. In Part *a*, which contains data for those who averaged fewer than three hours per week contacting each colleague, having frequent contacts and/or many colleagues paid off handsomely in terms of increased performance—note the predominance of pluses. However, for those who averaged three or more hours per week on communication with each colleague, the frequency with which they contacted colleagues and the number of their colleagues seemed unimportant—note that pluses and minuses were about equally present in Part *b*.

Thus it appeared that there were several different paths to effective interaction: spend much time on communication (in which case the other factors did not seem to matter), or spend little time but contact many people frequently. The situation to be avoided, apparently, was that of spending little time on infrequent contacts with few colleagues.

Colleague Contacts and Preferences or Requirements for Working with Others

Having discovered that contacts with colleagues did seem to enhance the performance of Ph.D's and engineers, and having specified some of the forms of contact which were most helpful, we wondered whether there were some subgroups or situations in which such contacts would not be useful. For example, would colleague contacts lose their usefulness to scientists who said they did not care to work with others?

To answer the question, scientists were separated into two groups according to their preferences for working with others.[6]

[6]Information from three items (found to be moderately intercorrelated among themselves) was added together to form an index of preference for working with others. These items were parts of Question 62 (quoted in Chapter 6) and inquired about the importance the respondent attached to: 62E—"working with colleagues of high technical competence"; 62F—"having congenial co-workers as colleagues"; and 62G—"working under chiefs of high technical competence."

TABLE 4 *These data indicate that contacts with colleagues tended to be positively related to scientific performance regardless of the scientist's preference for working with others (note predominance of +'s in Parts a and b).*

(A + or − sign appears when the correlation was at least ±.15; double signs appear when the correlation was statistically significant at the .05 level. P = papers for Ph.D's and patents for engineers; R = unpublished reports; C = scientific contribution; U = usefulness. All performance measures have been adjusted for length of experience. N's vary somewhat owing to missing data.)

		Frequency of contacting colleagues				Time spent contacting colleagues				Number of close colleagues in own group				Number of colleagues in other groups in org'n.				Scorecard	
a. Relationships for those with *low* preference for working with others		*P*	*R*	*C*	*U*	*P*	*R*	*C*	*U*	*P*	*R*	*C*	*U*	*P*	*R*	*C*	*U*	+'s	−'s
Ph.D's in dev. labs	(*N* = 34)	0	++	+	++	0	++	+	++	0	0	0	++	0	++	0	0	14	0
Ph.D's in res. labs	(*N* = 25)	++	++	+	+	+	+	+	+	0	++	++	+	+	++	++	++	22	0
Engineers	(*N* = 93)	0	0	0	++	0	++	0	0	++	0	0	++	++	0	0	+	11	0
																	Totals	47	0
b. Relationships for those with *high* preference for working with others																			
Ph.D's in dev. labs	(*N* = 37)	0	0	0	0	+	++	0	−	0	0	+	++	0	++	+	++	11	1
Ph.D's in res. labs	(*N* = 20)	−	+	+	+	0	+	+	+	+	+	0	+	0	0	+	0	10	1
Engineers	(*N* = 105)	0	0	0	+	0	0	0	0	++	0	0	++	+	++	++	++	12	0
																	Totals	33	2

The results, shown in Table 4, were surprising. The relationships between contacts with colleagues and performance were as positive among scientists who did *not* especially prefer to work with others (Part *a*) as they were for scientists who did prefer to work with others (Part *b*). (Among Ph.D's in research labs, the relationships were somewhat *stronger* for the former group.) Thus Table 4 suggests that contacts could be helpful even to the relatively unsocial scientist.

Two other analyses explored the effect of work requirements from a somewhat different viewpoint. It seemed possible that scientists who were motivated strongly from within might benefit less from colleague contacts than would men with low inner motivation. If scientists were divided into groups which differed in motivation, would the relationships between colleague contacts and performance be stronger in one group than in the other? Although the tables are not shown, the results from two separate analyses were straightforward: strength of motivation did not seem to affect the extent to which colleague contacts enhanced performance. They were just as useful to the highly inner-motivated scientists as to the less motivated men.[7]

Thus colleague contacts seemed useful for a wide range of Ph.D's and engineers. The positive relationship between contacts and performance appeared even for people who were not especially interested in working with others, and people who indicated strong sources of inner motivation.

SUMMARY AND IMPLICATIONS

To conclude the chapter, let us tune in on an imaginary conversation with a research director:

> You seem to have presented a pretty straightforward set of findings. Let me see if I've got the gist of them: contacts with colleagues went with higher performance.

Yes, that is the main point. It should be pointed out that this result appeared for three different groups: Ph.D's in research and development labs, and for engineers. However, they were not nearly so clear for two other groups, assistant scientists and non-Ph.D scientists.

> What surprised me was the way your findings seemed to hold up for everyone, even people I thought might be "immune" to the effects of contacting colleagues—those people who didn't want to work with others, for example, or those in labs where the work was done autonomously, or your highly motivated people.

[7] In one analysis, scientists were divided according to their scores on a question which inquired about their involvement in their work (the question is reproduced in Chapter 5). In the other analysis, the index "own ideas as a source of motivation" (described in Chapter 6) was used to divide the scientists.

At first we were surprised too. But as we began thinking about the kinds of benefits which might occur through interaction with colleagues, it became obvious that they could be helpful to scientists in many different situations, though perhaps for different reasons. Of course, one possible reason contacts may have enhanced performance was . . .

> Hold it! What makes you think that contacts "enhanced" performance? Might it not work the other way around—high performance increased a man's contacts?

Undoubtedly many scientists were sought out by others who wanted to learn something from the man they contacted. (An important teaching function may have been going on here.) So we would not want to rule out the possibility that some contacts occurred as a *result* of a man's previous high performance. But even teachers often claim they benefit from contacts with their students!

When we found that the positive relationship between colleague contacts and performance appeared even after we had taken into account differences in experience, in supervisory status, and in which person originated the contacts, then it looked as if contacts did enhance performance—at least sometimes.

> Okay, so perhaps I would do better science if I talked shop with my colleagues. Why should that be?

That is just what we were coming to. (But we would point out that contacts can occur in many ways; talking is just one.)

How can colleagues enhance performance? Well, one way, of course, is by providing new ideas—jostling a man out of his old ways of thinking about things. Several of the notions described in Chapter 4 on Diversity fit in here also. But colleagues may do much more. Sometimes a colleague may know something another man needs to know: "Hook it to the red terminal and wait ten minutes," or "Go see Fred, he knows all about it." And, of course, important coordination may occur: "Why not ask Ruth to run it for you; I'm not keeping her too busy right now."

Then there is the possibility of a colleague catching an error which the man himself is too engrossed to see: "You're crazy, Joe, the company couldn't possibly afford to produce it." Sometimes knowing that even one other person thinks a problem is worth working on may be all it takes to keep a man going in a new area: "Gee, Bill, it would be great if you could solve that one!"

Still another way colleague contacts may help a person is in keeping him on his toes—simple things, like putting in a good day's work, or running a test the way it should be done, or providing some friendly (but nevertheless real) competition for promotion or recognition. Thus there may be many ways you could benefit by talking shop with your colleagues. Of course they won't all happen at once and some may not be relevant for you.

Many of these benefits seem as relevant for the scientist in the autonomous lab, or the highly motivated man, or the man who is not eager to work with others, as for anybody else. So perhaps it was not too surprising that colleague contacts helped the performance of those men also.

In short, it may be a mistake to think of contacts with colleagues as providing only intellectual stimulation and new ideas. There may be a lot of error catching, coordination, and maybe even some needed relaxing ("Come on, John, you can't win them all; let's get a Coke").

> I recall that you mentioned that the highest-performing scientists had far more than average contact. This suggests that many of my men (who seem pretty much like the ones responding to your questionnaire) should be having more contact with their colleagues. Have you any suggestions as to what I could do?

Our data suggest several specific possibilities.

Contacts were especially useful if originated by the persons concerned (the man or his colleagues), but even meetings originated by the organization seemed to help some. Furthermore, we found that frequent contacts with many colleagues seemed more beneficial than frequent contacts with just a few colleagues. Similarly, having many colleagues both inside and outside one's own group seemed better than having many colleagues in one place and just a few in the other. So anything you can do to promote these forms of contact should be in the right direction. Set up teams, committees, evaluation groups, maybe lunch gatherings, but keep the situation loose. After all, the goal is good communication between individual men, not a complicated or rigid set of formal meetings.

One important thing that you can do is to make sure that men working in related areas are *aware* of each other's activities, interests, and problems. If this condition is met, your men can themselves seek the contacts which promise to be useful.

It may also be fruitful to take a more general view of your lab. What *opportunities* are there for good communication? What are the *risks?* Is Scientist A likely to lose credit for a bright idea if he tries it out on Scientist B? Or is he likely to be laughed at if what he thought was a bright idea turns out to be wrong? A tradition of mutual helpfulness and support may promote effective communication.

4

DIVERSITY[1]

In Both Research and Development, the More Effective Men Undertook Several Specialties or Technical Functions.

Ours has been called an age of specialization. What implications does this have for division of labor within an R & D laboratory? Should one scientist concentrate exclusively on research and another on development, or should each scientist keep one foot in each territory? Should he strive for "depth" within a few problem areas, or for "breadth" in several areas? Is either of these approaches better in certain kinds of laboratories? Is either of them to be preferred at certain stages in the individual's career?

There are other related questions. How much of his working day should the scientist or engineer spend on strictly technical tasks, and how much (if any) on administration, teaching, or communication? For that matter, is there an optimal total length to his working day? How many projects can a technical man work on profitably at one time?

This chapter reports several lines of analysis, all directed toward the question: What is the most fruitful way in which an R & D man can spend his time? We did not, of course, simply ask the scientists what they *preferred* in terms of depth or breadth for tasks, or amount of time they wanted to spend in research or teaching or administration. Rather we asked how they were actually spending their time, and we then examined how well scientists performed who spent their time in different ways.

As we explored the data, our conviction mounted that *diversity* was essential. In a number of ways, scientists who were highly specialized or one-sided were less effective than those with several interests.

[1]These analyses were largely supported by the U. S. Army Research Office (Durham). This chapter expands an article by D. C. Pelz and F. M. Andrews, "Diversity in Research," *International Science and Technology*, July 1964, pp. 21–36. Detailed data on percent of time spent in technical work appeared in F. M. Andrews, "Scientific Performance as Related to Time Spent on Technical Work, Teaching, or Administration," *Administrative Science Quarterly*, 1964, vol. 9, pp. 182–93.

We invite the reader to retrace in this chapter the steps by which we found ourselves reaching this conclusion; to provide a sense of direction, we offer the following road-map. Taking one path, Andrews asked how much time it was profitable for the technical man to spend on technical work as such, and discovered that those spending full time were less effective than those giving only three-quarters time. Pelz meanwhile examined the number of "areas of specialization" the scientist felt he possessed, and observed that the more of these the better. Following this hint, Pelz looked at various R & D functions which the individual might engage in, such as basic research, invention, etc.; again he generally observed that the more of these the better, in all types of laboratories. The chapter then reports data on how much time scientists or engineers spent on research as such, or on development or technical services as such, for best results; it then examines how the foregoing patterns varied for people at different career levels, and concludes with some practical implications.

Time in Research versus Administration versus Teaching

The questionnaire had asked how the scientist allocates his time to "technical work" (that is, research or development), or to "administration," or to "teaching or training" (see Question 5 box). It seemed plausible that some degree of teaching might enhance scientific achievement. Here

Question 5. Of your total work time, about what proportion do you normally spend on the following types of activities? (If it fluctuates, strike an average.) Enter nearest 5–10%.

	percent of time
A. Teaching and training	____%
B. Technical work, other than teaching	____%
[This was further broken down into:	
My own work	
Supervising technical work of others	
Collaborating with colleagues	
Consultation and technical service]	
C. Administrative and other nontechnical work	____%
[Including:	
Internal administration, expediting services	
Communicating with higher ups	
Communicating with outside groups or clients]	______
Total for all work (should add to 100%)	(____%)

we were following a trail blazed several years ago when Leo Meltzer, a colleague then at the Survey Research Center, was analyzing data from a nationwide survey of physiologists by himself and Seymour Lieberman.[2] Meltzer found that full-time researchers published less than those spending three-quarters time, and this held true within each academic rank. (Many of his respondents were university professors.)

Although some time spent in teaching might enhance scientific achievement, it seemed equally plausible that diversion into administration would hinder. But just how much administrative distraction could the scientist tolerate, we wondered, before his productivity suffered?

We had expected to find that the more time in technical work the better (except, perhaps, for those in university departments). What we weren't sure of was what amount of administration was clearly harmful.

We were therefore surprised to get the results plotted in Charts 1-A through 1-D. Chart 1-A shows data for the 100 Ph.D's who completed the long form questionnaire in development-oriented laboratories. It appeared that those Ph.D's who spent essentially full time in technical work, and therefore little if any time on administration, were *less* effective than those who spent about three-quarters time. Since little formal teaching was done in these labs, clearly the bulk of time not spent on technical work was being given to administrative tasks.

We see a similar trend in Chart 1-B for the 75 long-form Ph.D's in research-oriented labs. Again, those spending only half or three-quarters of their time on strictly research activities were more productive scientifically than those who spent full time. Meltzer's preliminary findings on physiologists were confirmed. Here, of course, the time spent on non-research activity might be given either to teaching or to administration or to both. We shall say a word later about the distinction between these two.

Chart 1-C shows a similar but modified picture for the 200 engineers. Again performance was low among those who spent full time, and optimal for those who spent only *half* time in strictly technical activities.

Chart 1-D for "assistant scientists" (in either research or development laboratories dominated by Ph.D's) shows a picture very similar to that for the Ph.D scientists. Among those spending full time in research, performance was lower than for those putting in three-quarters time. (Very few, as we might expect, spent less than three-quarters time on technical work. When they did, their scientific contribution was extremely

[2]Unpublished data. For other reports on this study see Leo Meltzer and James Salter, "Organizational Structure and Performance and Job Satisfaction of Scientists," *American Sociological Review*, 1962, vol. 27, pp. 351–62; R. W. Gerard, *Mirror to Physiology: A Self Survey of Physiological Science*, American Physiological Society, Washington, D.C., 1958.

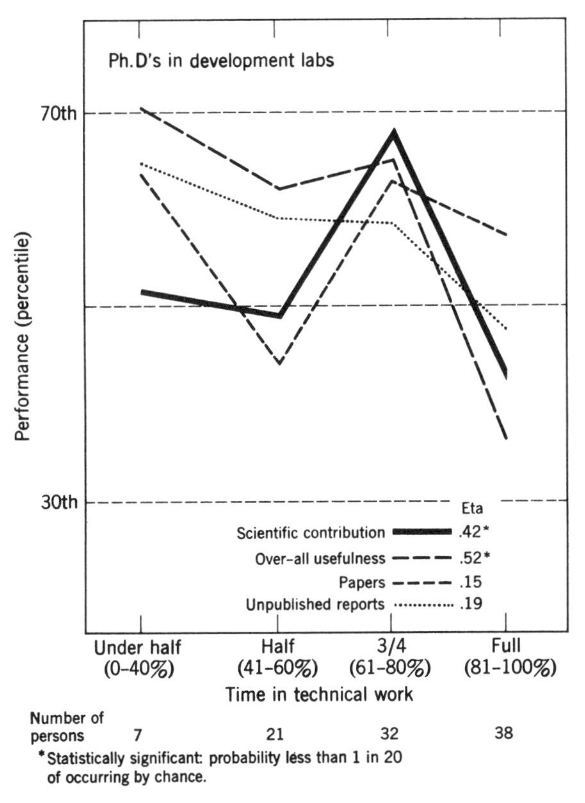

Chart 1-A. *For these Ph.D's in labs where executives valued development of better products more than scientific publication, performance was lower by most standards when the scientist spent all of his working time on technical activities, compared to three-quarters time.*

low, although output of reports was high. Perhaps they were administering a production line for grinding out technical memos.)

We haven't plotted the data for group E—the few nondoctoral scientists in research-oriented laboratories of the government not dominated by Ph.D's. Scientific contribution was higher when the person spent half time or less in technical work (although output of reports and papers fluctuated).

Some Further Explorations. What accounted for these trends? Were the people spending full time in technical work simply junior members who had not yet produced much? Were those spending three-quarters or half time senior members or supervisors who not only produced more, but also had administrative chores thrust on them? We checked this idea carefully, and found that it did not explain away the results. Among

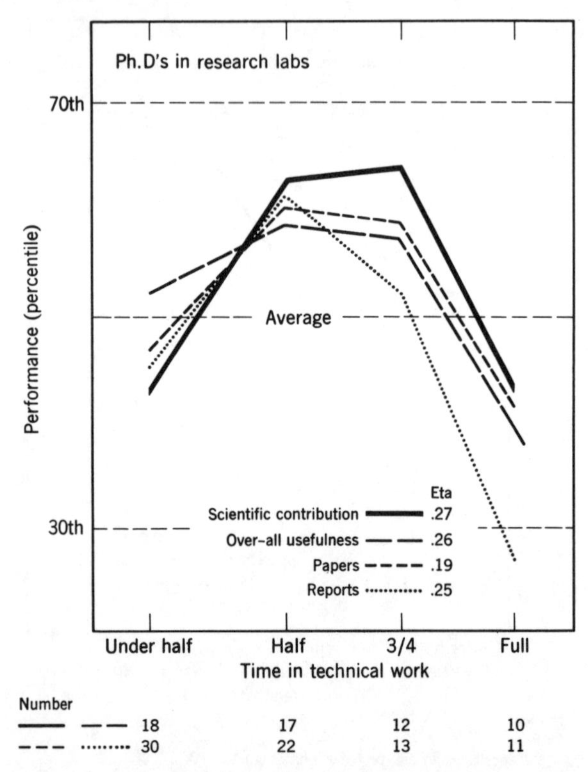

Chart 1-B. *For these Ph.D's in labs where executives placed higher value on scientific publication than on product development, all four performance measures stood higher when the scientist spent only half to three-quarters of his working time on strictly technical activities.*

mature nonsupervisory scientists and engineers at each career level, the effect held up sharply. Full-time researchers were simply less effective.

Another hunch was that those who were *teaching* or training in their spare time, rather than administering, were being stimulated. So we did some further breakdowns to see whether scientists did better if they spent their nontechnical time teaching instead of administering. In development labs, of course, little formal teaching was done, so we could only compare those who did some teaching with those who did none.

Among engineers, some time in teaching did indeed prove a slight advantage (although among Ph.D's in development there was no difference). But for the Ph.D's in research labs a most curious result appeared. Here we could compare scientists who spent their nonresearch time mainly in teaching with those who spent it mainly in administration. And among nonsupervisors in research, the part-time administrators were judged to be better scientists! (Among supervisors it made no difference.)

Was it possible that some administrative activity actually stimulated scientific achievement? In university settings (one can speculate), "administration" might be more than minor routine. It might include the seeking of research funds from sponsors, or serving with colleagues on admissions or salary committees where the goals of the department are debated.

We saw in Chapter 2 that staff members who allowed two or three "decision-making sources" to influence their technical goals were more effective. Possibly "administration" in mild doses provides a channel for these influences. It is not a mere exercise in pushing papers. Rather, it may expose the man to fruitful ideas outside his immediate problem area.

Whatever the interpretation, it can be said that the man who spent some time in administration or teaching had a certain diversity in his work.

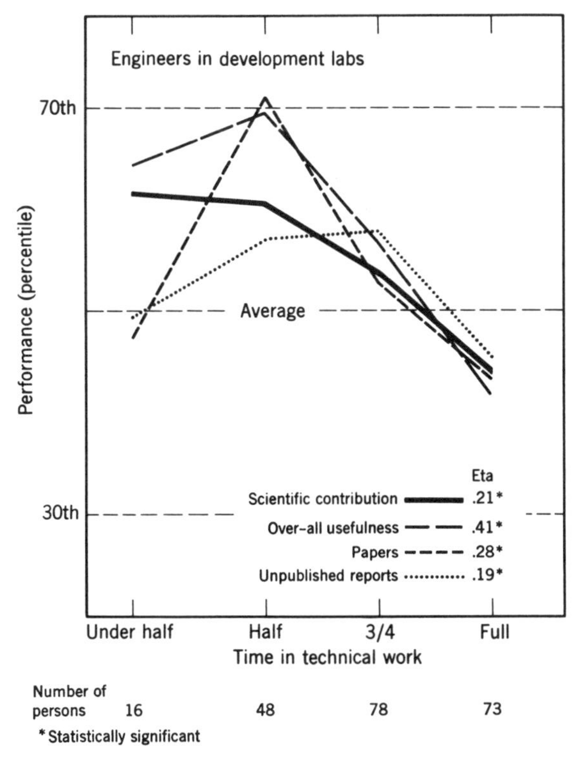

Chart 1-C. *Among these engineers in development-oriented labs, all four measures of performance were lower among those spending full time in technical work than those spending three-quarters or half time. Contribution and usefulness continued high among those with less than half time.*

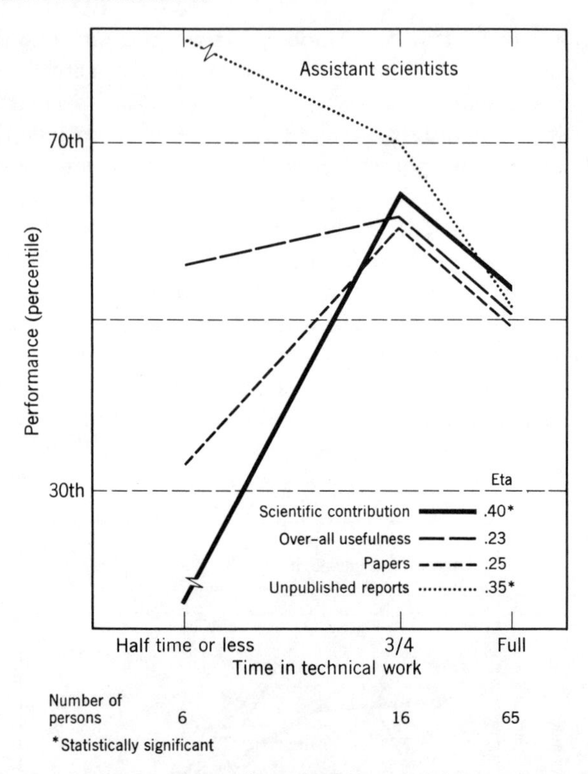

Chart 1-D. *Among "assistant scientists" in Ph.D-dominated laboratories, performance on all measures was again lower for those who spent full time on technical activities, compared to those spending three-quarters. The very few who spent half time or less produced many reports, of low scientific value.*

Number of Areas of Specialization

Another indicator of diversity was simply the *number* of research specialties a man felt he had. The following box shows the questionnaire item used to determine this, and the data are plotted in Charts 2-A through 2-D.

> *Question 2.* Within a discipline or field an individual may develop an area of specialization—a content area about which he knows a great deal. If you have such areas of specialization, please list them below in order of proficiency. (Limit to areas in which currently active.)
>
> A. Most proficient ____________
> B. Next most proficient ____________
> C. Third most proficient ____________
> Check here if no particular area of specialization ____

Chart 2-A shows the four measures of performance for Ph.D scientists in development-oriented laboratories. The more areas of specialization an individual reported, the higher was his scientific performance (with minor exceptions; more papers were published by those having two areas rather than three, and a few Ph.D's were moderately useful without any specialized areas).

Chart 2-B shows a similar trend for the Ph.D's in research-oriented labs. Except for a very few persons who reported no areas of specialization, we see the same trend: the more areas, the higher were several measures of performance.

The engineers in development laboratories are shown in Chart 2-C. Those with two or three areas of specialization again performed better than those with one.

Chart 2-D for "assistant scientists" indicates that scientific contribution and usefulness rose mildly but steadily with increasing number of areas; and, in general, so did publication of papers.

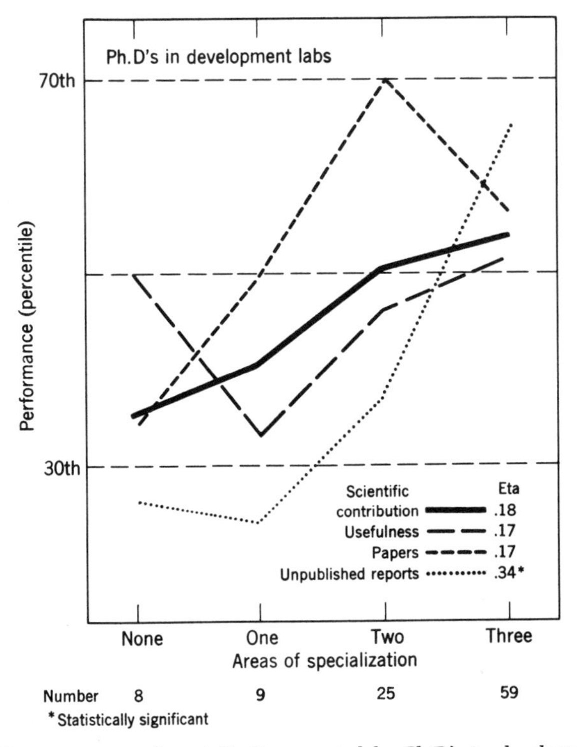

Chart 2-A. *The more areas of specialization reported by Ph.D's in development labs, the higher their performance tended to be (although more papers were published by those having two areas rather than three, and a few were moderately useful without any specialized areas.)*

For the final group of nondoctoral scientists in research labs (data not shown), a similar mild trend was noted: contribution, usefulness, and reports rose as number of areas increased from one to three (papers peaked at two areas).

Before we go further, the reader is sure to have questions. Might it not be, for example, that as the scientist matures he acquires more specialized areas, and also produces more? Might the relationship not be due simply to a rising level of competence and experience?

The data have already been adjusted, of course, to rule out length of time since the degree, and time in the organization. In addition we re-examined the results separately within each career level, and will show some details a little later. Here we can say simply that such an analysis did not wipe out the results; if anything, the trends at certain levels were strengthened.

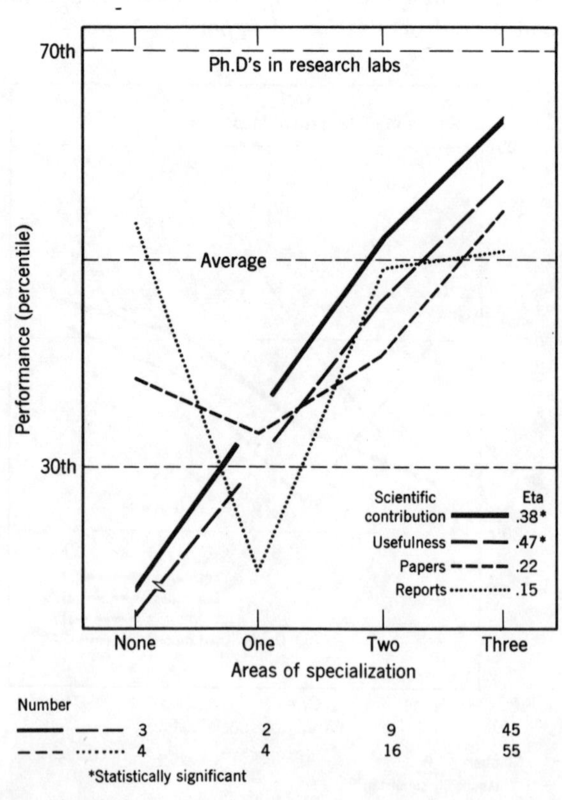

Chart 2-B. *The more areas of specialization among these research Ph.D's, the higher their performance on most measures; a few who named no specialty, however, wrote many reports of little scientific value.*

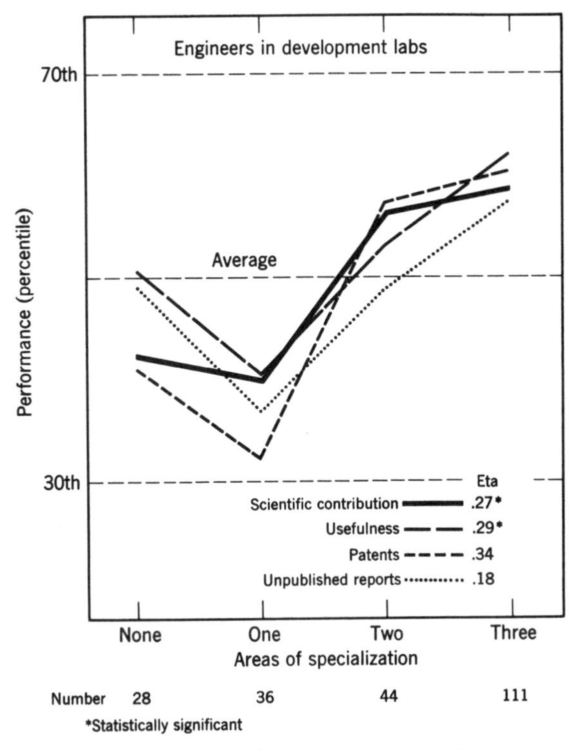

Chart 2-C. *Engineers performed better on all measures when they had two or three areas of specialization, although some with no specialty were moderately useful.*

As we pondered these results, we suspected that the phenomenon which accounted for them might be the same as that which led scientists spending less than full time on their technical work to perform well. *Diversity* in the R & D tasks of the scientist or engineer appeared to accompany high performance. But we wanted to check this further with other evidence.

Number of R & D Functions

In a third analysis we explored the question: How should the scientist allocate his time among various R & D functions such as basic research, applied research, development or technical services? To get an idea of how he currently divided his technical work, we asked the question shown on page 65.

Category IA for "general knowledge" was intended to represent basic research without using that ambiguous term, whereas category IB for "specific knowledge for solution of particular problems" was intended to

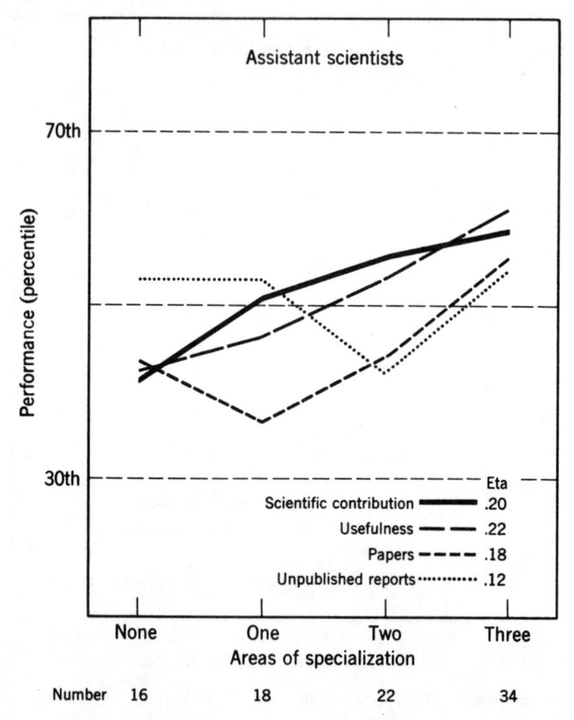

Chart 2-D. *In Ph.D-dominated laboratories, the scientific contribution, usefulness, and publication of these assistant scientists rose mildly with increasing number of areas, but report writing showed no trend.*

cover applied research. Note the distinction between IIC on "improvement of existing products or processes," and IID on "invention" of new ones—the former presumably more routine, the latter more creative.

Here we were following a clue suggested by Herbert A. Shepard's earlier study of twenty-one industrial laboratories.[3] One of his tentative findings was that in labs oriented toward pure research, teams doing applied work or development were rated lower by executives, whereas in relatively applied laboratories the reverse happened: teams doing more basic research were down-graded. It seemed reasonable, then, that in our research-oriented labs we might find that the more time scientists spent on research (to the exclusion of development) the better; and, conversely, in development labs we might find that the more time they spent on development the better.

[3]Unpublished data. See "Field Studies in the Organization and Management of Research," progress report, Sloan Research Fund Project #504, Massachusetts Institute of Technology, February 1954.

Question 14. The technical work of scientists and engineers covers a broad range of activities. At present, about what percent of your time (other than teaching) is directed toward each of the following purposes (either your own work, or work for which you are responsible)? Enter nearest 5–10%. FILL ALL SPACES.

	Percent of work
I. *Research* (discovery of new knowledge, either basic or applied)	____%
A. General knowledge relevant to a broad class of problems	(____%)
B. Specific knowledge for solution of particular problems	(____%)
II. *Development and invention* (design of particular products or processes; translating knowledge into useful form)	____%
C. Improvement of existing products or processes	(____%)
D. Invention of new products or processes	(____%)
III. Technical services to help other people or groups [including testing, analysis by standardized techniques, consultation, trouble-shooting]	____%
IV. Other purposes (specify) ____________	____%
Total time should add to 100%	(____%)

What we actually found will be shown a little later. First, let us see what happened when we followed up a hunch based on the previous results. By now we had discovered the pattern of areas of specialization. We had also seen (Chapter 2) that the more "decision-making sources" the person involved in deciding his goals, the better. Would something similar emerge if we looked at the number of different R & D functions carried on by the scientist?

We recorded for each scientist the number of R & D functions to which he devoted at least a slight amount of time (6% or more). A clear picture appeared, as shown in Charts 3-A through 3-D.

For Ph.D's in development labs (Chart 3-A) and for Ph.D's in research labs (Chart 3-B), the general tendency was clear: most measures of performance increased as the scientist engaged in more kinds of R & D functions, up to a maximum at four functions.

In research labs, though, the curves dropped at five functions. Apparently there is profit in diversification, but also danger. Carry it too

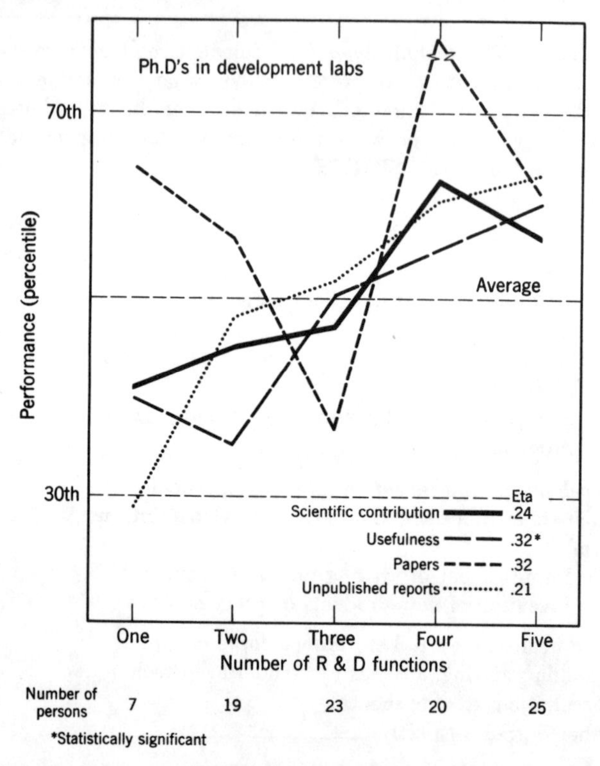

Chart 3-A. *In general, the more different R & D functions these development Ph.D's engaged in, the higher their performance, although a few published many papers (of minor scientific value) when concentrating on a single function.*

far and one becomes superficial. At the other extreme, a person tended to publish more papers if he specialized in a single R & D function, but his scientific contribution and report output were low.

For engineers (Chart 3-C), performance was above average when the person engaged in five different activities, below average when he had only one.

For the assistant scientists (Chart 3-D), performance tended to peak at three or four functions, and to drop at five. The parallels with data for both Ph.D groups were strong.

For the remaining small group of non-Ph.D's in government research labs, however, the picture was unclear (data not shown). A generally rising trend appeared for reports (maximum at five functions), but for contribution and published papers, the curves resembled a W, with maximum at either one or five functions. Perhaps this group found it necessary to pursue either the path of "specialist" or that of "generalist."

Again the reader might wonder whether experience alone could explain these results. Were senior scientists or supervisors not only performing more different kinds of R & D functions, but also producing more? As we shall see later, restricting the analysis to single career stages did not eliminate the effect, but in some cases sharpened it.

Once again, the more productive scientists were those with moderate diversity (several R & D functions) in the content of their work.

Optimal Allocation of Time among Various Functions

What, then, is the optimal time for an individual to spend in each kind of R & D function, for each kind of laboratory? Under three headings we shall consider results for time in research as such, development as such, and technical service.

Time in Research as Such. First, the next four charts show how per-

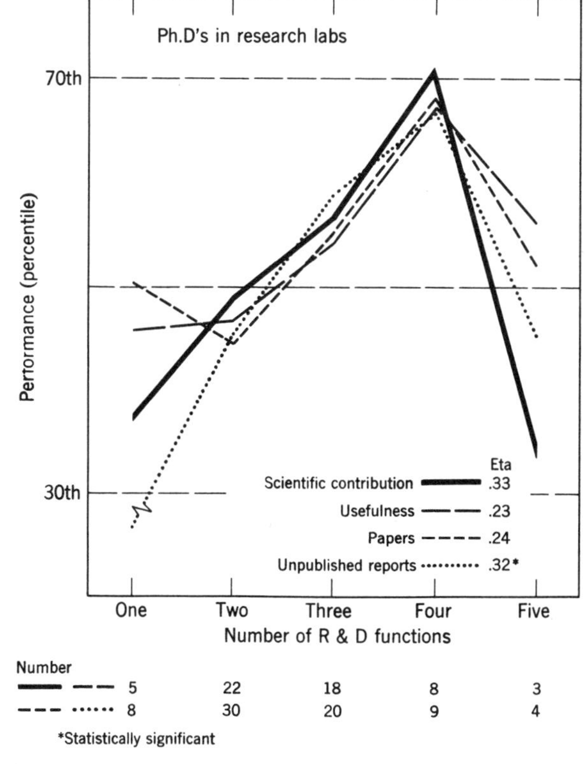

Chart 3-B. *The more R & D functions up to four performed by these research Ph.D's, the better their performance. The few who attempted all five functions may have spread themselves too thin.*

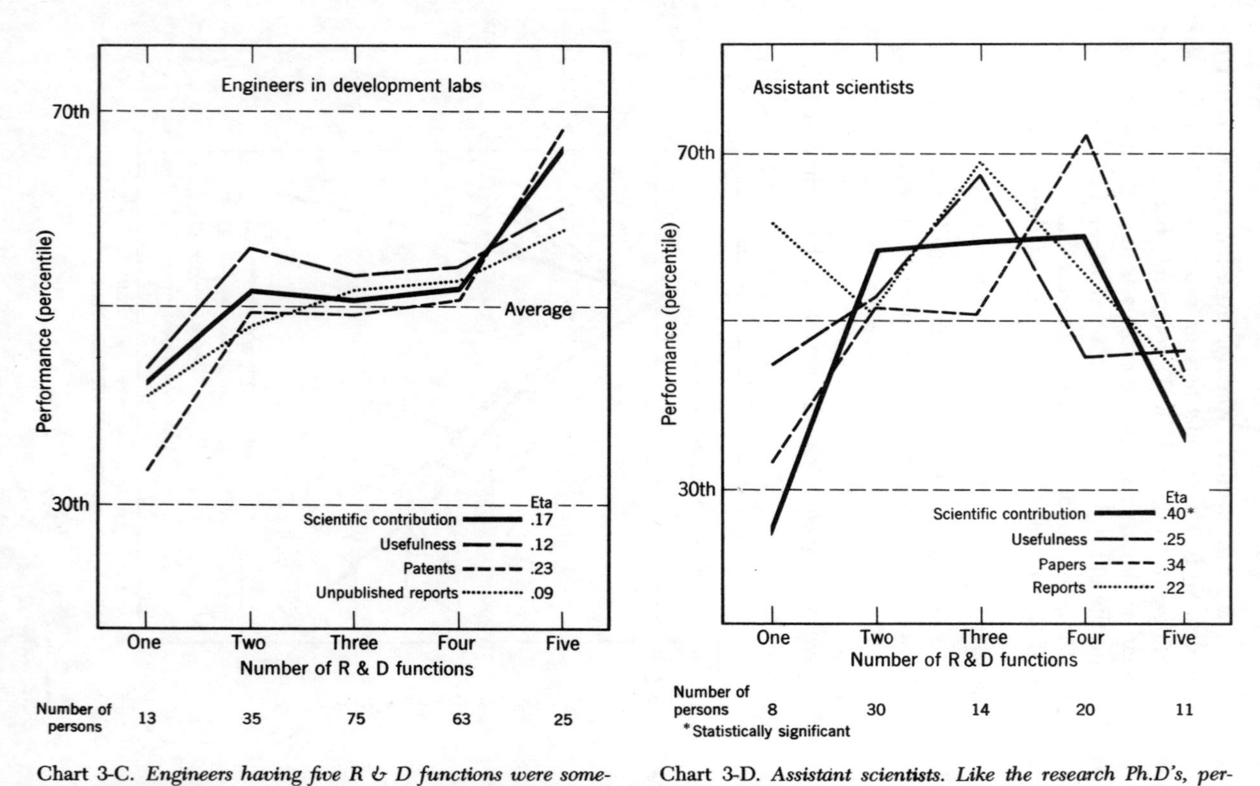

Chart 3-C. *Engineers having five R & D functions were somewhat above average, whereas those having a single function were a little below.*

Chart 3-D. *Assistant scientists. Like the research Ph.D's, performance dropped if they attempted all five.*

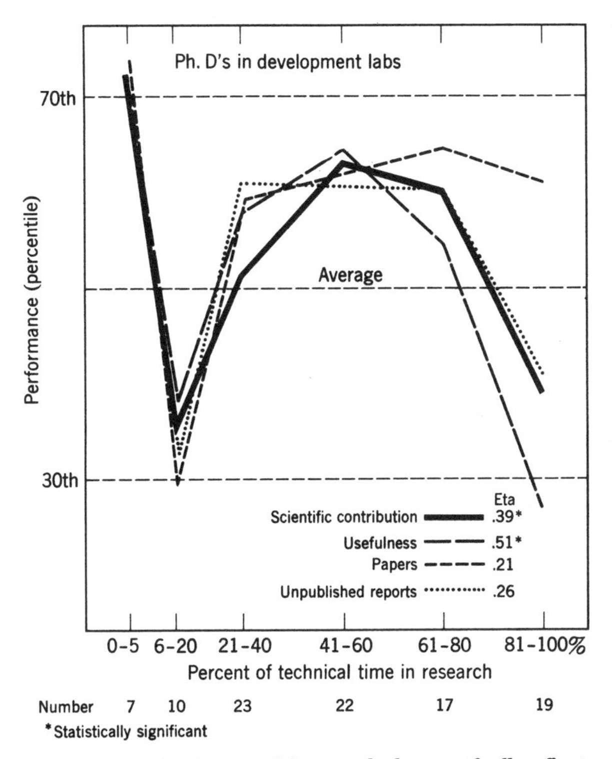

Chart 4-A. *Most Ph.D's in development labs were both scientifically effective and useful when spending about half of their technical time in research as such, although a few excelled when avoiding research entirely. Those who spent full time in research continued to publish, but their papers had minor scientific value and usefulness.*

formance varied according to the proportion of time spent in research ("discovery of new knowledge, either basic or applied")—Category I in the box. For simplicity we have combined time spent in pursuit of "general knowledge" and "specific knowledge for particular problems."

Among Ph.D's in development (Chart 4-A), a few individuals were highly productive when they spent no time in research. Among the remainder, however, the optimal proportion of time was about one-half to three-quarters. Less effective were scientists who devoted either very little time (one-eighth) or a great deal of time (more than four-fifths) to research.

A similar trend appeared among Ph.D's in research labs (Chart 4-B). Best performers were those spending one-half to three-quarters time in research as such.

For engineers, however, (Chart 4-C), the pattern was flat and ambiguous: engineers who spent most of their time on research were slightly better on scientific contribution, but slightly below par on patents and

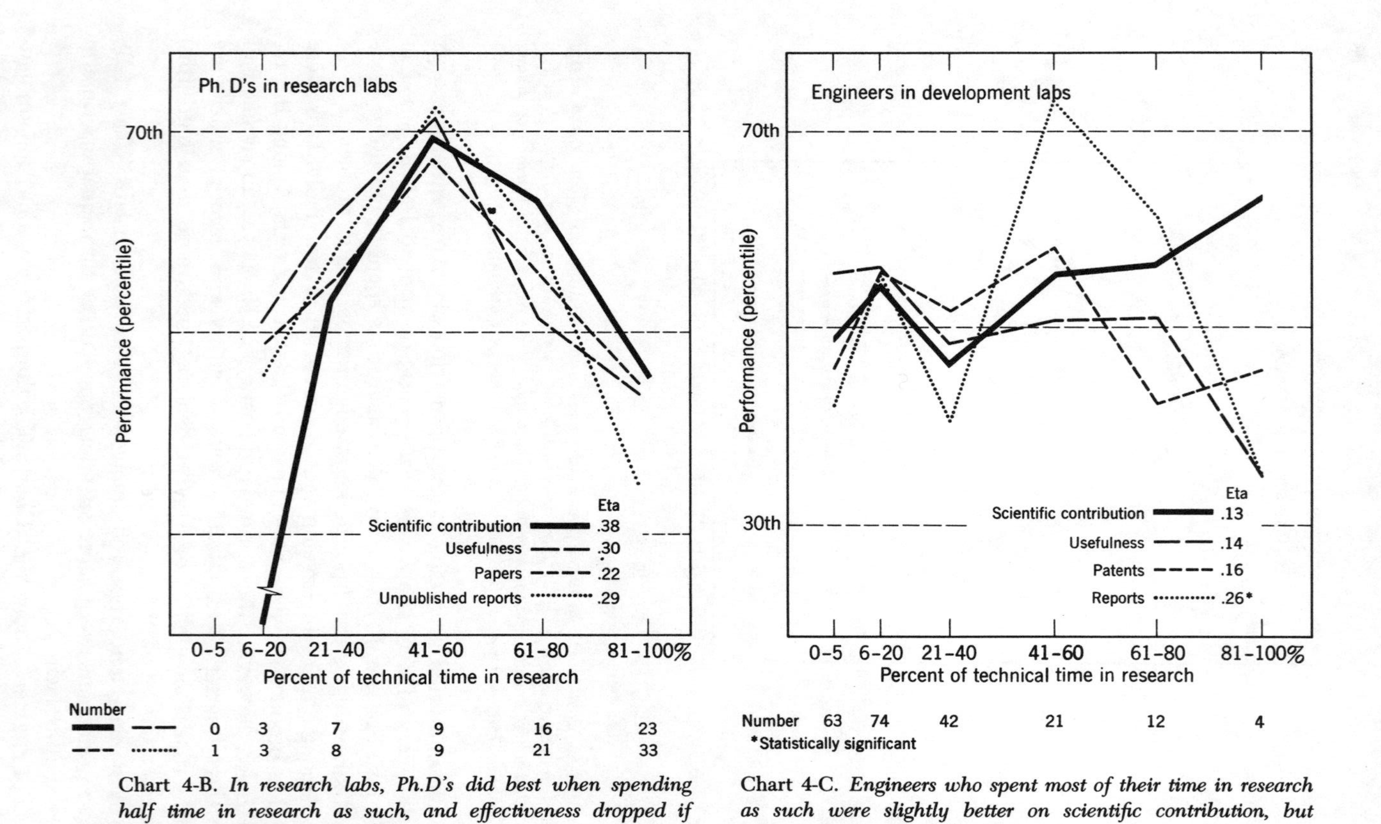

Chart 4-B. *In research labs, Ph.D's did best when spending half time in research as such, and effectiveness dropped if they gave full time to research.*

Chart 4-C. *Engineers who spent most of their time in research as such were slightly better on scientific contribution, but average or below on other measures.*

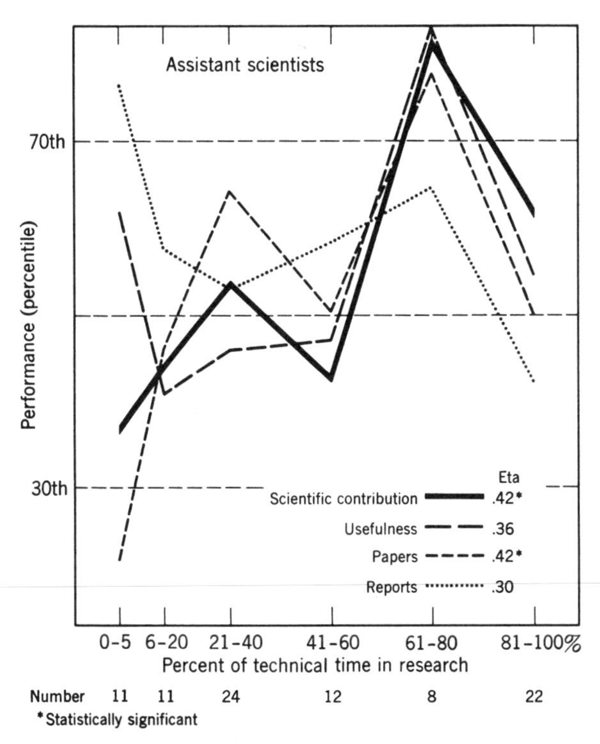

Chart 4-D. *An unsteady but rising trend appeared in the data for assistant scientists, some of whom were outstanding when they spent three-quarters time in research as such. Like Ph.D's in development, however, a few wrote many useful reports with no time spent in research.*

reports. For assistant scientists (Chart 4-D), the pattern resembled that for the Ph.D's—maximum performance when the person spent about three-quarters time in research as such. Among the final group of non-Ph.D scientists in research, the curves were flat.

Time in Development. The next charts show how performance varied with proportion of time spent in development (either improvement of existing products or processes, or invention of new ones)—Category II in the box.

For the bulk of Ph.D's in development (Chart 5-A), best performance occurred when they spent between one-eighth and one-half of their time in development activities.

Even among Ph.D's in research (data not plotted), we were surprised to find that performance was somewhat better if they spent an eighth to a third time on development, rather than *no* time or considerable time. (As one would expect, very few spent more than half time.)

Chart 5-C for engineers shows a modification of the same trend. Some

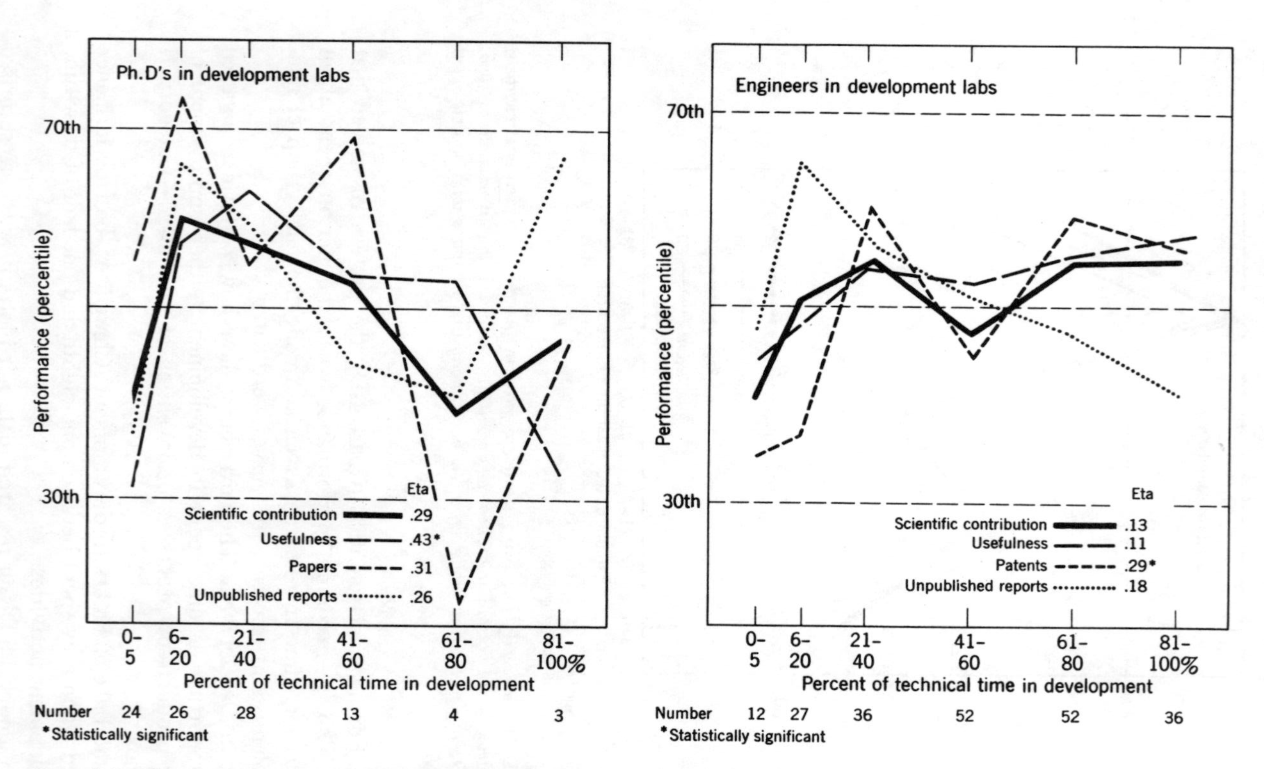

Chart 5-A. *These scientists generally contributed and published more when they spent one-eighth to one-half of their technical work time in development activities. A few who spent full time wrote many reports, of dubious usefulness.*

Chart 5-C. *Two humps appeared in this chart for engineers. Performance was a little better if the person spent either one-quarter time in development, or three-quarters time or more, rather than half.*

of these men tended to be productive if they spent about one-third time in development. But others contributed or patented more if they spent three-quarters to full time.

A similar pattern appeared among assistant scientists (not plotted). Their contribution was higher if they spent either modest time in development (one-eighth to one-third) or considerable time (three-quarters or more), but not simply half time. (For nondoctoral scientists, the trends were again ambiguous.)

There was a hint here that some nondoctoral engineers and assistant scientists could afford to specialize more than the Ph.D's, giving the bulk of their time either to research *or* to development.

Technical Services. In addition to "research" and "development," the main remaining category was "technical services." We shall summarize these results without plotting. Few Ph.D's spent more than a third time in technical services, but even here those who gave *some* time to technical services were judged to contribute slightly more (both in research and in development labs) than those who gave no time.

Among non-Ph.D's, the rating of scientific contribution generally declined as more time was spent in technical services (although the number of unpublished reports was high for about half time). Even so, those who gave a little time (about an eighth) to technical services were slightly more effective than those who gave none.

To Sum Up. Again we see that both in research and in development labs the more effective scientists and engineers (with some exceptions) did not concentrate exclusively on research, or on development, or on technical services. But neither did the more effective people totally avoid these categories. For all three kinds of functions, some time was better than none.

There is, of course, a link here with the preceding analysis (Charts 3-A through 3-D) on *number* of R & D functions. If the better scientists were those who spent close to half time on research, nearly as much on development, and a little on technical services, it followed necessarily that the better ones would turn out to be spending time in *several* R & D functions rather than concentrating in a few.

Career Levels

If it is true that scientists who diversify are more effective, does this hold true at all career levels? Are there certain stages at which the scientist ought to dig deep, or others at which he ought to broaden out?

Previously we had defined a set of career levels which would have analogous meaning across different kinds of laboratories. Each laboratory, of course, had its own ladder of job grades. By looking at job titles and salaries found at each grade, and typical characteristics of their occupants (education, length of service, number of subordinates, if any), we were able to establish rough parallels among job ladders in industry, government, and university.

We then defined three career levels for doctoral scientists and four for nondoctorals, mainly on the basis of job grade, but also setting limits of work experience. Details are given in Appendix E.

Level 4 or "supervisor" included members of industrial or government labs who headed a section or larger unit; and in the university, department chairman and other full professors. They had to have at least six years of experience since the Ph.D or B.S.

Level 3 or "senior" included mature nonsupervisory investigators with substantial responsibility who were usually in their late thirties (equivalent to government grades GS-12 and up, or university titles of associate or assistant professor).

Level 2 or "junior" included younger scientists or engineers. Some might be starting their first job, but typically they were in their early thirties, and about eight years beyond the B.S. (equivalent of government grades GS-9 or 11, or university title of instructor).

Level 1, which we have called "apprentice," was limited to nondoctorals at the bottom of the professional ladder (equivalent of government grades GS-5 and 7). Typically they were recent college graduates under 30, performing routine tasks, but a few were highly experienced non-B.S. technicians who had risen to professional status.

Within each of these career levels, we repeated our analyses. In Chart 6-A we see results for Ph.D's on number of areas of specialization. (Since the trends were similar in research and in development, we have combined the two groups here.) At the right, among scientists having at least three specialties, there was not a sharp difference among those at each career level. But as we move to the left on the chart, we see that the difference among the levels increased. Supervisors did not seem to be especially disadvantaged, but possession of fewer specialties was of greatest handicap to younger people. One gets the impression that lack of breadth was a marked drawback for the younger man, but not for those at older levels.

Somewhat the same picture is given by Chart 6-C for engineers. The apprentices stood apart from the others; specialization did not matter much, or may have helped. Breadth was especially important among junior and senior (nonsupervisory) engineers. Having a single specialty was most harmful to the juniors, moderately harmful to seniors, but possibly beneficial for the supervisors.

Another interesting picture appeared for number of R & D functions. Since the pattern was clearest among engineers, we show their data in

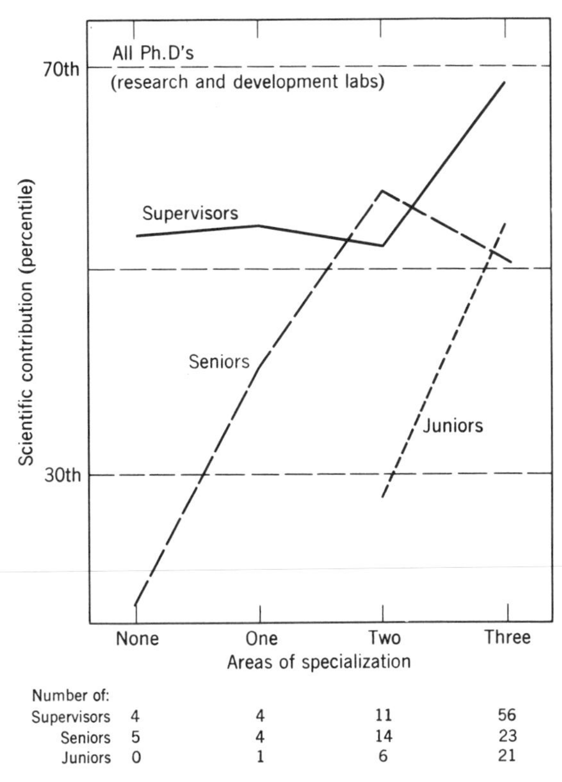

Number of:	None	One	Two	Three
Supervisors	4	4	11	56
Seniors	5	4	14	23
Juniors	0	1	6	21

Chart 6-A. *Among all Ph.D's in research and development labs combined (the curves for both were similar), lack of breadth was of most handicap to junior investigators, next for seniors, and least inhibiting for supervisors.*

Chart 7. Apprentices, as well as junior engineers, performed better when they limited their attention to two or three R & D functions. But there were some juniors who did well when covering five functions.

Among senior engineers, this bimodal effect was heightened. Best records were made either by those who concentrated on one major function, or those who diversified over four or five. We can almost see these men taking one of two routes to success: either that of a narrow specialist, or that of the broad generalist. For those who had become supervisors, however, all paths were equally successful—specialist, generalist, and mixed.

Number of Projects and Working Hours

At the beginning of the chapter we raised some other questions concerning number of projects and length of the working day.

For most of the groups, we found a very slight tendency for scientists

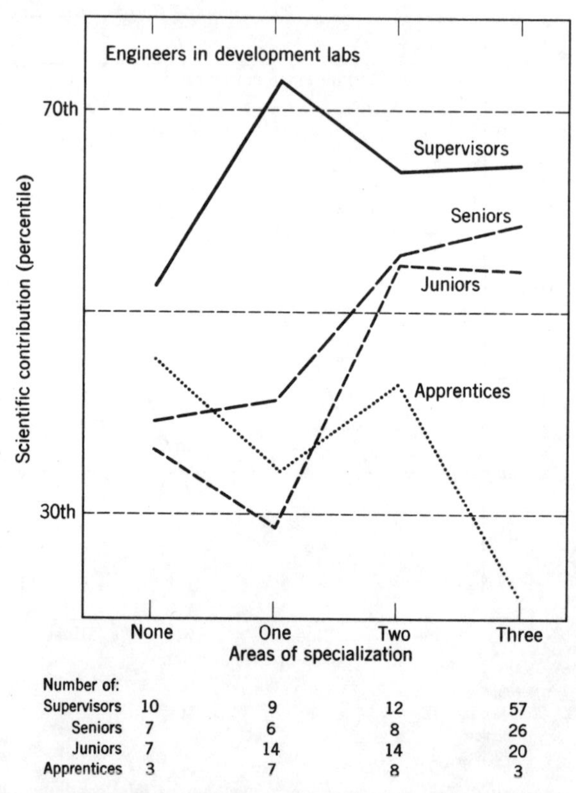

Number of:	None	One	Two	Three
Supervisors	10	9	12	57
Seniors	7	6	8	26
Juniors	7	14	14	20
Apprentices	3	7	8	3

Chart 6-C. *Among engineers, apprentices were not harmed by lack of specialties, but juniors and seniors performed better if they had two or three. Having a single specialty was possibly beneficial for supervisors.*

to perform better if they worked on two or three projects, rather than one or none. For Ph.D's in research, however, there was no relationship. The conclusion seemed to be that the scientist performed best when he utilized two or three different skills, and faced both scientific and applied problems in his work. It did not matter whether his work was organized around one or several projects, so long as it called for a mix of activities.

In almost all groups, the scientists performed less well if they worked only a standard eight-hour day or less. But it did not follow that the longer the hours the better the job done. Generally a nine- or ten-hour day, on the average, gave better results than an 11-hour day (but "assistant scientists" wrote more reports under this condition).

Once again there was a hint that excessive concentration was not healthy. All work and no diversity was making Jack a dull scientist.

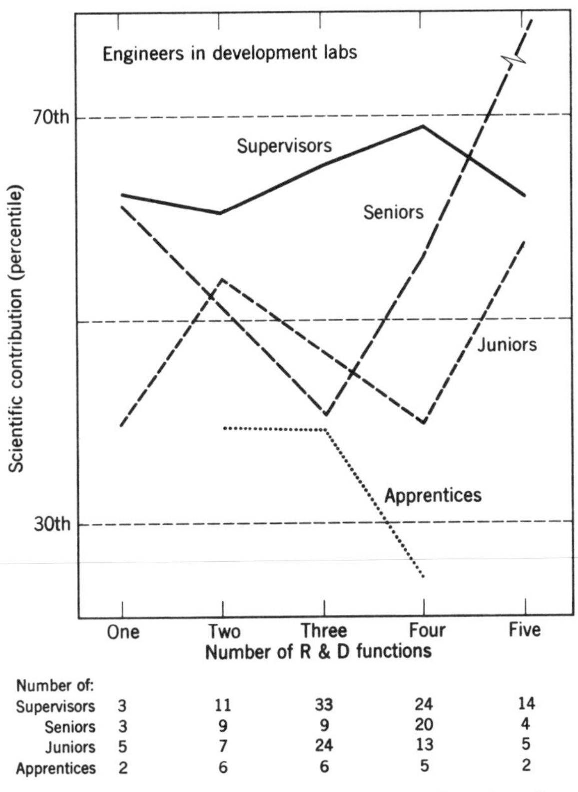

Number of:	One	Two	Three	Four	Five
Supervisors	3	11	33	24	14
Seniors	3	9	9	20	4
Juniors	5	7	24	13	5
Apprentices	2	6	6	5	2

Chart 7. *Among engineers in development labs, juniors and seniors showed a bimodal effect; either few R & D functions, or many, accompanied higher performance.*

SUMMARY AND IMPLICATIONS

At the risk of going beyond the data, let us sharpen the potential implications through the following conversation, based on typical questions from technical audiences.

What do these findings imply for administering research organizations?

In the first place, complete protection of the technical staff from administrative duties may be a mistake. We saw in Chapters 2 and 3 that a scientist may benefit if his colleagues are involved in setting his technical goals and if he maintains substantial contact with them. This chapter suggests that the scientist may also benefit from a mild exposure to administrative tasks. For example, he might participate in review conferences where each staff member periodically outlines his future directions. Serving on such review committees need not be limited to a few supervisors. If this responsibility were spread around among all the professional staff, it would add zest and breadth to the research climate.

Second, younger scientists should be stimulated to develop several specialized areas.

Finally, the research director should not assign some groups exclusively to research, others exclusively to development. Rather, he should encourage groups and individuals to tackle both "pure" and "applied" problems.

> I know you've not claimed outright that diversity stimulates creativity, but you've hinted at it pretty broadly. To me, the opposite makes more sense. Effective scientists are just naturally curious people, always interested in what other people are doing. The bright, productive ones are noticed and consulted about all sorts of problems, and that's why they develop more specialties and get into different kinds of R & D activities.

Maybe you're right. With survey data like ours you can never prove one causal hypothesis over another. But why do you insist that it's always the individual and not the environment that makes the difference? On this premise, the man is either inherently capable or he's not, and it's the outstanding ones who get drawn into the environments we find linked with achievement.

But maybe something else is happening. Maybe the best people are deliberately exposing themselves to the kinds of contacts that will stimulate them. If this is so, then isn't it possible for a good-but-not-outstanding scientist to learn a trick or two? If he put himself into the kinds of situations that typify the best scientists, or if his manager saw that this happened, might not the next-best man actually boost his own achievement?

> Suppose you're right. What does it mean for me? Say that I'm a research manager. Do I tell my section heads to assign each investigator to three more projects? You don't expect me to believe I can spoon-feed diversity?

No. Such a tactic would be artificial. Note that it doesn't matter how many *projects* the man works on. The important thing is the number of *skills or specialties* he brings to bear, and you don't develop those overnight.

What you *can* do, as a manager, is this. Next time you need to probe a certain specialized area—whether with a one-week review or a one-year pilot project—*don't* give it to the man (or the group) who already knows the most about it. Pick a man (or a small group) working in a related area, and try to interest him in exploring this one. He'll dig into the field with a new zest and excitement. He may uncover some fresh ideas that experts in the area would overlook.

Now we know this is not the most efficient way to run a laboratory. And it won't do for a crash program in a crisis. But when you can, try it. In the long run, we think, you will build breadth and flexibility into your organization because you will be building diversified skills. And these will open the door for creative advances all along the line.

5

DEDICATION[1]

Several Simple Questions Showed that High-Performing Scientists and Engineers Were Deeply Involved in Their Work.

In some laboratories the air seems to hum with excitement. Investigators are absorbed in what they are doing—their individual projects or the laboratory's mission. Coffee breaks buzz with shop talk instead of baseball; 5 o'clock quitting time is the exception instead of the norm. The morale, in one sense of that overworked term, is superb.

Can such an atmosphere of excitement—call it enthusiasm, perhaps, or zest, or dedication—be measured by checklists on questionnaires? And would it relate to scientific performance? We wanted to find out.

This chapter describes our interest in measuring *intensity* of work motivation—morale in its original sense. Several questions for doing so, particularly an item on one's feeling of "involvement" in the work, showed correlations with performance. Intensity of motivation was generally stronger among people who were self-directed. The chapter closes with some practical suggestions on strengthening the sense of involvement.

Morale

The concept of "morale" is an old one. Its origin is the "esprit de corps" of military units. The will to win—to persist in the face of danger, discouragement, even defeat—has always been recognized as essential to ultimate victory.

[1] Preliminary work on this topic was reported in D. C. Pelz, "Intensity of Work Motivation, as Related to Output," Analysis Memo #2, December 1960, available in Publication #1741 from the Survey Research Center, University of Michigan. This preliminary work was supported by a pharmaceutical company. Further work is reported in D. C. Pelz, "Self-Estimates of Motivation Strength, as Indicators of Scientific Performance," Analysis Memo #17, May 1962 (Publication #1922). The latter analysis and the present chapter were supported by a grant from the U. S. Army Research Office (Durham). We are indebted to S. S. West for the early analysis on intensity of motivation.

In the 1920's and 1930's the notion of "industrial morale" became popular, and morale surveys fashionable. But in the process, the original meaning of the term was distorted. To our mind, esprit de corps is fundamentally a question of motivation—the will to survive and to win. It is sustained, true, by the fighting man's trust in his leaders and comrades, and pride in his unit; but it is not generated by contentment. Yet the industrial morale surveys asked how well satisfied workers were with their pay, their working conditions, company benefits, and the like.

When social scientists began to study how the latter kinds of satisfaction were related to productivity of industrial employees, they often found little connection. Perhaps they should not have been surprised. Satisfaction is not equivalent to motivation; in fact, it may drop as motivation rises.

In designing our study of scientists, therefore, we avoided a focus on satisfaction as such, and attempted rather to probe into "morale" in its original sense of dedication to goals.

Could this intangible quality be measured? A basis for believing that it could came from a visit Pelz made to the research laboratories of a major oil company in 1955 where he had a chance to interview nineteen technical men. One impression which came through in these interviews was the extent to which different people were *intense or relaxed* about their job. It was not a question of satisfaction or enjoyment, but something more like involvement or dedication. Some were intensely interested in their work, others casual (but well satisfied with working conditions). From these subjective impressions, Pelz rated the nineteen people on intensity of motivation. And when he later compared these ratings with the over-all value of each man in the eyes of the company (as these men were ranked by supervisors), a clear relationship appeared—almost without exception, the more highly motivated were doing better work.

Could this elusive quality of enthusiasm or intensity of work motivation—of morale in its original sense—be captured by the dry checklists of a self-administered questionnaire?

More interviews were carried out in the next two years to catch the phrases that highly motivated scientists used to describe their work. From them we formed a number of questionnaire items. Those that turned out most promising are shown below. Note a contrast between these questions and some of the ones described in Chapter 6. In Chapter 6 we are concerned with tapping the *kinds* of motivations impelling the scientist, whether directed toward adding to science or climbing a ladder, whether originating from inner sources or external stimuli. Here, in contrast, we are trying to measure the *intensity* of a man's dedication to the job, regardless of the direction or origin of this motivation.

Question 53. Some individuals are completely involved in their technical work—absorbed by it night and day. For others, their work is simply one of several interests. How involved do you feel in your work?

[On a six-point scale, the respondent rated himself from "not much involved" to "completely; the most absorbing thing in my life."]

Question 57. How interesting or exciting do you find your present work?

[Six-point scale from "hardly interesting at all; pedestrian" to continually exciting."]

Question 11. One project may seem like a "labor of love" with which you feel strongly identified; it seems like a part of yourself, your own brainchild. Another project may be just an assignment about which you feel detached and impersonal. How do you feel toward each of your projects?

[Space was provided to list up to three projects, each rated on a six-point scale of identification from "nonexistent; I feel completely detached" to "intense; seems like a part of myself." The response for the main project was used in this analysis.]

Question 52. How important do you personally feel your present technical work is, that is, how much of a contribution is it likely to make, either to technical knowledge, or to practical application, or by other standards?

[Six-point scale from "of no importance" to "utmost; one of the most important of its type anywhere in the world today."]

Question 54. How challenging do you find your present work in the sense of demanding concentration, intelligence, or energy?

[Five-point scale from "hardly challenging at all" to "extremely challenging; demands all that I can offer," with a nonscale response of "completely over my head; I feel lost."]

How were we to know whether our attempt was successful—whether we really were able to capture the quality of enthusiasm? One test was to see whether several questions designed to tap this factor were, in fact, intercorrelated—whether they seemed to be measuring something in common. The five items listed did show a modest, though statistically significant, set of intercorrelations (large enough, that is, so that one can be confident that the correlations were not simply due to chance).

A second criterion was whether the responses differed among various groups in a meaningful way. Some information on this point will be given later in Table 3.

The third, and most important, test was to see whether the items related to scientific performance. If they did, this fact would not prove we had measured enthusiasm. But the lack of such evidence would at least show we had failed. The results are given in Table 1. For simplicity, the individual correlations are represented by symbols.

How the Measures of Intensity Related to Performance

The question on involvement was asked of all 1300 scientists, but the other four were asked of the 40% who filled out the long questionnaire. For consistency, only the latter are used in Table 1 (data on involvement for the total sample will be shown in Table 2). The performance scores used here have been adjusted, as described in Chapter 1, to rule out effects due to length of working experience, and to equalize means among university, government, and industrial settings.

What did we find? The results were gratifying. Although the correlations were mild, they were consistently positive across all five analysis groups. Particularly effective was the feeling of being involved in one's work; eight out of ten correlations with judged contribution and usefulness were large enough to be statistically significant. (When the same question was used later in a study of industrial employees, again it correlated in a consistently positive way with performance ratings, although *satisfaction* in the same study did not.) Here, then, was a promising candidate to measure "morale" in the sense of intense motivation or enthusiasm.

The weakest candidate was the feeling of the work's importance. This was significantly related to some performance measure in only three of the five groups (all the other measures were significantly related in at least four groups); and only four of the correlations were significant (the others had at least six). To maintain enthusiasm, then, it was more necessary that the work be interesting or challenging, than be seen as "important."

Toward a Motivation Index

We had hoped that by combining several indicators of intense motivation, we might arrive at a motivation index which would be more useful than any single item. In this, we were disappointed. One such measure was obtained by summing the individual's responses on the five items shown in Table 1. This over-all motivation index, shown at the bottom of Table 1, did correlate in a consistently positive way with performance, but not any more strongly than the single best item, involvement. It added nothing to what involvement already said.

Puzzled, we branched out in several directions, searching for a better total index. However, none of the efforts was especially successful.

TABLE 1 *This table is based on correlations (product-moment r's) between various items intended to measure motivation intensity and four criteria of performance. The following symbols are used: +, − = r is ±.10 or larger; double symbols indicate r is statistically significant.*[*] *The scorecard at the right sums the number of positive or negative symbols. Highly involved persons were consistently better performers in all five groups; the other four items also related positively but less strongly. A total motivation index worked about as well as the single item of involvement.*

	A. Ph.D's, devel.	B. Ph.D's, res.	C. Engineers	D. Ass't. scients.	E. Non-Ph.D's, res.	Scorecard +'s	Scorecard −'s
Involvement (Q. 53)							
Contribution	++	++	++	+	++	9	0
Usefulness	++	+	++	++	++	9	0
Papers/patents[†]	+	++	0	0	0	3	0
Reports	++	+	+	+	++	7	0
						28	0
Identification (Q. 11)							
Contribution	+	+	++	+	++	7	0
Usefulness	+	0	++	++	++	7	0
Papers/patents	0	0	+	0	0	1	0
Reports	++	+	+	0	++	6	0
						21	0
Interest (Q. 57)							
Contribution	0	++	++	+	++	7	0
Usefulness	+	+	++	+	++	7	0
Papers/patents	0	++	0	0	−	2	1
Reports	++	+	+	+	0	5	0
						21	1
Challenge (Q. 54)							
Contribution	+	+	++	+	++	7	0
Usefulness	+	+	++	+	++	7	0
Papers/patents	0	+	0	0	+	2	0
Reports	++	++	0	0	+	5	0
						21	0
Importance (Q. 52)							
Contribution	++	+	+	0	+	5	0
Usefulness	+	0	++	++	+	6	0
Papers/patents	+	+	+	−	0	3	1
Reports	++	0	+	+	+	5	0
						19	1
Motivation index (see text)							
Contribution	+	++	++	+	++	8	0
Usefulness	++	+	++	++	++	9	0
Papers/patents	+	++	+	0	0	4	0
Reports	++	+	+	+	++	7	0
						28	0
Number of persons	101	79	219	90	45		

[*] Probability is less than 1 in 20 that an *r* this large would occur by chance in a group of this size. In computing the correlations, data were weighted to compensate for sampling rates, but statistical significance was based on actual number of persons.

[†] For engineers, patents were used instead of papers.

Perhaps there are conditions under which motivation that is too strong will inhibit rather than stimulate. If a scientist wants a solution too keenly, he may fixate on what has worked before, and overlook creative detours. Chapter 12 explores certain hunches along these lines. Meanwhile let's recognize that we had made at least a promising start in capturing the elusive quality of dedication.

Involvement in Nine Settings

The reader may be curious to know how well our measure of involvement was working for us. Table 2 shows results for the full 1300 cases

TABLE 2 *The following is based on responses from all 1300 scientists and engineers. Part a shows percentages at the strong and weak ends of the involvement scale; the difference between them indicates that university Ph.D's were most strongly involved, whereas non-Ph.D's generally were less so. (**Boldface differences** indicate the strongest responses, italic differences the weakest. I = industry, G = government, U = university.)*

Part b is based on correlations of involvement with the four performance measures, using the same symbols as in Table 1. A highly consistent pattern appeared; except for Ph.D's in government research labs, the more involved scientists performed better by almost all standards.

	A. Ph.D's, devel.		B. Ph.D's, res.		C. Engineers		D. Ass't. scients.		E. Non-Ph.D's, res.	Scorecard	
	I	*G*	*U*	*G*	*I*	*G*	*I*	*G*		+'s	−'s
a. Level of involvement in each setting											
"Very strong, intense"	44%	37%	58%	42%	29%	26%	21%	27%	31%		
.	.	.	.	.	.	.	.	.	.		
.	.	.	.	.	.	.	.	.	.		
.	.	.	.	.	.	.	.	.	.		
"Low to moderate"	11	23	11	9	20	28	31	28	29		
Strong minus weak	33	14	**47**	33	9	−2	−*10*	−1	2		
b. Correlation of involvement with:											
Contribution	++	+	++	0	++	++	++	++	++	15	0
Usefulness	++	+	0	0	++	++	++	++	++	13	0
Papers/patents[*]	+	++	++	+	0	++	+	++	+	12	0
Reports	+	0	++	0	++	++	0	0	0	7	0
										47	0
Number of persons	106	77	140	65	401	140	118	116	121		

[*] For engineers, patents were used instead of papers.

within the nine different settings in university, government, and industrial locations.

An impressive pattern emerged in Part *b* of the table. In every group, with the exception of government Ph.D's in research labs, highly involved scientists were performing better by almost all standards. Part *a* also shows that involvement was most intense among university Ph.D's, and considerably milder among non-Ph.D's, with assistant scientists in industry slightly below everyone else. Except for the assistant scientist category, people in government seemed a little less involved than those in the university or industry.

Not a single negative correlation (of −.10 or larger) appeared between involvement and performance. The scorecard showed a total of 47 positive symbols. The correlations themselves were not unusually large—mostly in the .20's and .30's. But they were consistent and often big enough to be statistically significant.

Some Speculations

People who were thoroughly involved in their work—who did not put it aside at the end of an eight-hour day, but took it home and slept on it—were consistently able to turn out better performance.

But do not jump to the conclusion that these involved people worked constantly. We also looked at a measure of the total length of working day; Chapter 4 reported our findings briefly. True, the scientists who only put in a routine eight-hour day were consistently lower performers, whereas those who worked nine or ten hours on the average were somewhat better. But the longest hours did not necessarily go with the highest performance.

We also saw in Chapter 4 that the best performers did not put 100% of their time into technical work as such, nor did they concentrate exclusively in research or exclusively in development. They maintained a *mix* in their technical activities. There was even a suggestion that a mild dose of administration could be stimulating (certainly it did not keep the best performers from high achievement).

So although the highly involved person works an hour or two longer each day, he does not let it obsess him around the clock. He takes time out for other things. He backs away from his work and lets the diverse ingredients brew, perhaps at a subconscious level. Now maybe it is precisely the involved man who can do this. His absorption is a catalyst, perhaps, that permits interaction among disparate intellectual elements. During his relaxed periods, they can work on each other to form new structures.

Correlation of Involvement with Inner Motivation

The reader may wonder, as we did, to what extent involvement in the job is synonymous with being internally motivated. Were highly involved people the self-reliant ones who drew upon their own ideas and own previous work for stimulation, and rejected stimulation from a supervisor? Table 3 shows some pertinent data.

An index of stimulation from one's own ideas (based on Questions 13E, 13J, and 62L—see Chapter 6 for details) was indeed consistently correlated with the various dedication measures (mostly in the .20's and .30's—not large, but consistent). We are puzzled by the complete absence of correlation among Ph.D's in development. Apparently they

TABLE 3 *The index of motivation from one's own ideas as described in Chapter 6 (stimulation from one's own work and curiosity, and freedom to follow one's own ideas) was correlated with each intensity measure, and found to be related, except for Ph.D's in development. The independence index (own ideas index minus dependence on supervisor) correlated less well. A person could be challenged by his work without having to resist his supervisor. (Symbols are the same as in Table 1.)*

	A. Ph.D's, devel.	B. Ph.D's, res.	C. Engineers	D. Ass't. scients.	E. Non-Ph.D's, res.	Scorecard +'s	Scorecard –'s
Index of stimulation from own ideas as correlated with:							
Involvement	0	++	++	++	+	7	0
Identification	0	+	++	++	++	7	0
Interest	0	++	++	++	++	8	0
Challenge	0	0	++	++	+	5	0
Importance	0	+	++	++	+	6	0
Motivation index	0	+	++	++	+	6	0
Independence index as correlated with:							
Involvement	0	++	0	++	++	6	0
Identification	+	+	++	++	++	8	0
Interest	0	++	0	++	++	6	0
Challenge	–	0	0	++	0	2	1
Importance	0	+	0	+	+	3	0
Motivation index	0	+	+	++	+	5	0
Number of persons	101	79	219	90	45		

could be dedicated whether or not they drew upon their own ideas. But in most other groups, reliance upon one's own previous ideas was an important way of becoming absorbed in the work.

Note that the question inquiring about "challenge" showed the weakest relationship. A job could be seen as challenging—and therefore as exciting and stimulating—even though it might not stem from the man's own ideas. This point (and the odd result for Ph.D's in development) suggest that although dedication is more likely than not to be generated from within the person, it can also be generated from without.

The last point is sharpened by data from a related index on "desire for independence." This was constructed by taking the own ideas index, and subtracting an index of dependence on supervisor (also detailed in Chapter 6). The correlations of the new index with the intensity measures dropped a little, and especially so for challenge. One doesn't have to resist one's supervisor in order to become absorbed in or challenged by one's work. Presumably a skillful supervisor can, in fact, do much to generate a sense of challenge through active intervention.

SUMMARY AND IMPLICATIONS

What is the research supervisor or manager to make of these results? We have seen that the feeling of intense involvement in one's work—call it dedication or commitment, perhaps—was consistently found among high performers in numerous settings.

But how is the supervisor or manager to encourage a sense of involvement, if indeed he can do anything about it? We can offer several clues and speculations, inviting the reader to search his own experience for others. One of the questionnaire items which proved to be consistently associated with high involvement was being able to *influence* other people or groups who had substantial weight over the man's technical goals (see Chapter 2, which shows Question 31 and some data based upon it). Maybe involvement and influence are simply two expressions of the same subjective feeling. Maybe the man who (because of his own personality or background) is more absorbed in his work also feels he has more to say about the organization's decisions concerning him. Or perhaps if he is a dedicated worker, he is given more leeway and voice. But we suspect the reverse can happen, and does happen. Scientists who are given a genuine share in the decision-making process—and by this we mean not simply left alone to pursue whatever they like, but brought into genuine policy decisions—thereby become more involved in, and committed to, the technical goals that are being decided.

The supervisor can help in another important way, we think. Maybe it is true that a good scientist is his own best audience; perhaps he doesn't need appreciation by other people to know when he is getting somewhere. But the results in Chapter 7 will suggest otherwise. As will be shown, the best scientists distinctly felt that the organization was, in fact, appreciating them by giving not only opportunities to follow their own interests, but also material rewards and association with key people. The outstanding scientist is excited about his own work, but it also helps if other people are excited too. The supervisor who can see meaning and significance in the work his subordinates are doing thereby helps to reinforce their own enthusiasm.

Chapter 4 on diversity suggests other clues. Suppose a visitor comes to the lab to find out what is going on. The chore of talking to him is usually taken over by the supervisor; he doesn't want to intrude on the working time of his scientists. But why not spread this administrative distraction around? Suppose each scientist, now and then, took time out to explain to a visitor what he personally was up to, what his section was up to, how it fitted in with the total program of the organization, and, at the same time, he got ideas about the visitor's interests and problems. Might this not lend a new sense of perspective? Might not the scientist ask himself unasked questions, become curious about unsolved problems?

We suspect that a deliberate and mild amount of diversity in the work content can be one of several means to build a sense of personal involvement.

6

MOTIVATIONS[1]

Among Various Motives Characterizing High Performers, an Outstanding Trait Was Self-reliance.

What motivates scientists? And, in particular, what motivates the more effective ones? Do they strive toward scientific prestige, organizational status, financial returns, or just the fun of solving problems and learning new things?

These questions bear on our interest in a "stimulating research atmosphere." If we can find out what motivates high or low performers, we may gain some clues on how to build and maintain a vigorous climate.

Research directors are concerned with "incentives." Knowing what motivates high achievers, the director may be able to set up more effective systems of reward and recognition.

The chapter starts with a finding from an earlier study that the strength of orientation toward one's *discipline* typified high performers, but not orientation toward an *organizational career*. In addition to such measures of the direction of motivation, the present study sought clues as to different sources of ideas, and different styles of approach to the work. We again found that strength of science orientation modestly typified high performers. But some stronger motivational components also appeared, particularly reliance on one's own ideas as a source of stimulation, and an interest in breadth rather than specialization. Perhaps the earlier finding was due more to the need for self-direction than to the man's interest in scientific contribution as such. The chapter concludes with some practical suggestions for reward systems.

How We Started

In our earlier studies of the National Institutes of Health, Robert C. Davis located one cluster of job factors suggesting that the individual

[1]The analysis of this chapter was mainly supported by grants from the National Science Foundation.

aspired to a career in his *scientific field* (that is, he rated as important the chance to contribute to science, to use his knowledge and skills, and to have freedom to carry out his own ideas). From this set, Davis formed an index of "science orientation." A second cluster of items suggested aspiration toward a career within his *organization* (the individual placed importance on advancing to a more responsible and important job, and on associating with top executives), which generated an index of "institutional orientation."[2]

Davis found that among the NIH scientists, those who scored strong on science orientation (compared to those scoring moderate) were judged more effective by colleagues, whereas the strength of institutional orientation (aspiration for organizational status) was unrelated to performance.[3]

Yet when our Michigan staff compared notes with other investigators, such as Morris I. Stein (who had studied creativity of industrial chemists) and Herbert A. Shepard (creativity and productivity of industrial research teams), we found no confirmation. Their highly ranked chemists and teams appeared just as interested in monetary rewards and in promotion "up and out" of research as in research for its own sake. Clearly, further work was needed.

Three Types of Motives

A series of personal interviews with scientists, sandwiched between the NIH study and the current one, suggested three types of motives.

We tried to formulate questionnaire items to tap these, using language which scientists had employed in their interviews. Shown in the boxes are three general questions, each with multiple sub-items. Those which later proved useful are given. Question 13 aimed particularly at the *sources* or origins of work problems. Question 19 contained all of the items that we later used to measure *style of approach.* Question 62 contained many of the items used to measure *direction* (or orientation).

[2]These concepts stemmed from Robert Merton's distinction between "cosmopolitan" and "local" orientations. A number of related typologies offered by Reissman, Blau, Wilensky, Gouldner, Caplow and McGee, and Lazarsfeld and Thielens are summarized by William Kornhauser, *Scientists in Industry: Conflict and Accommodation,* University of California Press, Berkeley and Los Angeles, 1962, pp. 118–30. Citing these approaches, Barney G. Glaser has found evidence in data from a large government medical research organization that, "In contrast to previous discussions in the literature treating cosmopolitan and local as two distinct groups of scientists, this paper demonstrates the notion of cosmopolitan and local as a dual orientation of highly motivated scientists." See "The Local-Cosmopolitan Scientist," *American Journal of Sociology,* 1963, vol. 69, pp. 249–59.

[3]R. C. Davis, "Commitment to Professional Values as Related to the Role Performance of Research Scientists," doctoral dissertation, University of Michigan, 1956.

Question 13. Listed below are several sources from which projects can originate. . . .Regardless of the actual origins of your projects, what sources (in general) offer you the most stimulus to perform well?

[Ten sources were listed, and for each the respondent rated "amount of stimulus" on a five-point scale from "none" to "very strong." Some of the sources were:]

A. My immediate supervisor
B. Higher-up research supervisors
E. My own previous work or plans
F. The technical literature
G. Problems arising in practical applications
J. My own curiosity

Question 19. Scientists and engineers may differ widely in their *characteristic approach* to their work—both the kinds of problems that attract them, and the way they go about the task. How closely does each statement describe the approach you typically prefer to use?

[Nineteen statements followed. Respondent rated "how closely statement describes me" on a seven-point scale from "not at all" to "completely." Selected items were:]

B. I mainly prefer problems that will help to *build my professional reputation.*
C. I mainly prefer problems that will lead to *advancement in organizational status.*
D. I prefer areas where I can be fairly sure of *some acceptable results,* even though not spectacular.
E. I prefer to map out *broad features of important new areas,* leaving detailed study to others.
F. I prefer to probe *deeply and thoroughly* in selected areas, even though narrow.
I. I'm effective as a *"right-hand man,"* carrying the ball for a more experienced advisor.
J. I prefer to develop my ideas *"inside my head,"* before testing them against nature.
L. I prefer to spend enough time to find *general principles* that apply to many situations.
M. I prefer to find *immediate solutions* to specific problems.
N. I find it fruitful to utilize *abstract concepts* several steps removed from direct observation.
P. I like to bring about *order and simplicity* in chaotic or complex material.

Question 62. Listed below are different kinds of opportunities which a job might afford. If you were to seek a job, how much importance would you personally attach to each of these (disregarding whether or not your present job provides them)?

[Thirteen factors were listed, which respondent rated on "importance I would attach" using a five-point scale from "slight or none" to "utmost." For example:]

A. To make full use of my present knowledge and skills
B. To grow and learn new knowledge and skills
C. To earn a good salary
D. To advance in administrative authority and status
H. To associate with top executives in the organization
I. To build my professional reputation
J. To work on difficult and challenging problems
L. To have freedom to carry out my own ideas
M. To contribute to broad technical knowledge in my field

Direction of Motivation

Our analysis started with the matter of direction or orientation of motives. Procedural decisions confronted us. How many directions should we try to measure? Should each questionnaire item be used separately, or could we profitably combine related items and construct an index for each cluster?

Correlations among the items revealed several clusters, including two major ones of orientation toward science and status—a replication of the NIH result.

Andrews did a factor analysis on all of the university scientists, and a separate analysis on one-third of the industrial scientists and engineers, using 43 motivational items. Government data were not then available.[4] Among the basic dimensions found in each group, six had much the same meaning for both the university and industrial scientists. One corresponded closely to the NIH index of science orientation; a second closely resembled the index of institutional or status orientation.

[4]Reported in F. M. Andrews, "An Exploration of Scientist's Motives," Analysis Memo #8, March 1961, available in Publication #1825 from the Survey Research Center, University of Michigan. A factor analysis starts with a matrix of correlations between each pair of items, and then seeks a small number of underlying dimensions which can account for these correlations. For the statistically sophisticated reader, we may note that blind rotations were completed independently for the two factor analyses using the Varimax method of rotation. Within each analysis, eight factors accounted for almost all the common variance; six of these had similar meaning for both groups.

George F. Farris made a separate analysis among long-form scientists, using items from Question 62 only. (Chapter 7 gives more detail about his work.) He also obtained the same two clusters within research and development departments separately. (The science cluster he labeled "desire for self-actualization.")

Accordingly we constructed an index of science orientation and another of status orientation, using the items indicated in Table 1. Appendix G shows the intercorrelations among the items.

Differences among Groups. How valid were the questionnaire responses? One test was to see whether they differed meaningfully among scientists in various settings. Table 1 shows some results.

The index of science orientation was constructed to repeat as closely as possible the original NIH index. One of its components was desire to advance technical knowledge (Question 62M). Another component was desire for freedom (Question 62L), which appears later in Tables 3 and 4. In addition, in view of the high intercorrelations of these items with a number of other items, a broader index of "professional orientation" was also tried.

The separate cluster of status-oriented items is represented in the next index. The final entry is a separate item (Question 60—see box) in which the individual was asked to make a *choice* between science and status aspirations. Andrews found this item in the status factor among industrial, but not among university, scientists.

> *Question 60.* Do you think about your career more as: (A) a series of opportunities to engage in activities you like to do, or (B) a progression up one or more organizational ladders to a position in which you aspire to be?
>
> [Respondent chose among five alternatives from "almost entirely as (A)" to "almost entirely as (B)." For correlational analysis, the (B) or ladder preference was scored "high."

The construction of Table 1 is illustrated with the first entry. The index was arbitrarily divided into strong, moderate, and weak; the table shows the percentage of scientists falling in each. For an over-all indication of the atmosphere in each setting, we have subtracted the "weak" percentage from the "strong." Ph.D's in university research labs (as we might expect) overwhelmingly endorsed a science orientation; at the other extreme, engineers and assistant scientists in industrial laboratories were markedly nonoriented toward science.

For simplicity, the remaining entries show only the "strong" and "weak" ends of each measure and the percentage difference between them. (Because these categories were arbitrary, do not try to compare two scales.)

TABLE 1 *Direction of motivations in nine settings. Among Ph.D's in university labs, the percentage with strong science orientation far exceeded that with weak, but among assistant scientists in industrial laboratories, the reverse was true (Part a). For simplicity, Parts b, c, and d show only the strong and weak ends of each scale, and the difference. Ph.D's in research were strongly motivated toward scientific and professional goals; assistant scientists in industry were noticeably weak on these goals; engineers in industrial labs were strongly status-oriented.*

(I, G, and U refer to Industry, Government, and University. Differences in **boldface** *indicate the settings in which a given factor was strongest; differences in italics show where the factor was weakest.)*

	A. Ph.D's, devel.		B. Ph.D's, res.		C. Engi- neers		D. Ass't. scients.		E. Non-Ph.D's, res.
	I	*G*	*U*	*G*	*I*	*G*	*I*	*G*	
a. Science orientation* (Q. 62A + L + M)									
Strong	28%	36%	51%	34%	12%	16%	10%	21%	20%
Moderate	35	27	38	46	29	30	27	38	40
Weak	34	37	10	17	58	53	63	41	40
No answer	3	–	1	3	1	1	–	–	–
	100%	100%	100%	100%	100%	100%	100%	100%	100%
Strong minus weak	–6	–1	**41**	17	–46	–37	*–53*	–20	–20
b. Professional orientation (Q. 62A + B + I + J + L + M)									
Strong	31	35	45	43	23	24	15	26	30
⋮	⋮	⋮	⋮	⋮	⋮	⋮	⋮	⋮	⋮
Weak	25	26	11	12	37	33	50	30	30
Difference	6	9	**34**	**31**	–14	–9	*–35*	–4	0
c. Status orientation (Q. 19B + C + 62C + D + H)									
Strong	23	26	11	20	40	20	21	32	25
⋮	⋮	⋮	⋮	⋮	⋮	⋮	⋮	⋮	⋮
Weak	23	25	46	35	15	25	31	23	27
Difference	0	1	*–35*	–15	**25**	–5	–10	9	–2
d. Career as ladder versus activities liked (Q. 60)									
Mainly ladder, or both	26	26	15	20	61	33	31	43	30
⋮	⋮	⋮	⋮	⋮	⋮	⋮	⋮	⋮	⋮
Entirely activities	36	34	43	31	10	34	32	28	31
Difference	–10	–8	*–28*	–11	**51**	–1	–1	15	–1
Number of individuals	106	77	140	65	401	140	118	116	121

*The strong and weak ends of each scale are arbitrary; in Part *a*, for example, the weak end includes the bottom two steps of a four-point index, whereas in Part *c*, the weak end includes only the bottom step. Do not try to compare two scales, but do compare different groups on the same scale.

Ph.D's in research (especially in the university) said they were strongly interested in advancing science, compared with other groups; and they were somewhat higher on the over-all index of professional orientation.

University Ph.D's were conspicuously disinterested in higher status (according to their reports). In contrast, engineers in industrial laboratories were strongly interested. Faced with a choice (Question 60), they clearly saw their career as climbing a status ladder rather than engaging in "activities they liked."

These patterns made sense. Scientists in the various groups seemed to answer the questions in an honest and meaningful manner, not capriciously.

Relationship to Performance. How, then, did the direction-of-motivation measures relate to performance? A series of correlation coefficients (*r*'s) was obtained, between each measure of motivation and each criterion of performance, within the nine settings. A large matrix of *r*'s is confusing to the eye, so to clarify trends, the simplification in Table 2 is presented.

What did we observe? The index of science orientation, and the more inclusive version called professional orientation, were mildly but consistently positive in their relationship to various performance measures. (The single item of desire to "advance technical knowledge"—not shown here—was also mildly positive in its effects.) In general, strength of these measures affected output of *reports* more strongly than they affected scientific contribution or usefulness. Professional orientation, oddly, was least strongly related to papers (or patents for engineers).

Interest in status, on the other hand, was hardly related to performance, except perhaps to the writing of reports. The latter measure thus seemed responsive to several kinds of motives, and may reflect a general *energy* factor.

Note what happened in the final item where the individual had to choose between "activities liked" (science) and "organizational ladder" (status). Status-choosers lost out mildly in performance. Status orientation as such did not hinder, unless the ladder climber had to sacrifice research interest to do so.

Next, look at the data vertically, by group. Among Ph.D's in development labs and engineers, those in *government* seemed a little more responsive to science or professional motives than did those in *industry.* (Since the two measures overlapped, these were not independent findings.)

Note too that assistant scientists in both industry and government benefited the most—from both kinds of motives. We know that these groups are generally given little opportunity to create. Perhaps some individuals with intense motivation were able to overcome the obstacles with massive output of reports and papers.

The reader might want to know, in passing, of some other motivational directions that did *not* make much difference. One was an index of "interest in self-development," including desire to use abilities (Question 62A) and to learn new skills (Question 62B). Scientists in various settings claimed a similar interest in this goal, but it was only mildly correlated with performance. Interest in advancing "the nation's well-being" (Question 62K) was a little stronger in government non-Ph.D's than other groups, but largely unrelated to performance either there or elsewhere.

Source of Motivation

In terms of sources of motivation, we wanted to distinguish at least between "internal" and "external" sources, and if possible to distinguish among several external sources. Correlations among three items (see Appendix G) justified construction of an index of "own ideas as source," based on: own previous work as stimulus (Question 13E), own curiosity as stimulus (Question 13J), and desire for freedom to carry out own ideas (Question 62L). This index, and two of the component items, are shown in the following tables.

Another cluster included three items in which the supervisor was seen as the important source of motivation: stimulus received from immediate supervisor (Question 13A), as well as higher technical supervisors (Question 13B), and respondent views himself as an effective "right-hand man" for an experienced advisor (Question 19I).

It turned out that the indices of own ideas and of supervisor as sources were negatively correlated. Therefore an over-all index of "need for independence" was constructed by subtracting the two, that is, reversing the sign of the supervisor index and adding it to the own ideas index. A person scored high on need for independence if he insisted on self-direction and denied receiving ideas from a chief or advisor.

In Andrews' factor analysis, both university and industrial matrices yielded another factor which might be called "caution" (Andrews suggested "motive to avoid risks"). It included Question 19D (prefers to be fairly sure of acceptable results, even though not spectacular), which is shown separately in Table 3.

Differences among Groups. According to Table 3, Ph.D's in research labs (especially in the university) drew heavily on their own ideas and wanted freedom to carry them out; all the Ph.D's (especially in university research) denied dependence on a supervisor. Engineers did not insist on freedom for their own ideas. Assistant scientists were cautious and clearly dependent on their supervisors for stimulation. These differences are meaningful, and lend confidence in the validity of the answers.

Relationships to Performance. Table 4 goes on to show, by means of correlation coefficients, how well scientists performed when they relied on these various sources.

TABLE 2 *The following is based on correlations (r's) between the direction-of-motivation measures listed in Table 1, and four criteria of performance (adjusted for length of experience). I, G, and U mean Industry, Government, and University, respectively.*

+ = *r is mildly positive (+.10 or more)*
++ = *r is positive and statistically significant*°
− = *r is mildly negative (−.10 or less)*
−− = *r is negative and significant*°

For a general picture, the scorecard at the right sums the number of positive or negative symbols. Scientific and professional orientations showed mildly positive relationships, especially to unpublished reports; status orientations were mostly unrelated to performance, or mildly negative.

	A. Ph.D's, devel.		B. Ph.D's, res.		C. Engi-neers		D. Ass't. scients.		E. Non-Ph.D's, res.	Scorecard		
	I	*G*	*U*	*G*	*I*	*G*	*I*	*G*		+'s	−'s	Net
a. Science orientation												
Contribution	0	+	+	+	0	0	0	0	0	3	0	+ 3
Usefulness	0	+	0	+	0	+	0	++	0	5	0	+ 5
Papers/patents†	+	0	0	+	0	+	++	−	0	5	1	+ 4
Reports	0	++	0	++	0	+	+	++	+	9	0	+ 9
										22	1	+21
b. Professional orientation												
Contribution	0	+	++	+	0	+	0	0	0	5	0	+ 5
Usefulness	0	0	+	+	0	+	0	++	0	5	0	+ 5
Papers/patents	0	0	0	0	0	0	+	0	0	1	0	+ 1
Reports	0	+	0	+	0	++	++	++	++	10	0	+10
										21	0	+21

c. Status orientation												
Contribution	−	0	+	0	0	0	0	0	0	1	1	0
Usefulness	0	0	+	0	0	0	0	0	0	1	0	+1
Papers/patents	0	−	0	0	0	+	++	+	0	4	1	+3
Reports	0	0	+	−	0	+	++	+	+	6	1	+5
										12	3	+9
d. Career as ladder versus activities liked												
Contribution	0	+	−	0	−−	0	0	−	−	1	5	−4
Usefulness	+	0	−−	0	−−	0	0	−	0	1	5	−4
Papers/patents	0	0	0	+	0	0	0	0	−	1	1	0
Reports	0	0	0	0	0	0	+	+	+	3	0	+3
										6	11	−5
Column totals +'s	2	7	7	9	0	9	11	11	5			
Column totals −'s	−1	−1	−3	−1	−4	0	0	−3	−2			
Number of individuals	106	77	140	65	401	140	118	116	121			

*By "statistically significant" is meant that correlations of this size would not arise by chance more than five times in 100, if the true r were zero.
†For engineers (group C), patents were used instead of papers.

TABLE 3 *This shows how nine settings compared on internal or external sources of ideas, using the same system as Table 1. Ph.D's (especially in the university) emphasized inner sources, whereas assistant scientists were cautious and dependent on their chief for stimulation.* (**Boldface:** *factor was strong;* *italics: weak.*)

	A. Ph.D's, devel.		B. Ph.D's, res.		C. Engi-neers		D. Ass't. scients.		E. Non-Ph.D's, res.
	I	*G*	*U*	*G*	*I*	*G*	*I*	*G*	
a. Own ideas as source (index; Q. 13E + J + 62L)									
Strong	18%	27%	46%	35%	15%	19%	8%	22%	26%
⋮	⋮	⋮	⋮	⋮	⋮	⋮	⋮	⋮	⋮
Weak	18	20	8	17	26	27	42	26	20
Difference	0	7	**38**	18	−11	−8	*−34*	−4	6
b. Own previous work as stimulus (Q. 13E)									
"Very strong"	20	36	41	38	21	23	14	26	28
⋮	⋮	⋮	⋮	⋮	⋮	⋮	⋮	⋮	⋮
"None to moderate"	27	20	18	18	32	29	46	29	29
Difference	−7	16	23	20	−11	−6	−32	−3	−1
c. Freedom to carry out own ideas (Q. 62L)									
"Utmost importance"	32	27	64	42	19	19	27	22	22
⋮	⋮	⋮	⋮	⋮	⋮	⋮	⋮	⋮	⋮
"Slight to considerable"	24	33	7	18	40	41	35	38	27
Difference	8	−6	**57**	24	*−21*	−22	−8	−16	−5

d. Supervisor as source (index; Q. 13A + B + 19I)									
Strong	12	16	4	10	28	34	52	48	20
. . .	. . .	. . .	. . .	. . .	. . .	. . .	. . .	. . .	. . .
Weak	46	35	69	48	21	18	9	11	26
Difference	−34	−19	*−65*	−38	7	16	**43**	**37**	−6
e. Need for independence (index *a* minus *d*)									
Strong	37	32	64	43	11	9	6	8	15
. . .	. . .	. . .	. . .	. . .	. . .	. . .	. . .	. . .	. . .
Weak	14	15	6	12	21	22	46	33	14
Difference	23	17	**58**	31	−10	−13	*−40*	−25	1
f. Caution: want acceptable results (Q. 19D)									
Strong	27	42	20	32	29	32	42	46	35
. . .	. . .	. . .	. . .	. . .	. . .	. . .	. . .	. . .	. . .
Weak	44	36	53	34	41	39	28	22	32
Difference	−17	6	*−33*	−2	−12	−7	14	**24**	3

TABLE 4 *The following correlations between the inner and outer sources of motivation listed in Table 3 and four criteria of performance. As in Table 2, "+" and "−" mean positive or negative r's of .10 or larger, and the double symbols refer to statistically significant r's.*

Scientists who relied on their own ideas were effective, whereas those who relied on a supervisor were not. The stronger the need for independence and self-reliance, the higher was performance on several criteria.

	A. Ph.D's, devel.		B. Ph.D's, res.		C. Engi-neers		D. Ass't. scients.		E. Non-Ph.D's, res.	Scorecard		
	I	*G*	*U*	*G*	*I*	*G*	*I*	*G*		+'s	−'s	Net
a. Own ideas index												
Contribution	+	++	++	+	0	++	0	+	+	10	0	+10
Usefulness	0	+	0	+	0	++	+	++	+	8	0	+ 8
Papers/patents*	+	0	++	++	++	+	0	0	0	8	0	+ 8
Reports	0	+	+	++	0	+	++	++	+	10	0	+10
										36	0	+36
b. Own previous work												
Contribution	0	+	+	0	0	++	+	+	+	7	0	+ 7
Usefulness	0	0	0	0	0	++	++	0	+	5	0	+ 5
Papers/patents	0	0	+	+	++	0	0	+	0	5	0	+ 5
Reports	0	0	0	++	0	+	++	++	0	7	0	+ 7
										24	0	+24
c. Freedom for own ideas												
Contribution	+	++	++	++	0	++	0	0	+	10	0	+10
Usefulness	0	+	+	++	0	++	0	+	0	7	0	+ 7
Papers/patents	0	++	0	++	++	++	+	0	0	9	0	+ 9
Reports	0	+	0	++	0	+	0	++	0	6	0	+ 6
										32	0	+32

d. Supervisor index												
Contribution	− −	0	− −	−	0	0	0	0	− −	0	7	−7
Usefulness	−	0	−	−	0	0	0	−	− −	0	6	−6
Papers/patents	0	−	−	−	−	−	0	−	0	0	6	−6
Reports	0	0	0	0	0	0	0	− −	0	0	2	−2
										0	21	−21
e. Need for independence (own ideas index minus supervisor)												
Contribution	+ +	+	+ +	+	0	+ +	+	+	+ +	12	0	+12
Usefulness	+	0	+	+ +	0	+	+	+ +	+ +	10	0	+10
Papers/patents	0	0	+ +	+ +	+ +	+ +	0	0	0	8	0	+ 8
Reports	0	+	0	+	0	0	+ +	+ +	+	7	0	+ 7
										37	0	+37
f. Caution: want acceptable results												
Contribution	− −	−	− −	−	0	−	0	0	−	0	8	−8
Usefulness	− −	−	0	0	0	− −	+	0	0	1	5	−4
Papers/patents	0	0	−	−	0	0	+	+	0	2	2	0
Reports	0	0	−	−	−	0	0	−	0	0	4	−4
										3	19	−16
Column totals +'s	6	13	15	23	8	23	15	18	11			
Column totals −'s	−7	−3	−8	−6	−2	−4	0	−5	−5			

* For engineers, patents were used instead of papers.

The index of own ideas as source showed a strong and consistent positive trend: more effective scientists, on all four measures and in most of the settings, had strong inner sources of motivation. The component item of desire for freedom to follow own ideas showed especially strong trends for most Ph.D's. Another component, stimulation from own previous work, was also definitely positive in its effects. According to the scorecard tally, inner source of motivation was more essential to high performance than was science orientation (Table 2).

On the next index, scientists who considered a supervisor or advisor as a useful source of stimulation were rather consistently *less* productive. And when we combined the indices of inner source and supervisor as source (reversing direction of the latter) into an over-all index of need for independence, the same trends were emphasized. Independent and self-reliant scientists and engineers were substantially more effective; dependent individuals were below average.

Also, those who wanted to be safe and sure, to achieve results which would be "acceptable even though not spectacular," were generally less effective, particularly in scientific contribution. Oddly, though, caution did not interfere with publishing or patenting.

Other Sources. In addition to the measures shown previously, we attempted to measure several other kinds of external sources. One index we called "isolation as a source," based on "I'm rather a lone wolf; prefer to work by myself" (Question 19G), and "I'm [not] a strong team man; work best in collaboration with colleagues" (Question 19H, scale reversed). But it failed to correlate with performance. The person who simply withdrew was neither a high nor a low performer. "Inner motivation" does not mean isolation from people, but an independence of *thought*—confidence in one's own judgment.

A contrasting index was a desire to have "competent colleagues" as a source of motivation. It was based on importance attached to working with colleagues of high technical competence (Question 62E), having congenial co-workers as colleagues (Question 62F), and working under chiefs of high technical competence (Question 62G). This also made little difference to performance.

Another index was "client or practical problems as source," based on: problems arising in practical applications seen as a stimulus (Question 13G), and client or sponsor seen as stimulus (Question 13I). A final cluster was "technical literature as source," based on: stimulation from colleagues elsewhere (Question 13D), technical literature (Question 13F), and a prominent authority or teacher (Question 13H). Neither correlated consistently with performance.

Conclusion. Thus it appeared that effective scientists and engineers might report stimulation from a variety of sources—from practical prob-

TABLE 5 *This contrasts scientists in various settings on "style of approach," using the same system as Table 1. Ph.D's in research labs (especially in the university) were abstract in their approach, and disinterested in solving immediate problems; assistant scientists in industrial labs showed the opposite pattern. Engineers in industry emphasized breadth of approach; assistant scientists in government stressed depth.*

	A. Ph.D's, devel.		B. Ph.D's, res.		C. Engineers		D. Ass't. scients.		E. Non-Ph.D's, res.
	I	*G*	*U*	*G*	*I*	*G*	*I*	*G*	
a. Broad versus deep (Q. 19E – F)									
Mainly broad	40%	39%	33%	26%	51%	39%	23%	16%	32%
⋮	⋮	⋮	⋮	⋮	⋮	⋮	⋮	⋮	⋮
Mainly deep	30	40	39	51	20	32	39	54	43
Difference	10	–1	–6	–25	**31**	7	–16	–38	–11
b. Abstract (Q. 19J + L + N + P)									
Strong	18	31	41	38	20	31	12	21	27
⋮	⋮	⋮	⋮	⋮	⋮	⋮	⋮	⋮	⋮
Weak	30	22	20	22	23	26	32	26	19
Difference	–12	9	**21**	16	–3	5	*–20*	–5	8
c. Solve immediate problems (Q. 19M)									
Strong	45	48	20	26	47	39	55	44	39
⋮	⋮	⋮	⋮	⋮	⋮	⋮	⋮	⋮	⋮
Weak	16	14	40	25	12	19	9	18	18
Difference	29	34	*–20*	1	35	20	**46**	26	20

TABLE 6 *This depicts correlations between several measures of "style of approach" and four criteria of performance. The same symbols are used as in Tables 2 and 4.*

Scientists who maintained a broad perspective were more effective than those who wanted to probe deeply, especially on over-all usefulness. Interest in solving immediate problems was both a help and a hindrance.

	A. Ph.D's, devel.		B. Ph.D's, res.		C. Engi-neers		D. Ass't. scients.		E. Non-Ph.D's, res.	Scorecard		
	I	*G*	*U*	*G*	*I*	*G*	*I*	*G*	res.	+'s	−'s	Net
a. Broad versus deep index												
Contribution	0	0	+	+	0	+	0	+	0	4	0	+ 4
Usefulness	+	+	0	++	++	++	+	++	++	13	0	+13
Papers/patents*	−	0	++	+	0	0	0	0	0	3	1	+ 2
Reports	0	+	++	+	++	+	0	0	+	8	0	+ 8
										28	1	+27
b. Broad (Q. 19E)												
Contribution	−	+	+	++	0	++	0	0	0	6	1	+5
Usefulness	0	+	0	++	0	++	+	0	++	8	0	+8
Papers/patents	0	+	++	0	0	+	+	0	0	5	0	+5
Reports	0	0	++	+	0	+	+	+	++	8	0	+8
										27	1	+26

c. Deep (Q. 19F)												
Contribution	0	0	0	–	0	+	0	0	0	1	1	0
Usefulness	–	–	0	0	–	0	0	0	– –	0	5	–5
Papers/patents	+	0	– –	–	0	0	0	0	0	1	3	–2
Reports	+	0	–	–	0	0	0	0	– –	1	4	–3
										3	13	–10
d. Abstract index												
Contribution	0	+ +	0	+	0	+	0	0	+	5	0	+5
Usefulness	0	+	0	+	0	0	0	0	0	2	0	+2
Papers/patents	0	0	0	0	0	0	0	–	0	0	1	–1
Reports	0	+	+	+	0	0	+	+	0	5	0	+5
										12	1	+11
e. Immediate problems												
Contribution	–	+	–	–	0	0	+	–	– –	2	6	–4
Usefulness	0	0	–	0	0	0	+ +	0	0	2	1	+1
Papers/patents	0	+	– –	0	0	0	0	0	0	1	2	–1
Reports	0	+ +	–	–	0	+	0	– –	0	3	4	–1
										8	13	–5
Column totals +'s	3	13	11	13	4	13	8	5	8			
Column totals –'s	–4	–1	–8	–5	–1	0	0	–4	–6			

° For engineers, patents were used instead of papers.

lems or from the technical literature, from professional colleagues or from isolated study. The critical element was not the specific source, but an underlying factor of *intellectual self-reliance*—confidence in one's own ideas.

Style of Approach to the Work

We examined several measures of the style with which individuals approached their work. Table 5 shows how a few of these differed among the nine settings.

An index of broad versus deep perspective was based on the items "I prefer to map out broad features of important new areas" (Question 19E), and "I prefer to probe deeply and thoroughly in selected areas, even though narrow" (Question 19F, sign reversed). An index of interest in abstract approach was based on Questions 19J, L, N, P. In Andrews' factor analysis, a "motive to use abstract concepts," had emerged as a separate dimension in both university and industrial populations.

Engineers in industry especially claimed an interest in breadth rather than depth, whereas assistant scientists in government did just the opposite.

Ph.D's (especially those in university departments) indicated a preference for abstract concepts, and a distaste for finding "immediate solutions to specific problems" (Question 19M). Assistant scientists in industrial laboratories showed the opposite pattern—strong preference for solving immediate problems, and a disinterest in abstract approaches.

Style as Related to Performance. Table 6 shows how various measures of style correlated with four criteria of performance.

Among most of the groups shown in Table 6, interest in a broad, rather than deep, approach was a distinct advantage, especially in terms of over-all usefulness. People who were most indispensable to the organization had a wide grasp of major new developments, and did not become sidetracked in narrow specialties.

Interest in abstract concepts, however, was only slightly (though positively) related to contributions and reports, mainly among Ph.D's in government. Interest in immediate solutions of specific problems showed an interesting pattern, sometimes definitely helping, sometimes hindering.

In general, scientists in *government development* labs (both Ph.D's and engineers) were more strongly affected by these motivations than those in industry.

SUMMARY AND IMPLICATIONS

What have we learned from this sizable collection of data?

Briefly, this: We started vigorously along one promising path—the contrast in *direction* between a career in science or in the institution—

and found it moderately instructive. In terms of actual performance, the first of these was mildly helpful to performance, but the second was unrelated.

In the process of search, however, we came upon two other avenues of considerable importance. When we turned our attention to the *source* of motivation, we clearly found that those who relied on inner sources (their own ideas) were highly effective, whereas those who relied on supervisors for stimulation were below par. These results bear close comparison with our discussion in Chapter 2 on the effects of decision-making by the chief alone.

This finding, perhaps, casts light on the earlier NIH result. The measure of "science orientation" used there had three components: desire to use one's skills, interest in scientific contribution, and need for freedom to follow one's own ideas. The first component did not work in the present study. The second worked only mildly. Perhaps, then, the last component of self-reliance was the essential core that had generated the earlier result.

Finally when we examined the *style* of approach to the work, an interest in "broad mapping of new areas" was clearly an aid, whereas desire to "probe deeply in a narrow area" tended to handicap. These results are relevant to our discussion in Chapter 4 on the importance of diversity.

Practical Implications

What is the research director to conclude from all this? What has he learned that might guide him in building a more vital atmosphere?

We feel the results have significant implications for incentive plans and systems of recognition.

The findings point to the vital importance of self-reliance and independence. Yet systems of organizational rewards, as they typically operate, create dependence. Consider the usual performance review. A single supervisor rates the scientist and recommends promotions, and on this evaluation hangs the scientist's future. (Upper echelons also enter the evaluation, but as a rule they rely heavily on the judgment of the immediate supervisor.) The immediate chief also assigns and coordinates tasks.

This is all in keeping with orthodox management theory. "Lines of authority and responsibility must be clear! Never bypass the supervisor!" Yet if you deliberately wanted to stamp out independent thought in the subordinate, could you design a better system?

"But," the research director may ask, "what else can the organization do? Someone needs to coordinate and to evaluate. Isn't that what supervisors are for?"

We suspect there are many alternatives, if management really wants to be creative in discovering them. How do you build the individual's pride in his own work? How do you encourage him to map his own plans rather than lean on the hierarchy?

- Make sure he has a chance two or three times a year to tell a gathering of colleagues what he is up to, where he has come from, and where he plans to go. Let him meet face-to-face with higher executives or research users who can point out the mountains that need to be climbed—and then turn him loose to climb. In meetings to review progress on designs, let the engineer who did the work explain it, not the section head.

- If he has no recent report or paper or patent in which he can take pride of authorship, *prod* him to produce one every so often that bears his own name (with not more than one or two co-authors). Then see that the contribution is featured in the company newsletter or, even better, in newspapers and trade journals.

- If output takes the form of designs rather than of papers or reports, let these be signed by the actual designers. If necessary, identify the subparts to which each individual has contributed.

- Where letters or memos are handled by a specific individual, let him sign them personally (the boss can co-sign if necessary).

- Base monetary and status rewards not just on supervisory judgment, but give major weight to evaluations by colleague panels of actual work accomplished. Let the individual demonstrate his claim to their respect by reports, papers, or designs he has authored, or by his presentations at colleague seminars.

We were dismayed at the number of nonproducers of various scientific outputs in certain groups. Two out of five assistant scientists, and the same proportion of non-Ph.D's in research labs, had published not a single paper in five years; among engineers, the figure was four out of five. Half the engineers had not a single patent to their credit in the past five years. One out of five industrial engineers and non-Ph.D's in research, and two out of five assistant scientists (government plus industry), had not even produced a single report in the past five years for which they claimed co-authorship.

Is it unreasonable for a scientist to deprecate his own ideas when there is no tangible example of them to which he can point with fatherly pride?

The reader should not infer that we consider monetary rewards unimportant. Pay and title must jibe with achievement, or the man will feel unappreciated, may quit. In fact, we examined some data two years later in one government organization, to compare a few men who had left with those who stayed. They turned out to be similar in many respects, including colleague evaluation. (One of the fears was that better people were leaving; our data said not.) But there *was* an interesting difference (though the number of instances was too small to say for sure). Those who had left seemed to have a larger discrepancy between their organizational status (and pay) and their evaluated level than did those who stayed. In short, the leavers were *under-recognized* by the organization, relative to achievement in the eyes of colleagues.

Pay and status must be commensurate with achievement, and the research manager is justified in giving careful attention to reward systems. (Further evidence on this point will appear in the next chapter.) But at the same time, it is dangerous to rely on creating *motivation* toward status as the major incentive. Only among assistant scientists, those at the bottom of the prestige ladder, did strong striving for status benefit performance.

7

SATISFACTION[1]

Effective Scientists Reported Good Opportunities for Professional Growth and Higher Status, but Were Not Necessarily More Satisfied.

Typical morale surveys ask whether salary is "adequate," whether supervision is "effective," whether promotional opportunities are "satisfactory." In our study we avoided this format, and focused rather on motivation. A strongly motivated individual is not necessarily satisfied; a stimulating environment may not score as adequate in the eyes of an impatient investigator.

Nevertheless, administrators are necessarily concerned with problems of satisfaction or dissatisfaction, since squeaking wheels do require grease.

What can be said, then, about the level of satisfaction of scientists in our study, regarding various aspects of their jobs?

And what was the connection, if any, between satisfaction and performance? Were the more effective scientists happy? Unhappy? Neither?

The chapter opens with the concept of satisfaction, defined as the extent to which job factors desired by the individual are actually provided, and discusses a related concept of congruence between an individual's personal interests and those of the organization. Scientists whose personal interests were in perfect congruence (or agreement) with those perceived for the organization wrote many reports, but work of better scientific value and usefulness was done by scientists who disagreed moderately with the organization.

The chapter then takes up satisfaction regarding "self-actualization" and status advancement (each scored as the *difference* between the amount desired and the amount provided). These satisfaction scores showed moderately positive relationships to performance, but the simple ratings on provision of these factors related even better. A total provision score

[1]This analysis was supported largely by grants from the National Science Foundation. Preliminary analyses on which it draws were supported by a grant from a pharmaceutical company.

across all job factors correlated well with contribution and usefulness, and better than total satisfaction.

Closing on a practical note, the chapter points out the paradox that a research manager cannot rely on scientists' desire for status as an incentive for achievement, but that when achievement occurs it should be rewarded commensurately (both by extrinsic recognition such as pay, and by intrinsic rewards such as challenging assignments).

Satisfaction Has Two Components

Industrial morale surveys over many years have asked employees to rate the degree of "adequacy" or of "satisfaction" regarding various features of the working situation. Answers to such questions have not shown consistent relationships to productivity. In one early Survey Research Center study in an insurance office, for example, members of high-producing sections were somewhat more satisfied with regard to their supervision, but *less* satisfied concerning promotional opportunities. One enthusiastic headline writer was moved to exclaim: "Best workers gripe the most!" Actually the finding made good sense if you think of more productive employees as motivated by ambition, and resentful of obstacles in the way of promotion.

Satisfaction is a relative matter. If neighbor Jones buys a color TV set, our own black-and-white becomes inadequate. Around the world the "revolution of rising expectations" means that although conditions are no worse than before, aspirations have increased; the result is explosive frustration.

It has been obvious to numerous investigators that satisfaction should be broken into two components. On the one hand, there is the strength of desire for some factor. (The terms "aspiration" and "expectation" have also been used; these are not quite equivalent, but they have the same effect.) On the other hand, there is the degree to which the desired factor is actually present. The agreement or discrepancy between the two generates the feeling of satisfaction or of deprivation, of success or of failure.[2]

In the present study we did not ask directly about satisfaction. But we did ask questions trying to measure the two components: the strength of *desire* for (or importance of) various factors in the job, and the extent to which each of these was being *provided.* From the discrepancy between the two responses, a satisfaction score could be obtained.

[2]Martin Patchen has explored how these phenomena govern workers' satisfaction toward their wages in *The Choice of Wage Comparisons,* Prentice-Hall, Englewood Cliffs, New Jersey, 1961. Also, "A Conceptual Framework and Some Empirical Data Regarding Comparisons of Social Rewards," *Sociometry,* 1961, vol. 24, pp. 136–56.

Congruence between Interests of the Individual and the Organization

Satisfaction depends on agreement between what is desired by the individual and what the situation provides. A closely related concept deals with harmony or conflict between what is desired by the individual and what the organization *wants* of him. Does the individual want autonomy while the organization requires coordination? Lack of fit between these two can generate feelings of frustration akin to the feeling of dissatisfaction.

Notions of congruence are central in the writings of Chris Argyris about individuals in organizations.[3] These ideas stimulated George F. Farris to spend several months analyzing our data,[4] guided by statements such as the following: "Proposition I. *There is a lack of congruency between the needs of healthy individuals and the demands of the formal organization.*" If one thinks of mature individuals as desiring "relative independence, activeness, [and] use of important abilities," said Argyris, then a disturbance arises since these "needs . . . of healthy individuals are not congruent with the requirements of formal organizations, which tend to require the agents to work in situations where they are dependent, passive, and use few and unimportant abilities."[5] Because of such noncongruence, Argyris posited that, "if the agents are predisposed to a healthy, more mature self-actualization," they will experience "frustration . . . failure . . . and short time perspective."

Measuring Congruence

First let us consider congruence between the desires of the individual and the requirements of the organization.

The following box shows one effort to measure these two sets of interests. We listed several experiences a scientist might have; we asked the individual to rate how strongly each would give him a feeling of personal accomplishment, and then how much each would help him to "get ahead" in his organization. If the things he enjoyed doing were the same things that the organization rewarded him for, then his personal desires were congruent with his perception of organizational requirements. (For simplicity, the term "interests" may be used to refer to either response.)

[3]C. Argyris, *Personality and Organization*, Harper, New York, 1957.

[4]G. F. Farris, "Congruency of Scientists' Motives with their Organizations' Provisions for Satisfying Them: Its Relationship to Motivation, Affective Job Experiences, Style of Work, and Performance," Department of Psychology, University of Michigan, November 1962, 50 pp.

[5]*Personality and Organization*, p. 233.

More accurately, we should speak of "perceived congruence." The extent to which an individual perceived the organization's interests correctly is, of course, open to question; but consistently our analyses have indicated that responses were reasonable and realistic.

Question 55. Among the following experiences, how strong a *feeling of technical "success" or "accomplishment"* in your field could each one give you?

[Respondent rated each of ten experiences on "potential feeling of accomplishment" on a five-point scale from "none" to "utmost." Items were:]

A. Contributing to a product with high commercial success
B. Contributing to a product of distinctly superior quality
C. Contributing to something of value to the nation's well being
D. Publishing a paper which adds significantly to the technical literature
E. Securing a patent for an ingenious new device, process, or material
F. Executing an assignment rapidly and efficiently
G. Helping technical personnel to grow and develop
H. Solving a problem to the satisfaction of a sponsor or client
I. Coming up with highly original or creative ideas
J. Turning out a thoroughly sound and careful piece of work

Question 56. To what extent do you feel that each experience (if it occurred) would help you to *get ahead in your technical organization?* [Respondent rated the same ten experiences in terms of their "help in getting ahead," again on a five-point scale from "none" to "utmost."]

Using such data, how could a measure of congruence between individual and organizational interests be obtained? For each individual, we recorded the correlation (r) between his ten ratings from the respective standpoints of personal desire and organizational requirement. Perfect congruence between a scientist's personal and organizational ratings would generate an r of $+1.00$, complete disparity an r of -1.00.[6]

The average degree of congruence (or at least of perceived congruence) proved to vary considerably among settings, as shown in Table 1. Highest congruence, as one might expect, was reported by Ph.D's in research-oriented laboratories, where almost three-quarters had correlations of .50 or higher between personal and organizational interests. Non-Ph.D's in research laboratories were next, with correlations typically ranging in the

[6]We are indebted to Morris I. Stein for the format of the questions and method of measuring congruence. See M. I. Stein and R. Rodgers, "Creativity and/or Success?", paper delivered at American Psychological Association Convention, September 1957; see also "Creativity in American Life," selections from the writings and research of M. I. Stein, *University of Chicago Magazine*, December 1957, vol. 50, pp. 11–15.

TABLE 1 *Congruence between personal desires and perceived requirements of the organization was measured for each individual by a correlation coefficient between his ratings of ten experiences in these two frames of reference. Average congruence was highest for Ph.D's in research labs, moderate for non-Ph.D's in research labs, and relatively low for people in development laboratories.* (*Within each horizontal row, relatively large percentages are given in* **boldface,** *small percentages in* italic.)

Congruence	A. Ph.D's, devel.	B. Ph.D's, res.	C. Engi- neers	D. Ass't. scients.	E. Non- Ph.D's, res.
Low ($r = .29$ or less)	47%	*19%*	47%	40%	24%
Medium ($r = .30$ to .49)	23	9	21	21	**37**
High ($r = .50$ and up)	30	**72**	32	39	39
	100%	100%	100%	100%	100%

.30's and .40's. Development laboratories showed lowest congruence (both for Ph.D's and for non-Ph.D engineers), with nearly half the staff showing correlations below .30. For all respondents, the median correlation was about .40. Individual scores ranged widely; one person in five had +.70 or above, one person in seven a mildly negative correlation, but seldom below −.30. Very few were sharply in opposition to their organization; they probably would not have remained long.

How Congruence Related to Performance

We did a preliminary analysis using output of scientific products, before the performance evaluations were available.[7] A curious result appeared: for papers and patents, output seemed best when there was only mild congruence (r's in the .40's), whereas reports were most numerous with almost complete congruence (r's of .70 or up). Would the same trends appear when we examined scores of scientific contribution and usefulness? The results appear in the next four charts.

For Ph.D's the curvilinear effect tended to reappear, although the effects were not strong enough to attain statistical significance (Charts 1-A and 1-B). Although curves for the different criteria varied somewhat, in general they were above average when the individual's interests were only moderately congruent with those of the organization (r's in the .30's to .60's), and dropped when there was full congruence. The perfectly "adjusted" Ph.D scientist was not especially creative.

[7] D. C. Pelz, "Congruence between Personal and Organizational Values, as Related to Output," Analysis Memo #4, December 1960, available in Publication #1741 from the Survey Research Center, University of Michigan.

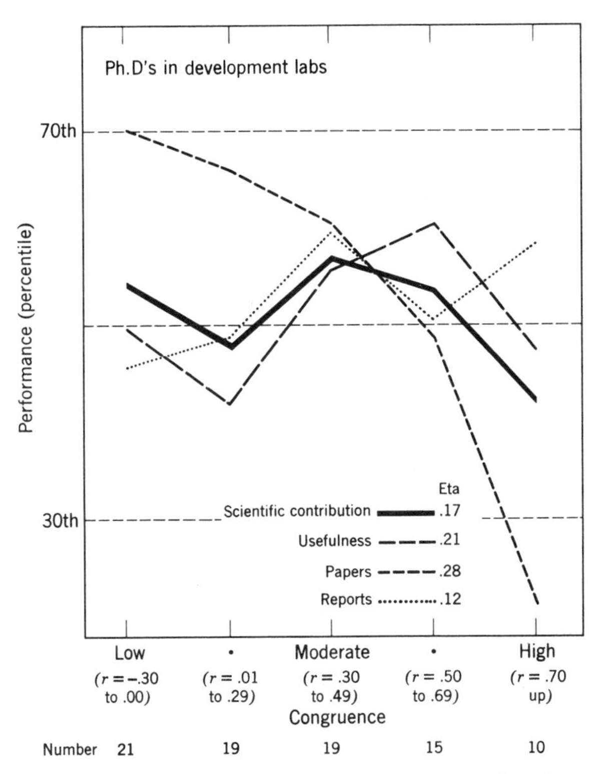

Chart 1-A. *Ph.D's in development who had high congruence—whose personal interests agreed fully with those of the organization—performed less well on three measures than did scientists with only moderate congruence; prolific publishers had the lowest congruence of interest.*

One striking curve is that for paper publication among Ph.D's in development (Chart 1-A): those with least congruence published the most, those with most congruence published the least. On a moment's reflection, this curve makes sense. By definition, a development-oriented lab values new products and not publication of knowledge. Active publishers, then, were likely to see a conflict between what they personally preferred to do, and what the organization rewarded. No such conflict appeared in research-oriented labs (Chart 1-B).

What about nondoctorals? Engineers (Chart 1-C) showed a very mild curvilinearity: three of the four performance measures were above average somewhere in the middle range of congruence, but not at the extremes. Only the output of unpublished reports showed a mild exception. Both among Ph.D's and engineers in development, many reports were written under maximum congruence.

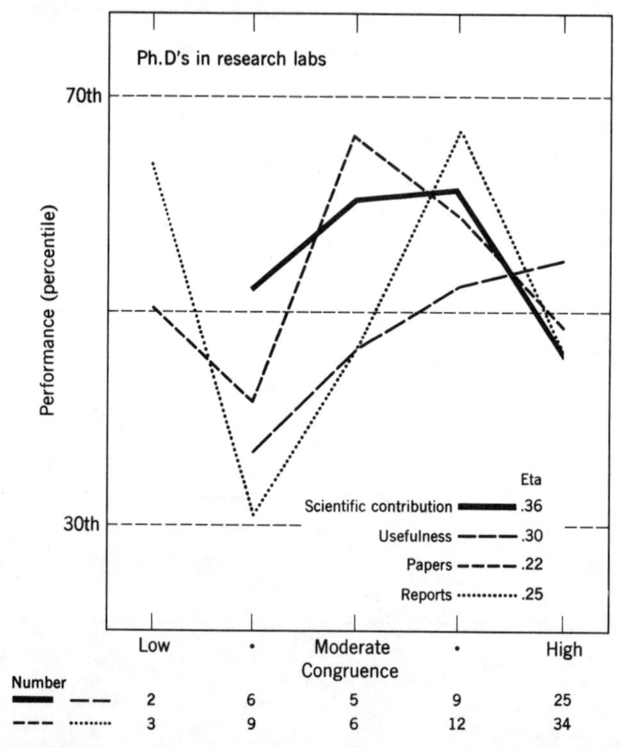

Chart 1-B. *On three measures, Ph.D's in research labs whose personal interests were perfectly congruent with those of the organization performed less well than Ph.D's with only moderately high congruence. A few individuals produced many reports when personal interests clearly disagreed with those of the organization.*

Most of the trends for assistant scientists were relatively flat, and are not shown. Again there was a mild tendency for contribution to peak with moderate congruence, although usefulness was maximum with high congruence.

Data for non-Ph.D's in research labs (Chart 1-E) showed some stronger trends than for the other groups. Again contribution and usefulness peaked at moderate congruence (statistically significant). But paper publication and report writing were maximum at high congruence.

In short: there were several hints that many *reports* were turned out by scientists whose personal interests agreed with those of the organization. But work of greater *scientific value* (as well as greater usefulness to the organization) was more likely to be done by scientists who did not fully see eye to eye with the organization.

What to make of these trends? Do they simply mean that any capable scientist is bound to challenge the bureaucracy now and then? Or is it possible that an organization can be too agreeable for the scientist's own good, that a mild degree of tension is stimulating?

Possibly both. With a cross section survey, analysis cannot prove either interpretation. But we did look at the data to see what kinds of personal and organizational interests were stressed by scientists at each end of the congruence scale. Low-congruence individuals tended to stress research goals (such as items D and I in Question 55), although seeing the organization as valuing development activities (such as items A, E, H). High-congruence individuals tended to report themselves and their organization as both stressing research, and both de-emphasizing development.

Does not the latter condition seem ideal for research? The individual sees the organization as supporting his own commitment to publication and originality. This stress-free environment was productive of report writing. But it did not accompany work of scientific significance. Conceivably a mild degree of tension may have been invigorating. Conceiv-

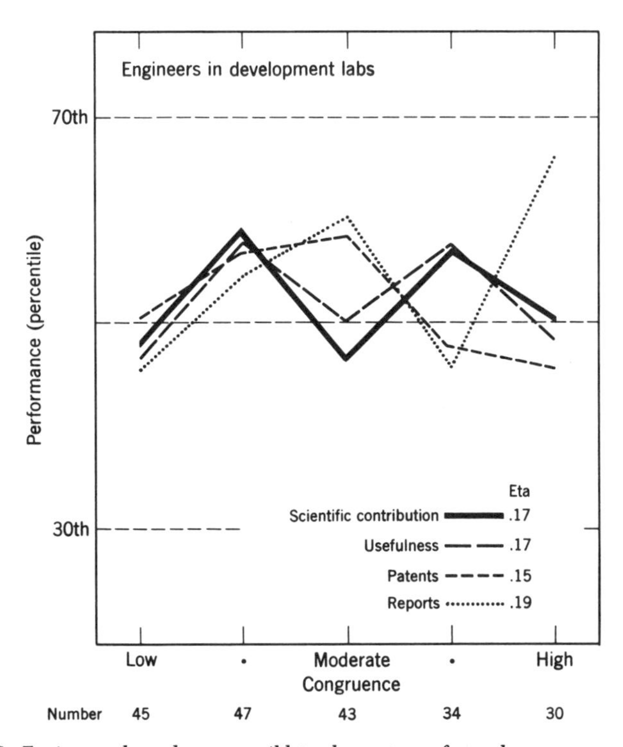

Chart 1-C. *Engineers showed a very mild tendency to perform above average when their congruence scores were neither very high nor very low; but they wrote most reports when they saw eye to eye with the organization.*

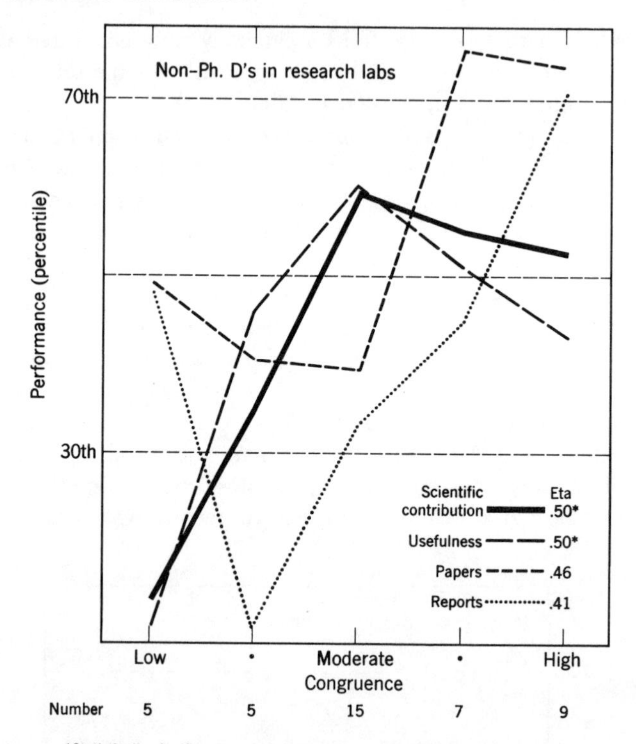

Chart 1-E. *Non-Ph.D's in research labs did work of scientific value as well as usefulness when personal interests agreed moderately well with those of the organization. But they wrote many reports and published many papers when they agreed fully with the organization.*

ably the organization's interest in product improvements may have stimulated broader scientific probing.

We now leave the concept of congruence and turn to the related concept of satisfaction. Congruence will come up again in Chapter 12 in which we shall examine the utility of various motivations in situations which differed in the stringency of their demands on the individual.

Satisfaction: Discrepancy between Desire and Provision

"A man's reach should exceed his grasp," mused Browning, "or what's a heaven for?" The questions we wanted to examine were: how close were our scientists to heaven; and how did they perform, the more closely they approached it?

In our questionnaire we attempted to measure "reach" and "grasp" separately by the methods shown in the box. (Question 62 was also used in Chapter 6; we repeat it here for convenience.)

Question 62. Listed below are different kinds of opportunities which a job might afford. If you were to seek a job, how much importance would you personally attach to each of these (disregarding whether or not your present job provides them)?

[Respondent rated each of 13 job factors on "importance I would attach" on a five-point scale from "slight or none" to "utmost"]:

A. To make full use of my present knowledge and skills
B. To grow and learn new knowledge and skills
C. To earn a good salary
D. To advance in administrative authority and status
E. To work with colleagues of high technical competence
F. To have congenial co-workers or colleagues
G. To work under chiefs of high technical competence
H. To associate with top executives in the organization
I. To build my professional reputation
J. To work on difficult and challenging problems
K. To work on problems of value to the nation's well being
L. To have freedom to carry out my own ideas
M. To contribute to broad technical knowledge in my field

Question 63. Now, to what extent does your present job actually provide an opportunity for each of these factors?

[The same 13 factors were rated on degree of "provision in present job" on a five-point scale from "slight or none" to "complete."]

To the extent that "grasp" or provision of a factor approached "reach" or desire for the factor, the individual was scored as satisfied on that factor. Dissatisfaction could arise from various combinations, such as intense need when provision was (say) moderate, or from moderate need when provision was nil. The individual would score as satisfied if both need and provision were strong, or if both were weak. The latter is more correctly a condition of apathy. These nuances may be important. But to consider them all simultaneously would make the analysis extremely complex. For simplicity, let us see what happens when we look separately at the two components of desire and of provision, and then at the degree of closeness or discrepancy between them.

It is true, of course, that these ratings of desire (or importance) and of provision, respectively, are not absolute. They are bound to be influenced by the respondent's idea of what is normal, just as the satisfaction ratings are. But note an important difference between our "provision" wording, and the typical satisfaction question. The latter might ask "how adequate is factor X?", with responses ranging from "not nearly enough" to "fully adequate." The respondent who knows that he has a good deal of X, but feels he needs more, is forced to check "less than adequate." Our format, on the other hand, permits him both to recognize the high level and to say that he wants still more.

Self-actualization and Status Advancement in Five Settings

On the basis of intercorrelations among the items in Questions 62 and 63, George F. Farris found a distinct cluster which in terms of Argyris' concepts he called a desire for "self-actualization" (items 62A, B, J, L). A second cluster indicated a desire to "advance in status" (consisting of items 62C, D, H). For each cluster he formed an index of desire and another of provision by adding the respective ratings.

Let us first see how scientists in various settings differed in their typical desire for self-actualization and status advancement, and their perception of opportunities for realizing these desires.

According to Chart 2, Ph.D's in research-oriented laboratories had the strongest need for self-actualization (use of abilities and learning of new ones, freedom to carry out their own ideas and to solve challenging problems). Assistant scientists stood lowest.

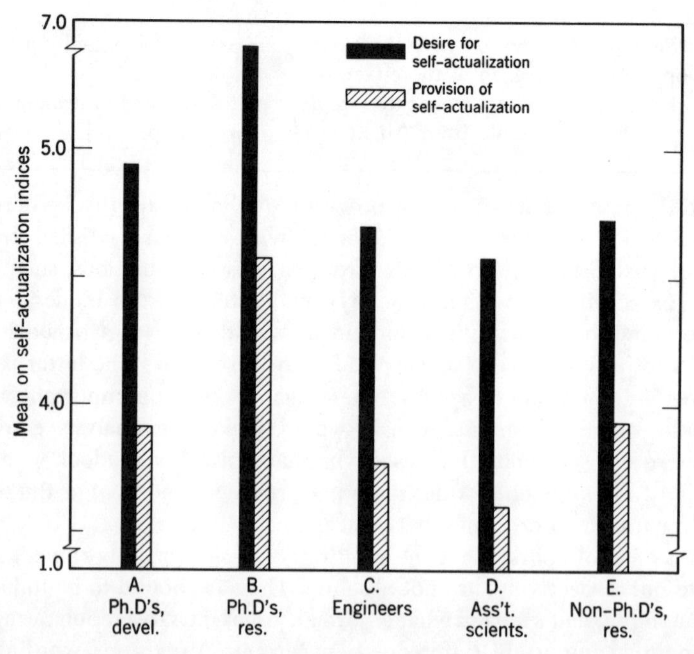

Chart 2. *Strength of desire for self-actualization, and provision of this factor, were based on four items from Questions 62 and 63, respectively. Research Ph.D's were highest in both their reported desire and provision, whereas assistant scientists were lowest in both respects. In terms of discrepancy between the two scores, Ph.D's in research counted as most satisfied with self-actualization.*

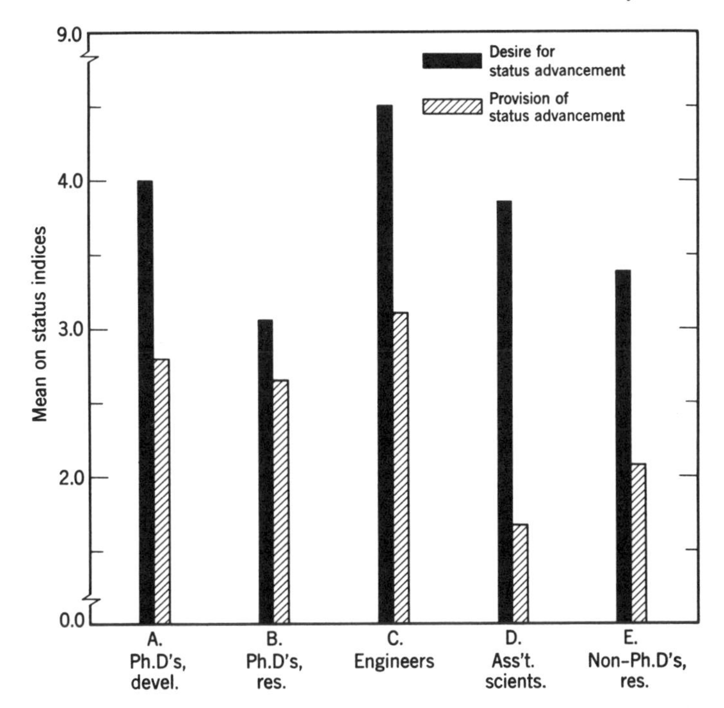

Chart 3. *Desire for and provision of status advancement were scored by a similar process. Engineers reported the best provision of status, but because of their very high aspiration, they scored as only moderately satisfied. Ph.D's in research had modest aspirations; and with high provision, they were well satisfied. Least satisfied were assistant scientists. (These scales are not equivalent with the scales for self-actualization; do not compare scores on the two charts.)*

At the same time, the provision of opportunities followed a similar profile: highest among research Ph.D's, lowest among assistant scientists. This is a familiar phenomenon in laboratory experiments on level of aspiration. The more you have, the more you want. As individuals perform better, their aspiration inches up; the "reach" stays a little ahead of the "grasp." If an individual does poorly, he adjusts his aspiration downward, so that he will not experience an unbearable sense of failure. (This sketch is over-simplified, and describes only the general tendencies.)

Our questionnaire ratings showed the same tendency. In general there was a mild positive correlation (typically in the .20's) between ratings of desire and of provision on most of the job factors. People who wanted more, had more, or vice versa.

In Chart 3, however, the parallels were not so clear. Here research Ph.D's reported the weakest need for higher status, and engineers the strongest. At the same time, research Ph.D's had about all the status they wanted; they were satisfied. Assistant scientists fared worst of all. Their status needs were moderately strong, but provision was lowest of any group. They keenly felt their deprived position at the bottom of the laboratory hierarchy.

Note the interesting contrast between engineers (group C) and non-Ph.D's in research (group E). Neither group was subordinate to a body of Ph.D's in the department. But the engineers as a whole felt relatively well off—better than any other group, in fact—in terms of the marks of prestige. The non-Ph.D's in research labs, however, felt almost as deprived as assistant scientists in Ph.D-dominated labs.

In general these profiles made reasonable sense; they were about what one might expect. Research Ph.D's, one might argue, were perhaps not quite telling the truth in their disdain of status aspiration; but no one would doubt that they were the most privileged of the five groups in this respect, as shown in their small discrepancy between desire and provision.

How Desire, Provision, and Satisfaction Related to Performance

Now we turn to the important question of how these scores related to various measures of performance. Were effective scientists generally happy, or unhappy, or neither?

For simplicity we have used correlation coefficients. A positive correlation indicates that as a given measure of desire or provision or satisfaction rose, performance in general went up. A negative correlation indicates that the higher such an item, the lower in general was performance.

The correlation coefficient has the weakness of failing to recognize curvilinear relationships—better performance at a moderate degree of a certain factor than at either extreme. Such curvilinear effects might lurk behind a zero correlation. But correlation coefficients have the great virtue of compactness: more information can be given in a brief space than would be possible with plots such as Charts 1-A to 1-E.

The data on *desire* for self-actualization and for status advancement to some extent cover territory already traveled. These are measures of motivation, drawing on some of the items whose relationship to performance was discussed in Chapter 6. The differences are that the data shown here were obtained only from the 500 long-form respondents (weighted), whereas Chapter 6 drew on all 1300 (unweighted); and the self-actualization and status indices consisted of somewhat different items compared to the previous indices of science, professional, and status orientation.

As shown in Table 2, desire for self-actualization showed a mild tendency to go with high performance (contribution and usefulness); but desire for higher status was not consistently related. These trends were parallel to those reported for indices of science and status orientation in Chapter 6.

The ratings of the level of *provision* on both factors, however, were clearly associated with performance, especially for the measure of usefulness (significant positive correlations in four out of five settings for both indices). Those scientists who were esteemed by colleagues for their scientific contributions, and particularly for usefulness to the organization, reported excellent opportunities to use their skills and carry out their own ideas, and for promotion and a good salary. Actual provision of such conditions was more strongly linked to achievement than was the desire to have them.

Some Interpretations

How should these results be interpreted? (a) Perhaps they simply mean that the organizations we surveyed were good at recognizing high performance. They rewarded achievers with more freedom, interesting work, better pay, and contact with executives. (b) Alternatively, it might be argued that the provision of such factors—giving the individual a chance to follow his own ideas, to tackle challenging problems, to be well paid and appreciated—actually helped to build a stimulating atmosphere in which individuals could do their best work.

It is impossible to establish or disprove either one or the other of these interpretations. Perhaps both are partially correct. Later we shall see that controlling on career level (job status) did not eliminate the results. It was not simply that high performers were those at higher levels who enjoyed more rewards and opportunities. The relationships existed even within career levels, strengthening interpretation (b) that such conditions, in fact, stimulated performance.

Regardless of interpretation, the data suggest a parodoxical comment about the reward system—whether rewards are *intrinsic* such as a chance to do what one likes, or *extrinsic* such as pay or title. First, the reward system is an essential feature of an effective organization. Management must give it careful attention; rewards must be commensurate with achievement. But second, as stressed at the conclusion of the preceding chapter, it would be shortsighted to rely on arousing status motivation as the primary *incentive* to accomplish. As scientists achieve, be sure to pay them well, but don't count on the desire for money or promotion as the main reason for achieving.

This interpretation may serve to reconcile an apparent discrepancy between the early results from our NIH study, and a study by Morris I. Stein of industrial chemists. He asked them to indicate how they would like to be rewarded for developing a major product by rank-ordering 12 rewards. Those judged by colleagues to be most creative ranked a cash bonus significantly higher than those judged to be less creative; the latter preferred a paid trip to a scientific meeting.[8] (The other eleven rewards received almost identical rankings by the creative and less creative groups.)

[8]M. I. Stein, "Creativity and the Scientist," Chapter 21 in Bernard Barber and Walter Hirsch (eds.), *The Sociology of Science,* The Free Press of Glencoe, New York, 1962, pp. 329–43.

TABLE 2 *The following shows by means of product-moment correlations, the relationships of various self-actualization and status measures to four criteria of performance. The following symbols are used:*

+ = *r is mildly positive (+.10 or more)*
++ = *r is positive and statistically significant**
– = *r is mildly negative (–.10 or less)*
–– = *r is negative and significant**

As in the previous chapter, desire for self-actualization had a mildly positive effect on performance, but status desire was inconsistent. Actual provision of both factors was clearly associated with higher performance, but satisfaction scores only moderately so.

	A. Ph.D's, devel.	B. Ph.D's, res.	C. Engi-neers	D. Ass't. scients.	E. Non-Ph.D's, res.	Scorecard +'s	Scorecard –'s	Scorecard Net
Self-actualization:								
Desire								
Contribution	0	++	0	0	++	4	0	+4
Usefulness	0	+	0	+	++	4	0	+4
Papers/patents†	0	0	0	0	–	0	1	–1
Reports	+	0	–	++	0	3	1	+2
						11	2	+9
Provision								
Contribution	0	++	++	+	++	7	0	+7
Usefulness	++	++	++	+	++	9	0	+9
Papers/patents	––	+	0	0	0	1	2	–1
Reports	+	+	+	0	0	3	0	+3
						20	2	+18
Satisfaction (provision minus desire)								
Contribution	0	+	++	+	0	4	0	+4
Usefulness	+	+	++	0	0	4	0	+4
Papers/patents	–	0	0	+	0	1	1	0
Reports	–	+	++	–	0	3	2	+1
						12	3	+9

	A. Ph.D's, devel.	B. Ph.D's, res.	C. Engi-neers	D. Ass't. scients.	E. Non-Ph.D's, res.	Scorecard +'s	−'s	Net
Status advancement:								
Desire								
Contribution	+	+	0	−	−	2	2	0
Usefulness	+	+	0	−	0	2	1	+1
Papers/patents	+	+	0	0	− −	2	2	0
Reports	0	0	0	+ +	−	2	1	+1
						8	6	+2
Provision								
Contribution	+	+ +	+ +	0	+	6	0	+6
Usefulness	+ +	+ +	+ +	+ +	+	9	0	+9
Papers/patents	−	0	+	0	−	1	2	−1
Reports	+	+	+ +	0	0	4	0	+4
						20	2	+18
Satisfaction (provision minus desire)								
Contribution	0	0	+ +	+ +	+	5	0	+5
Usefulness	+ +	+	+ +	+ +	0	7	0	+7
Papers/patents	−	0	0	0	+	1	1	0
Reports	0	+	+	− −	+	3	2	+1
						16	3	+13
Number of persons	101	79	219	90	45			

* Meaning that the probability of an r this large occurring by chance is less than 1 in 20, if the true r is zero. In computing correlations the data were weighted to compensate for sampling rates, but significance is based on actual number of cases.

† For engineers, patents were used instead of papers.

Perhaps Stein's able chemists were simply saying that when they achieved, they expected material recognition for doing so. This is not equivalent to saying that they achieved *in order* to receive a bonus.

Some Comments on Satisfaction Scores

Note that certain necessities follow when satisfaction is scored as the difference between what is desired and what is provided. (Assume throughout this discussion that desire and provision are not strongly related.) Suppose that (a) desire is *uncorrelated* with performance, but *provision* is strongly so. An illustration is the data for self-actualization among engineers. People high in provision will score as satisfied; therefore satisfaction will tend to behave in the same way as provision; both will show a positive correlation with performance.

Suppose instead that (b) *desire* is related to performance but provision is not; for example, among assistant scientists, the positive correlation of desire for self-actualization with reports. People with strong desire will score as dissatisfied; therefore satisfaction will tend to behave opposite to desire; satisfaction will correlate negatively with performance. "Best workers gripe the most." Those scientists who work hard because they are eager and ambitious will appear as frustrated.

Sometimes (c) both desire *and* provision are positively linked with performance (for non-Ph.D's in research, note that both desire and provision for self-actualization were correlated with contribution and usefulness). These two effects will cancel out in terms of the satisfaction score, which is likely to have *no* correlation with performance.

These considerations suggest that satisfaction (as we have scored it) is unreliable as a key to understanding scientific achievement. It is likely to correlate meaningfully only if the strength of *desire* is unrelated to performance although provision is related; and this is a dubious assumption. It would seem wiser for the investigator to measure the two components separately.

Further Measures of Desire, Provision, and Satisfaction

Further analyses reinforced the conclusion that ratings of *provision* as such were more promising indicators of high performance than were scores of *satisfaction.*

(a) One additional measure concerned satisfaction with the individual's autonomy or self-determination. In Chapter 2 we examined the weight

Question 29. Consider the choice of *goals or objectives* of the various technical activities for which you are responsible . . . Who has weight in deciding on these goals and objectives? . . . .

	Percent of weight in deciding goals
Myself	____%
Subordinates	____%
etc.	

Question 30. How much weight would you *prefer* different persons to have in deciding the goals of your technical work? Fill in percents below indicating the situation which you feel *would be most helpful* or stimulating to you in performing your job well.

	Preferred weight in deciding goals
Myself	____%
etc.	

exerted by various persons and groups in deciding the individual's technical goals. As a reminder, Question 29 is partially reproduced in the preceding box. Question 30 then went on to ask how much weight the respondent would *prefer* each of the decision-making sources to have. By taking the discrepancy between the weight a person assigned to himself, and his preferred degree of weight for himself, we generated a measure of "satisfaction with own weight in goals" (the larger the discrepancy, the lower the satisfaction).

The results in Table 3 (Part *a*) show that this measure of satisfaction was ambiguous in relation to performance. Among engineers, those who were satisfied (had as much own weight or autonomy as they thought they should) were significantly higher in contribution and usefulness. But in the other groups, individuals who were dissatisfied with their degree of autonomy published more papers. Perhaps these individuals had strong needs for independence; we have seen in Chapter 6 that self-reliant scientists were productive ones. Or perhaps they occupied subordinate roles, and published often in order to earn more autonomy. (Some trends in Chapter 12 suggest this view.) In any event, degree of satisfaction on this issue proved interesting but ambiguous.

(b) Another approach was to follow a typical procedure in an attitude survey in which a number of satisfaction scores are added to generate a measure of "total job satisfaction" or (loosely) "total job morale." We wondered how such a measure would relate to performance in our own data.

For each of the 13 job factors in Questions 62 and 63, we collapsed the importance and provision scales into three categories: moderate or less, considerable, and great or more. For each item, the individual was scored dissatisfied (score = 1) if his score on provision were less than what he preferred; neutral (2) if the provision were the same as preferred; or satisfied (3) if the provision were greater than preferred. The resulting discrepancy scores were summed across the 13 items for a "total satisfaction score." A neutral score on all items would generate an index of 26. It turned out that three out of five respondents were below this point, that is, generally dissatisfied, wanting more than they received; only two out of five received as much as they wanted, or a little more.[9]

When we correlated this over-all satisfaction index against the four performance criteria (Table 3, Part *b*), the relationships were surprisingly positive, especially for usefulness. Satisfied assistant scientists, however, wrote fewer reports.

[9]Preliminary results were reported in D. C. Pelz, "Satisfaction with the Work Situation, as Related to Output," Analysis Memo #5, December 1960, available in Publication #1741 from the Survey Research Center, University of Michigan.

TABLE 3 *The scientist who exercised as much weight in determining his goals as he preferred was scored as satisfied; but this measure (shown in Part a) had no consistent relationship to performance. A total satisfaction score across 13 job factors did correlate in a consistently positive way with performance, especially usefulness (Part b). A total score simply of the provision of the same 13 factors (Part c) correlated even better. Same symbols as in Table 2.*

	A. Ph.D's, devel.	B. Ph.D's, res.	C. Engi-neers	D. Ass't. scients.	E. Non-Ph.D's, res.	Scorecard +'s	Scorecard −'s	Scorecard Net
a. Satisfaction with own weight in goals								
Contribution	0	0	++	−	+	3	1	+2
Usefulness	+	0	++	0	+	4	0	+4
Papers/patents*	−−	−	0	−−	−	0	6	−6
Reports	0	−	0	0	−	0	2	−2
						7	9	−2
b. Total satisfaction score on 13 items								
Contribution	0	+	++	+	++	6	0	+6
Usefulness	++	++	++	++	+	9	0	+9
Papers/patents	−	+	0	0	+	2	1	+1
Reports	0	++	++	−−	+	5	2	+3
						22	3	+19
c. Total score on job provision								
Contribution	+	++	++	0	++	7	0	+7
Usefulness	++	++	++	++	++	10	0	+10
Papers/patents	−	+	0	0	0	1	1	0
Reports	+	+	++	0	0	4	0	+4
						22	1	+21
d. Total score on job desires								
Contribution	+	+	0	−	+	3	1	+2
Usefulness	+	0	0	+	++	4	0	+4
Papers/patents	0	0	0	0	−	0	1	−1
Reports	+	0	0	++	0	3	0	+3
						10	2	+8

*For engineers, patents were used instead of papers.

What to make of this? Our hunch was that on many of the items, outside of the self-actualizing area, the strength of desire had little effect on performance. Some evidence on this hunch was given in Chapter 6 with items such as desire "to work with colleagues of high technical competence," "to have congenial co-workers as colleagues," and "to work on problems of value to the nation's well being." Generally, they did not relate to performance. But the *provision* of various facilitative elements might accompany achievement.

(c) This idea was checked by constructing a total index of provision to see whether it correlated even better than total satisfaction. Using the same three-point scoring system on the 13 items employed for satisfaction, we constructed for each individual a score of "total job provision"—the total of his ratings on extent to which various job factors were provided.

(d) For good measure, we also summed the three-point scores on *desire* for the 13 job factors to yield a score of "total job desires." The correlations of these scores with performance are reported in Parts *c* and *d* of Table 3.

We were interested to observe that total job desires correlated positively with total job provision (*r*'s ranging from .29 to .53, not shown). Again, individuals with stronger wants concerning the job generally had greater opportunities also.

As it turned out, the score of total job *desires* (Part *d* of the table) correlated only mildly with performance (a net of eight positive symbols in the scorecard). The score of total *provision* (Part *c*), however, did even better than total satisfaction—a net total of 21 positive symbols. Note that total provision was significantly correlated with usefulness in each of the five groups. Note also that publication of papers (or patents, for engineers) was completely unrelated. People who published had neither more nor less than the average share of rewards.

The total provision score related nicely to performance. Across the five analysis groups its average correlation with usefulness was .38, and with scientific contribution .28 (corresponding average *r*'s for total satisfaction were .27 and .22). These relationships might reflect either the fact that good performers were recognized and knew they were recognized, or that substantial recognition in turn stimulated achievement, or both.

(e) For the record, we would like to report one other scoring system. We still felt it should be useful to take into account the level of desire as well as provision. But by computing the *difference* between provision and desire (as satisfaction was scored), one disparaged the provision of the very conditions which were most important. Instead of subtracting, why not *multiply* provision by importance? Let a high or low level of provision count for more on those factors which were most wanted by the individual.

Accordingly, each individual's provision of a given item was multiplied by his rating of its importance, and the products were summed to yield a total score of "provision-weighted-by-importance."

As it turned out, the total weighted provision score was almost identical with total (unweighted) provision score; correlations between the two were in the .90's. Therefore correlations between total weighted provision and performance were almost identical with those for total provision alone—possibly a little better, but not enough so to justify the complex scoring system. (Why were the measures so close? Partly, perhaps, because desire tended to parallel provision; what was provided was thus weighted up, what was not provided was weighted down, so that the end result by either method was similar.)

We returned to the conclusion that the simple score of *total provision of job factors*—ignoring the relative strength of *desire* for these factors—was about as strongly correlated with performance as any measure we have examined in the chapter.

Career Levels

Now let us return to an objection that was raised earlier. "Isn't it possible," the reader might ask, "that these relationships can all be explained quite simply by status level? As the scientist achieves, he becomes recognized; he rises on the organizational ladder, he gets more leeway to do what he wants, as well as better pay and the other marks of prestige. Since the senior scientists receive more gratifications than the junior, and also enjoy higher scientific reputations, this might completely account for your results."

This objection is partially answered by the adjustments in performance measures (described in Chapter 1) to rule out effects due to length of experience since the degree, and to time in the organization. However, even after adjustment, the scores tend to be higher for those who occupy higher levels.

Therefore the severe step was taken of repeating some of the foregoing analyses separately within each of four career levels. These are described in Appendix E: level 1 or "apprentice" for non-Ph.D's at the bottom of the professional ladder (mostly recent college graduates); level 2 or "junior" investigator for new Ph.D's (or for B.S. people with equivalent experience); level 3 or "senior" for mature nonsupervisory investigators with several years of experience; and level 4 or "supervisor" for section heads or up in government or industry, or university department chairmen and other full professors.

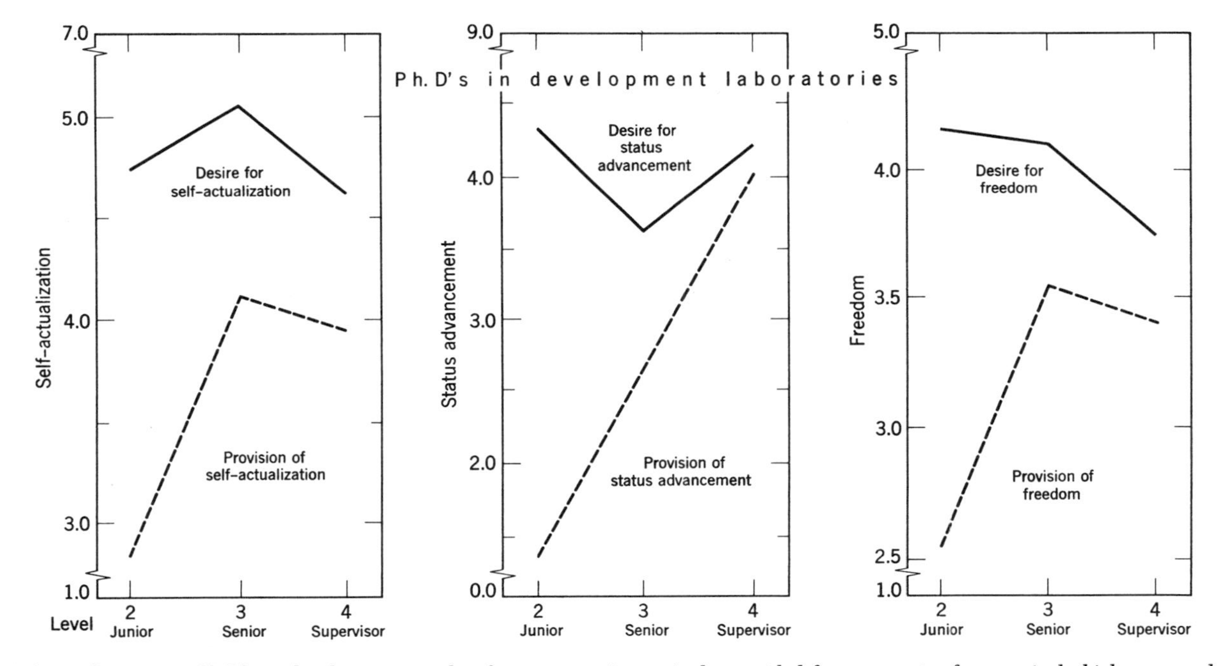

Chart 4 *shows that among Ph.D's in development, marks of status were increasingly provided for occupants of successively higher career levels. A mild rise in self-actualizing opportunity (including freedom to follow own ideas) occurred. But desire for these factors changed only slightly by career level. Number of persons at the three levels, respectively, were 17, 23, and 52.*

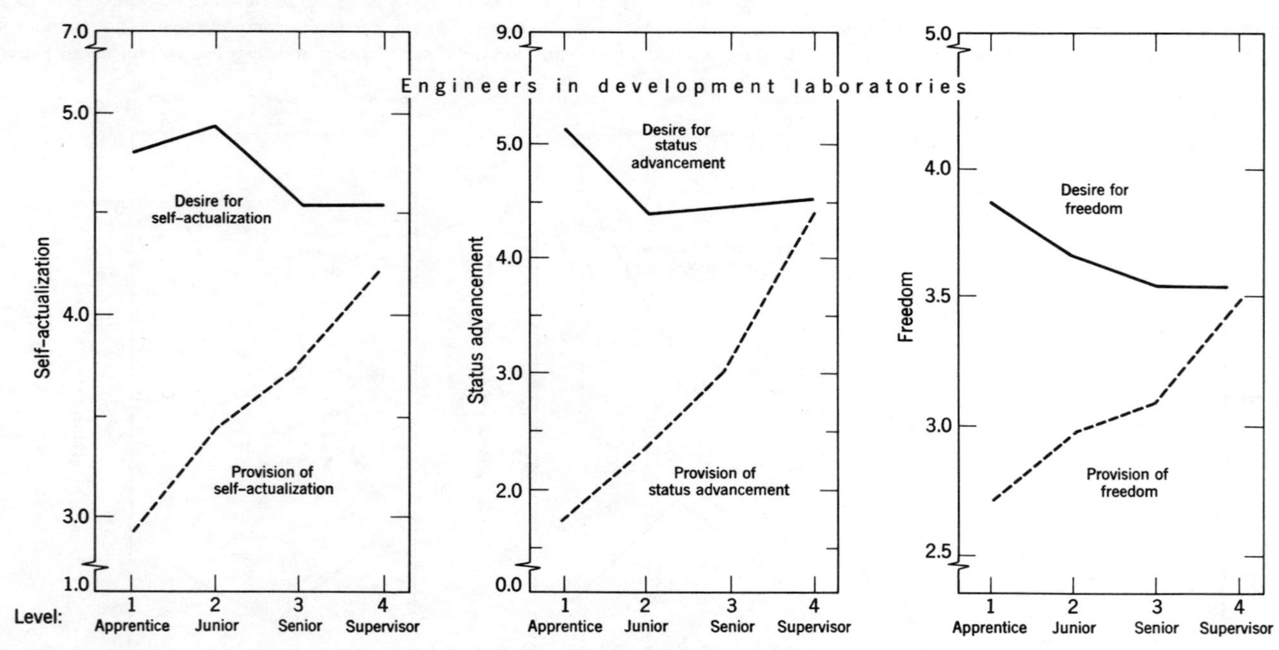

Chart 5. *Among engineers, provision of both status rewards and self-actualizing opportunity increased steadily at successively higher career levels, but desire for these remained constant or declined somewhat. Number of persons at the four levels, respectively, were 22, 55, 48, and 86.*

We looked first to see whether desire for, or provision of, various factors rose as career level increased (see Charts 4 and 5 for illustrations); and second, whether within each career level separately, relationships of job factors to performance had diminished or disappeared (see Tables 4 and 5).

Some results for Ph.D's in development laboratories are given in Chart 4. Provision of status rewards increased markedly among the three career levels. Opportunity for self-actualization rose distinctly between junior and senior levels, then flattened for supervisors. (Because of its importance as an indicator of self-reliance, we also show the component item of freedom for own ideas.) The strength of *desire* for status or self-actualization, however, showed no regular change with career level. (Senior investigators seemed a little more concerned with self-actualization, and less concerned with status, than were the other two levels; whether this difference was meaningful is hard to say.) There was an interesting trend for the freedom item, which *declined* slightly; supervisors seemed to want freedom a little less urgently than either junior or senior investigators.

The picture for engineers in Chart 5 was similar. Status rewards rose markedly among the four levels, and self-actualization opportunity (including the feeling of freedom) rose mildly and continued up for supervisors. At the same time, the desire for these factors was flat, or even declined mildly; note the slight drop in need for freedom.

Data for assistant scientists (not plotted) showed similar trends: increase in both provision measures with rising level, no change in desire or a slight drop. Ph.D's in research (not shown) also reported greater status provision at higher levels, but no further increase in opportunity for self-actualization, which was already very high; again needs for self-actualization and freedom declined somewhat.

We also plotted "total satisfaction score" and found that this rose steadily from level to level in all groups.

Now let us examine correlations between these factors and performance within each career level separately (Tables 4 and 5).

The symbols in Tables 4 and 5 are similar to those used previously. But since there were fewer individuals at each career level, and therefore the fluctuations in correlations due to chance would be larger, we set a higher criterion for a positive or negative trend: an r of $\pm.20$ or larger (rather than $\pm.10$ as previously). Double symbols again indicate correlations large enough for statistical significance, given the number of cases. (Because of the limited cases within non-Ph.D's in research, and for level 4 among assistant scientists, data for these groups are not shown.)

The results in Table 4 were surprising and gratifying. Correlations

TABLE 4 *This shows correlations between provision and satisfaction scores separately within each career level. Level 1 = "apprentice"; 2 = "junior"; 3 = "senior"; 4 = "supervisor." Because of the limited numbers of cases at each level, a stiffer criterion was set for showing trends:*

+, − = *r is* ±.20 *or larger*
++, −− = *r is statistically significant*

Provision of both self-actualization and status were still associated positively with performance, except for supervisory engineers.

	A. Ph.D's, devel.			B. Ph.D's, res.			C. Engineers				D. Ass't. scients.			Scorecard		
Level:	2	3	4	2	3	4	1	2	3	4	1	2	3	+'s	−'s	Net
Self-actualization:																
Provision																
Contribution	0	0	0	+	++	+	+	++	0	0	0	+	−	8	1	+ 7
Usefulness	+	+	0	+	++	0	++	++	++	0	0	+	+	13	0	+13
Papers/patents*	−	−−	0	+	0	0	−	0	0	0	0	0	0	1	4	− 3
Reports	++	−	+	+	0	0	0	++	+	0	0	0	+	8	1	+ 7
														30	6	+24
Status advancement:																
Provision																
Contribution	+	0	0	+	0	+	+	++	0	0	−	++	−−	8	3	+ 5
Usefulness	++	++	0	+	0	+	0	++	++	0	0	++	0	12	0	+12
Papers/patents	−	−	0	0	+	0	−	0	0	0	0	−	0	1	4	− 3
Reports	+	0	++	+	0	−	0	0	++	0	0	0	0	6	1	+ 5
														27	8	+19
Number of persons	17	24	53	20	26	32	22	55	48	88	33	30	22			

*For engineers, patents were used instead of papers.

TABLE 5 *The following relates the single item of freedom to carry out one's own ideas (a component of self-actualization) to four criteria of performance. High performers at each career level both wanted and experienced the feeling of freedom.*

+, − = *r is* ±.20 *or larger*
++, −− = *r is statistically significant*

	A. Ph.D's, devel.			B. Ph.D's, res.			C. Engineers				D. Ass't. scients.			Scorecard		
Level:	2	3	4	2	3	4	1	2	3	4	1	2	3	+'s	−'s	Net
Freedom for own ideas																
Desire																
Contribution	++	−	++	+	+	++	−	+	0	0	0	−	+	10	3	+7
Usefulness	++	0	0	+	0	++	+	0	+	0	0	0	+	8	0	+8
Papers/patents*	+	−	++	0	0	0	−	+	0	0	0	−	+	5	3	+2
Reports	+	−	0	0	0	+	0	0	+	0	0	0	+	4	1	+3
														27	7	+20
Provision																
Contribution	0	0	0	+	++	0	+	++	0	0	0	++	0	8	0	+ 8
Usefulness	0	0	0	+	++	0	++	0	++	0	−	++	++	11	1	+10
Papers/patents	−	−	0	+	+	0	0	0	0	0	0	+	+	4	2	+ 2
Reports	+	0	0	+	0	−	+	+	0	0	0	0	+	5	1	+ 4
														28	4	+24

*For engineers, patents were used instead of papers.

between provision of self-actualization and status, respectively, seemed about as strong within the various career levels as we had found when career levels were ignored. Despite the use of a stiffer criterion, positive trends were numerous, and many of these were statistically significant (almost one in five, not quite as large a proportion as previously).

We think the reader will be intrigued by Table 5, which shows a final set of results for the item of "freedom to carry out one's own ideas" (a component in self-actualization). When we related this to performance in each of the five primary analysis groups (data not shown), the results in the scorecard were feeble—only eight net positive trends, only one statistically significant. But when we looked within each career level, as in Table 5, we were amazed to note a net scorecard result of 20 positive trends, six of them significant. Holding level constant had increased rather than diminished the strength of the relationships.

The data in Charts 4 and 5 tell why. Instead of increasing with level, desire for freedom, if anything, dropped slightly. This meant that those at higher levels, who were also performing better, expressed less need for freedom. The result was a "suppressor effect" (as the statisticians would say), serving to hide the true relationship between desire for freedom and performance. When level was held constant, as in Table 5, the underlying positive trend came out sharply.

The measure of provision of freedom also held up. In the original five-group analysis, the scorecard showed 18 net positive trends, seven of them significant (data not presented). In Table 5, the scorecard for provision of freedom showed 24 net positive trends, eight of them significant.

Again the data confirmed the central importance of need for independence as a basic condition for scientific achievement, as discussed in Chapter 6.

SUMMARY AND IMPLICATIONS

What practical value can the research director gain from these results?

Chapter 6 concerned the motivations underlying high performance. We found there that a need for independence or self-reliance was an essential ingredient for scientific achievement, whereas ambition to rise in status was a dubious foundation on which to build.

Chapter 7 continued to examine such motivations, but also examined the actual opportunities provided by the organization to fulfill them. We considered a number of things the organization can do to reward achievement, falling into two broad categories: rewards *intrinsic* to the work itself (such as opportunity to use skills, to gain new knowledge, to deal with challenging problems, and to have freedom to follow up one's own ideas), and those *extrinsic* to the technical content (a good salary, higher administrative authority, association with top executives).

In this chapter we saw again that ambition for status was a shaky basis on which to spur achievement among staff members. But oddly, the actual provision of status rewards was definitely associated with achievement, even when we limited our attention to people at a single career level. Provision of intrinsic rewards (self-actualizing opportunities) was just as clearly associated, or more so.

The implication is, we feel, that the research director must give close attention to the whole system of rewards—both intrinsic and extrinsic. He must live with the paradox that extrinsic rewards cannot be relied on to motivate achievement, but that when achievement occurs, the extrinsic rewards should be consistent. And possibly the very provision of them will stimulate further achievement.

Some support for the last speculation can be cited from laboratory experiments on "dissonance theory." Workers who were paid more than they thought they deserved worked harder! [10]

This chapter also concerned the degree of congruence between activities which the individual valued personally, and those which he felt would help him progress in the organization. The best scientists (particularly among Ph.D's) saw only moderate congruence between their personal interests and those of the organization. What is the implication? Perhaps a laboratory remains vigorous when it encourages a certain tension between what the members want, and what they think the organization wants. The manager should be worried if he finds his staff perfectly willing to pursue those activities which he thinks his organization must stress.

In a similar way, we found that although satisfaction generally characterized high performers, there were several instances of the opposite. A certain amount of dissatisfaction, stemming from eager impatience, is perhaps inevitable in a healthy research atmosphere.

[10] Personal communication from Jack W. Brehm. See Brehm and Arthur R. Cohen, *Explorations in Dissonance Theory*, John Wiley and Sons, New York, 1962. Also see J. S. Adams, "Toward an Understanding of Inequity," *Journal of Abnormal and Social Psychology*, 1963, vol. 67, pp. 422–36.

8

SIMILARITY[1]

Colleagues of High Performers Disagreed with Them on Strategy and Approach but Drew Stimulation from Similar Sources.

Should a scientist work with colleagues who are similar to himself in their strategies for tackling technical problems? Or is contrast in their technical strategies helpful?

There are many other ways in which a scientist can resemble his co-workers: their respective previous experience in university, government, or industrial settings; their career goals; the sources from which they derive stimulation. Is similarity or dissimilarity along these lines beneficial, or inhibiting, or neither? The same questions may be asked regarding the scientist's similarity to his supervisor.

The answers to such questions, by helping to identify the mixtures of people which encourage innovation, may be of value in organizing research laboratories so as to stimulate creativity.

This chapter follows a lead from an earlier study which suggested that having dissimilar colleagues may (for some scientists) be a source of constructive "dither." The chapter describes six different ways of measuring similarity. It reports that for three of them, scientists with dissimilar colleagues were moderately better performers, although on one measure scientists with *similar* colleagues did slightly better. Two measures, however, and various ways of measuring similarity to the immediate chief, seemed to make no difference. The chapter concludes with some practical implications.

Previous Results

In a previous investigation at the National Institutes of Health, Mellinger and Pelz found that similarity to colleagues was related to performance, but in a complex way: the relationships depended on frequency of contact with colleagues. Senior scientists (GS-12 and up) who had *daily*

[1]This analysis was supported by grants from the National Science Foundation.

contact with their five main colleagues, and who were *dissimilar* to them in career orientations and past experience, performed at higher levels than those whose colleagues were similar to themselves. However, for senior scientists who contacted colleagues *weekly*, the relationship was just the opposite: better performance if colleagues were similar.

This finding suggested the concept of "uncertainty" or "dither." Having colleagues who think differently from oneself may be one source of intellectual jostling needed for innovation.[2]

Another source of such "dither" may be diversity in tasks and skills. Chapter 4 showed that scientists tended to perform better when they performed several R & D functions rather than one or two, when they spent less than full time on research, and when they had several areas of specialization.

In this chapter we shall consider *people* as a potential source of "dither." A case could be made that both similarity and dissimilarity are advantageous. Similar colleagues, it might be argued, provide easy friendships, good coordination, and the security of familiar surroundings. However, contacts limited to similar people could lead a scientist into comfortable but unproductive intellectual ruts. Conversely, dissimilar people might be intellectually stimulating but hard to work with.

Before we test these issues with our data, however, additional questions need to be raised. What kind of similarity are we speaking of? It is conceivable that scientists benefit if they are similar to colleagues in some respects, and different in others. We therefore constructed a variety of different measures of similarity, and examined performance in relation to each.

Second, it is quite possible that some types of individuals will benefit from dissimilar colleagues and other types from similar ones. In the earlier NIH study we found just such a difference. Accordingly, we studied various subgroups within our population, and found an interesting contrast to report.

Measures of Similarity

Of the various measures of similarity to colleagues which were constructed, four related to performance. These four will be discussed first.

[2] D. C. Pelz, "Motivation of the Engineering and Research Specialist," *General Management Series*, No. 186, American Management Association, New York, 1957, pp. 25–46. Warren Weaver suggested the term "dither" in an editorial in *Science* (August 7, 1959, vol. 30, p. 301.) During the war, his British colleagues built into antiaircraft computing devices a "small eccentric or vibrating member which kept the whole mechanism in a constant state of minor but rapid vibration. This they called the 'dither.' . . . We need a certain amount of dither in our mental mechanisms. We need to have our ideas jostled about a bit so that we do not become intellectually sluggish."

Respondents themselves identified their colleagues. Each scientist had named the colleagues with "the greatest significance for [his] work"—excluding supervisors above him and subprofessionals. (The questionnaire provided space to list up to five such colleagues; 3.7 were actually named, on the average.) For some measures of similarity we examined what the scientist told us about his colleagues. For other measures we went to the questionnaires which the listed colleagues had answered, and examined what they had said about *themselves.*

Similarity in Technical Strategy. This first measure was based on the scientist's *perception* of his similarity to each of his colleagues. Scores came from a question asking the respondent how similar he thought he was to each person in terms of the technical strategies they used in tackling problems. The question is shown in the following box. To measure similarity between the scientist and his five colleagues, we simply averaged his answers for these individuals.

> *Question 45.* The "approach" or "strategy" in tackling technical problems will vary—at what point you start, what concepts or methods you use, what sequence of steps you follow, etc. When you are talking over a technical problem with each person, to what extent do you find yourself adopting similar or different "technical strategies"?
>
> [With respect to each of five colleagues, respondents rated their respective strategies from "almost completely different" to "almost completely similar."]

Similarity in Style of Approach to Problems. In Chapter 6 we described our efforts to measure several broad categories of motivations. One category concerned "style of approach." Six scores of style were obtained for each participant: preference for an abstract approach, a concrete approach, broad mapping of new areas, deep probing of narrow areas, immediate solutions to problems, and long-range planning.[3]

Each distribution of scores was divided roughly into quartiles. To measure dissimilarity between the respondent and any colleague, we simply took the difference between his own quartile on a given score and the quartile in which the colleague fell (based on the *colleague's* response), and summed across all six scores (maximum difference, therefore, was

[3]Most of the items from which each component was scored are shown in Chapter 6. The components were as follows: abstract = Questions 19J + L + N + P; concrete = 19-O; broad = 19E; deep = 19F; immediate solutions = 19M; long-range = 19R + S. The items not appearing in Chapter 6 read as follows. In item 19-O the respondent indicated how closely the following statement described him: "I prefer to find out all I can by observation before trying to generalize." In 19R the statement was: "I prefer to plan a long-range series of related tasks which I follow more or less systematically." The statement for 19S was: "I prefer to alter my direction from week to week as new directions arise."

18). To measure similarity between the scientist and his group of colleagues, we averaged the individual similarity scores for all colleagues.

This was an "objective" measure, in the sense that *we* determined the agreement between the scientist and each of his co-workers, rather than using the scientist's own perceptions of similarity, as in the previous measure.

Similarity in Career Orientations. A second broad category of motives described in Chapter 6 concerned the *directions* or purposes toward which the scientist felt he was striving. Two important directions described there were orientation toward (a) contributing to science, and (b) advancing within the organization (that is, status).

Some scientists strive toward both purposes, others toward just one, still others apparently toward neither. These orientations have been identified in a number of other studies.[4]

By computations similar to those used for the style index, we determined the similarity between the scientist and each of his co-workers with respect to these career orientations.[5] This measure again was "objective": it was based on the answers of both the scientist and the colleagues with whom he was being compared.

Similarity in Sources of Motivation. Our fourth measure was based on six *sources* from which motivation or stimulation might originate. (These were also discussed in Chapter 6.) They were: (a) the index of "own ideas" as a source (own previous work or plans, own curiosity, desire for freedom); (b) clients or practical problems; (c) supervisors; (d) the technical literature; (e) isolation (working alone); and (f) competent colleagues.[6] Like the previous two measures, this one also was objective in method of construction. As before, it was computed by summing (across the six components) the quartile differences between the scientist and the person with whom he was being compared.

How Similarity to Colleagues Related to Performance

When we looked at the various measures of average similarity to colleagues, we found that the four previously described did relate to scientific performance. The correlations shown in Table 1 were not very

[4]Footnote 2 in Chapter 6 cites various studies in which these and related concepts have been explored.

[5]The science and status components of this measure were based on the following items, which appear in Chapter 6: science orientation = 62A + L + M; status orientation = 19B + C + 62C + D + H. Their intercorrelations appear in Appendix G.

[6]The components of this measure were based on the following items: own ideas = Questions 13E + J + 62L; client = 13G + I; supervisor = 13A + B + 19I; technical literature = 13D + F + H; isolation = 19G + H; competent colleagues = 62E + F + G. All of these items are quoted in Chapter 6 in either the boxes or under "other sources."

TABLE 1 *The following is based on correlations (r's) among similarity to colleagues in several respects and four criteria of performance (adjusted for length of experience).*

+ = *r is mildly positive (+.10 or more)*
++ = *r is positive and statistically significant*[*]
− = *r is mildly negative (−.10 or less)*
−− = *r is negative and significant*[*]

For a general picture, the scorecard at the right sums the number of positive or negative symbols. Scientists who were dissimilar to colleagues with respect to strategies, style, and orientations tended to perform at higher levels than those who were similar. With respect to sources of motivations, however, those who had similar colleagues did better.

	A. Ph.D's, devel.	B. Ph.D's, res.	C. Engineers	D. Ass't. scients.	E. Non-Ph.D's, res.	Scorecard +'s	−'s	Net
Similarity with respect to:								
a. Technical strategy								
Contribution	0	−	0	−	0	0	2	−2
Usefulness	0	0	0	−	−	0	2	−2
Papers/patents[†]	+	−	0	0	−	1	2	−1
Reports	0	−	0	0	−	0	2	−2
						1	8	−7
b. Style of approach								
Contribution	0	0	0	0	−	0	1	−1
Usefulness	0	0	−	−	−−	0	4	−4
Papers/patents	−	−	0	+	0	1	2	−1
Reports	−	−	0	0	−	0	3	−3
						1	10	−9
c. Orientations								
Contribution	+	0	−	−	+	2	2	0
Usefulness	0	0	−−	0	0	0	2	−2
Papers/patents	0	−	−	−	0	0	3	−3
Reports	−	0	−	0	−	0	3	−3
						2	10	−8
d. Motivation sources								
Contribution	+	++	0	0	+	4	0	+4
Usefulness	+	+	0	0	+	3	0	+3
Papers/patents	−	0	0	0	−	0	2	−2
Reports	+	0	0	0	+	2	0	+2
						9	2	+7
Number of persons	101	79	219	90	45			

[*] By "statistically significant" is meant that correlations of this size would not arise by chance more than 5 times in 100, if true *r* were zero.

[†] For engineers, patents were used instead of papers.

strong, and they did not appear everywhere we looked, but they occurred consistently enough to be intriguing.

We had, of course, expected to find negative relationships, that is, to find better performance among scientists who were *dis*similar to colleagues. These generally did appear on three of the measures, but on the fourth—to our surprise—the opposite trend occurred.

Part *a* of Table 1 shows that scientists in research laboratories (both Ph.D's and non-Ph.D's), and assistant scientists performed slightly better if they perceived themselves as dissimilar to colleagues with respect to *technical strategy.* The performance of engineers, in general, was not affected by this type of similarity (although as will be shown later, performance *was* affected for a certain subgroup of engineers).

As shown in Part *c* of Table 1, however, engineers performed better on all four criteria if their colleagues were dissimilar in *career orientations* (that is, if a science-oriented engineer had status-oriented colleagues, or vice versa).

The table shows also that most groups of scientists performed somewhat better if they differed from colleagues in *style of approach* (abstract versus concrete, broad versus deep, immediate solutions versus long-range)—Part *b* of the table.

That dissimilarity in at least three respects seemed to enhance performance could be viewed as another instance of the diversity phenomenon discussed in Chapter 4. Here we see that scientists and engineers who experienced some diversity among their group of colleagues—with respect to strategies, styles of approach, and career goals—seemed to benefit.

Perhaps colleagues who were dissimilar in these respects helped a scientist see problems in new perspectives or find solutions in unsuspected areas. We did much exploring to find particular conditions under which dissimilarity was especially helpful. These findings will be reported shortly.

First, however, we were intrigued to find that not all forms of dissimilarity were helpful. Among Ph.D's in research and development, and among non-Ph.D's in research, performance was a little better when colleagues were predominantly *similar* in *sources of motivation,* that is, when the individual and his co-workers were both self-directed, or both stimulated by practical problems, etc. These data appear in Table 1, Part *d.* (Although the data in Table 1 indicate that similarity in source of motivation made little difference for the performance of engineers and assistant scientists, as we pushed further we found some relationships here too.)

Thus it appeared that some combination of similar and dissimilar characteristics in one's colleagues might be best. It seemed helpful if sets of close colleagues shared a common enthusiasm for similar kinds of problems (derived from the technical literature, perhaps, or from practical problems), and preferred similar social relations (either isolation or close interaction, either self-reliance or dependence on a common chief). But diversity in one's colleagues seemed desirable with respect to intellectual strategy or approach in solving the problems.

Perhaps scientists who were similar to their colleagues in the type of problems they enjoyed and the social relations they desired experienced a kind of emotional security or support. But dissimilarity to colleagues in strategy or style of approach may have provided the intellectual jostling or "dither" required for innovation.[7]

Similarity Mattered More for Innovators

In a further effort to understand the nature of the relationships which appeared in Table 1, we examined a large number of environmental and social conditions to see whether the relationships would be stronger in some situations than in others.

One factor that made a difference concerned the scientist's research role. We wanted a way to separate scientists who did primarily creative work from those performing more routine activities. We had included no direct question on this matter, but we did have data on the extent to which each scientist believed he was useful to others in providing "original ideas." We used these answers to form two groups: respondents who thought their original ideas were among their own most useful functions, and those who thought that their other attributes were more useful. We then repeated the analysis shown in Table 1 for each such group separately.

The results (shown in Table 2) were intriguing. It appeared that our previous findings were applicable primarily to scientists who said they were useful for their original ideas. For convenience let us call these people "innovators." It was primarily such innovators who were helped by dissimilarity in style and in technical strategy—and at the same time were helped by similarity in motivational sources. But among other scientists who thought themselves not especially innovative, similarity or dissimilarity to colleagues had inconsistent effects.

Separating scientists in this way allowed some relationships to emerge which had been masked in Table 1. In the first analysis, dissimilarity in strategy or style seemed hardly to affect the performance of engineers. Table 2, however, shows that such dissimilarities tended to help the more innovative engineers, but not the less innovative ones. Once we made

[7]Dissimilarity in career orientation does not seem to fit the second category, although its effects for engineers were the same. See discussion at end of chapter.

this separation, the findings for the engineers became consistent with those for other groups.

(With respect to similarity in orientations, however, whether the individual was an innovator made little consistent difference; dissimilarity tended to help both the more innovative and the less.)

Thus dissimilarity to colleagues seemed most relevant for scientists who claimed to be innovative. This finding made good sense. It supported our hunch that colleagues who are intellectually dissimilar may provide new perspectives on problems that require creative solutions. Where we expected new perspectives to be most helpful (that is, for the innovators), we found the relationship most clearly.

At the same time, it is interesting that those innovators also benefited from colleagues who were *similar* in motivational sources. If it is true that innovators are especially liable to the "enormous anxieties that accompany creative thinking" (to use a phrase by W. J. J. Gordon[8]), then they may especially need extra sources of emotional support.

These tendencies have some fascinating implications for the role of the innovator in a research or development lab. They suggest that the more effective innovators do "stand apart" from their colleagues in some respects: they adopt different approaches for tackling problems. And yet the productive innovators are not isolated; they share their colleagues' enthusiasm for certain sources of problems, and they prefer similar kinds of social relations.

Some Further Explorations

Type of Job. We wanted to check on the scientist's report about the usefulness of his original ideas. Did self-described innovators, for example, in fact work at jobs requiring high creativity? In a subsequent setting (a large industrial laboratory whose data have not generally been included in this book), additional questions were asked about the kind of job.

We found that the perceived usefulness of scientists' original ideas was indeed related to replies about their jobs. Among the 350 scientists in the subsequent study, there was a significant positive correlation ($r =$ .35) between self-rated usefulness of one's original ideas and the degree to which the man's job consisted of "pulling together ideas from apparently unrelated areas and forming useful new combinations of them." In addition, there was mild but significant negative correlation ($r = -.18$) between usefulness of own original ideas and engaging in standardized data collection: "systematic observation and recording of facts about the properties of my subject matter."

Thus we gained some confidence that scientists who reported them-

[8]Quoted in S. Burry, "The Question of Creativity," *Industrial Design*, January 1957, p. 32.

TABLE 2 *This is again based on correlations (r's) among similarity to colleagues in several respects and four criteria of performance (adjusted for length of experience). Correlations were computed separately for "innovators," that is, those who believed their original ideas were highly useful (H), and those who ranked their original ideas of lower usefulness than other functions (L). Other symbols are as in Table 1.*

Among the innovators, higher performers tended to be those with colleagues who differed in strategies, style, and orientation, but who had similar sources of motivation. But for scientists low on innovation, these types of similarity showed inconsistent patterns.

	A. Ph.D's, devel.		B. Ph.D's, res.		C. Engi-neers		D. Ass't. scients.		E. Non-Ph.D's, res.		Scorecard Net totals	
	L	*H*	*L*	*H*	*L*	*H*	*L*	*H*	*L*	*H*	*L*	*H*
Similarity with respect to:												
a. Technical strategy												
Contribution	0	+	–	–	+	0	+	0	0	–	+1	–1
Usefulness	+	+	0	–	0	–	0	0	–	–	0	–2
Papers/patents*	+	0	0	–	+	–	0	–	0	+	+2	–2
Reports	+	0	0	0	+	0	0	–	–	–	+1	–2
											+4	–7
b. Style of approach												
Contribution	0	–	0	–	0	0	0	+	–	0	–1	–1
Usefulness	0	– –	+	–	–	–	0	0	–	–	–1	–5
Papers/patents	–	+	+ +	– –	0	0	+ +	– –	–	–	+2	–4
Reports	–	0	0	–	–	–	+	0	–	–	–2	–3
											–2	–13

c. Orientations												
Contribution	0	–	–	0	0	– –	0	–	0	0	–1	–4
Usefulness	0	–	0	0	0	– –	0	0	– –	+	–2	–2
Papers/patents	++	0	–	–	–	–	–	– –	–	+	–2	–3
Reports	++	0	–	0	– –	–	+	–	–	+	–1	–1
											–6	–10
d. Motivation sources												
Contribution	0	– –	–	0	–	0	–	+	–	0	–4	–1
Usefulness	0	+	0	0	0	+	–	0	–	++	–2	+4
Papers/patents	+	0	+	+	0	+	–	0	–	+	0	+3
Reports	+	0	0	+	0	0	0	+	0	++	+1	+4
											–5	+10
Number of persons	25	50	27	24	80	110	40	30	25	14		

* For engineers, patents were used instead of papers.

selves as "innovators" probably performed jobs requiring some creativity.

Different Kinds of Similarity. We also wondered whether it was correct to speak of different kinds of similarity, or whether scientists similar to colleagues in one respect would tend to be similar in other respects. The fact that similarity in sources of motivation showed different effects from the other kinds was one clue suggesting the wisdom of measuring each separately. Further confirmation occurred when we examined the correlations among the various measures (data not shown). We found that each was almost independent of the others, with one exception. Not surprisingly, people who were similar to colleagues with respect to career orientations also showed a slight tendency to be similar with respect to sources of motivation (mean r across five analysis groups = .22). Despite this positive relationship, the two measures had shown opposite effects!

In short, it made sense to distinguish different kinds of similarity. Scientists who resembled their colleagues in one respect might or might not do so in other respects.

Results for Main Colleague. We wondered whether we would get the same results if we looked at similarity to just the *main* colleague (the one listed first in order of significance). Our interest stemmed from an odd result in the earlier NIH study. There, scientists performed better if they had daily contact with several dissimilar colleagues, but if they were *similar* to their main colleague (and saw him daily).

This result had originally suggested the "security" concept; perhaps one's main colleague should be similar for emotional support, the others different for dither.

We repeated the analysis shown in Table 1 using just the main colleague, but found nothing new. The results for the main colleague were much the same as for the colleagues combined.

Although the importance of security did appear in other results (for example, Table 1), it did not re-appear in the form that had first suggested it.

Results for Supervisor. We also looked at similarity between the scientist and his administrative chief. These steps produced nothing. So far as we could tell, it did not matter whether a scientist was similar to his chief.

Mixtures of Similarity and Dissimilarity. As another follow-up on the earlier results, we wondered whether it might be desirable (on any given measure of similarity) to have a mixture among various co-workers, for example, to have a chief who was similar and a main colleague dissimilar; or to have these two individuals similar and the remaining four colleagues dissimilar. Numerous patterns of this sort were explored for each measure, but none of them proved fruitful.

Two Other Similarity Measures. Up to this point we have described four aspects in which a scientist might be similar to his colleagues, his main colleague, and his supervisor. All of the analyses described previously included two additional similarity measures: similarity in *past work experience* and similarity in *research role.* The first considered the type of institutions (industrial, university, government, etc.) the respondent and the person with whom he was being compared had worked in previously. The second compared people according to the research roles for which they thought they were most useful—providing technical know-how, critical evaluation, etc. Neither of these similarity measures related to performance.

SUMMARY AND IMPLICATIONS

The question-and-answer format will be used to explore some possible meanings of these results.

> Would you start by summarizing the major findings in a few words?

Certainly. These findings were not strong in magnitude, but they were reasonably consistent. Scientists tended to perform better if they named as colleagues individuals from whom they differed in the strategy of tackling technical problems, and in the style of approach to the work—abstract versus concrete, broad versus deep, etc. At the same time, scientists did a little better if they named as colleagues individuals similar to themselves in their sources of motivation—kinds of problems or types of social relations preferred. The findings were sharper for "innovative" scientists, those who rated themselves useful to colleagues for having original ideas. But similarity or dissimilarity to the immediate chief did not seem to matter.

> Isn't there an inconsistency there? Can you reconcile the opposite results for different measures of similarity?

Actually, the two findings make good sense. Differences between the scientist and his colleagues in their technical strategy or approach to the work may provide the intellectual jostling or "dither" which is needed for really creative work. On the other hand, similarity to colleagues in motivational sources—types of problems and social relations—may supply emotional "security" necessary to sustain the anxieties of creative activity.

> What about your results for dissimilarity in career orientations—whether the man is mainly driving toward scientific contribution or toward higher status in the organization? Don't you have some trouble fitting these results with the others?

We're somewhat puzzled. Similarity in career interests, on the surface, would seem to denote common interests in much the same way as similarity in motivational sources. And we found, in fact, a mild positive correlation between the two measures. Yet the effects on performance were opposite. One point to consider is that this measure seemed to work mainly for engineers. Perhaps among engineers these measures convey genuine intellectual differences. The science-oriented engineer may take an abstract or long-range approach; the status-oriented engineer, concerned no doubt with practical pay-off, may take a concrete or short-run approach; and each may benefit by this conflict in ideas.[9]

> What do you make of the fact that several of the results from your earlier study were not confirmed in your present data?

We were disappointed, naturally. But, of course, one purpose of the present study was to see what findings might show up consistently across several settings. And even if the early results were not strictly confirmed, they were valuable. They led to the formulation of general concepts such as "dither" and "security"; and these *have* appeared in a number of other results, though not in the exact form which originally suggested them.

We'd like to emphasize a parallel between these results and some to be presented in Chapter 13 on group age. It will be shown there that in effective older groups the members respected one another, but they maintained an atmosphere of "intellectual tension." Similarly, William Evan obtained data suggesting that effective industrial groups had personal harmony or liking, but considerable intellectual conflict.[10]

All these results point to the conclusion that scientists can benefit from those colleagues with whom they are personally compatible (in the sense of sharing common enthusiasms and working habits), but at the same time are intellectually competitive. Maybe that's not a very profound statement. But it does help to correct an oversimplified assumption that is sometimes made—that to work together, members of a group must be "cooperative" and "compatible" in all respects, and that the man who is "different" has to be isolated, or else he'll ruin the morale of the others.

> Suppose that all you say is right. What difference does it make to me, as a research manager, in the way I run my lab? Are you suggesting that I can change the interests and motivations of my people to make them more compatible or more competitive?

[9]Checking back to our data, we found, in fact, that among engineers similarity in orientation correlated modestly ($r = .20$) with similarity in style, although for other groups there was no relation.

[10]W. M. Evan, "Conflict and Performance in R & D Organizations: Some Preliminary Findings," *Industrial Management Review*, 1965, vol. 7, no. 1, pp. 37–46.

You might be able to do more to stimulate new interests than seems at first feasible—see some of the suggestions at the end of Chapter 11, for example. But let's consider some of the easier steps you could take. From time to time you will be setting up new sections, reorganizing departments, and pulling together committees to coordinate new projects. Usually you will consider mainly the technical content of what each man brings to the new group. According to our results, you should give weight to motivations as well.

With some thoughtful interviewing, you can probably select a group who will respect one another, share common enthusiasms, and prefer similar methods of working together. Maybe you already do this—you try to see that they will be compatible. But you should also try to put into the same group individuals with different modes of intellectual functioning: one man who works with his "head," another more with his "hands"; one individual who likes to map out a long-range, systematic approach, another who enjoys wrestling with short-range puzzles. Then name as coordinator someone who will see to it that these people exchange ideas vigorously.

There is another point that will make your job easier. Groups of diverse composition are mainly important in areas where you are hoping for really creative approaches. For other parts of your laboratory that deal more with standardized tasks, diversity in group composition won't matter so much.

And finally, there are certain symptoms that you can watch for. If you find your staff members arguing hotly, and coming back the next day to argue some more, good! If you find them nodding politely to each other, or even enjoying the same social activities, without shop-talk—not so good! They are not gaining the full benefit of intellectual jostling. The potential dither may be there, but it is dormant.

How can you awaken it? That may call for some creative management. One possible way: set up periodic seminars or conferences on current unresolved problems, and invite several individuals to discuss how they would approach these—where you suspect that each individual will use a rather different approach. This may set off sufficient debate to spill over into other areas of their work.

9

CREATIVITY[1]

Creative Ability Enhanced Performance on New Projects with Free Communication but Seemed to Impair Performance in Less Flexible Situations.

Scientists, like other people, vary in their ability to be creative. Although some jobs in science are of a purely routine nature, it is commonly believed that creative ability is a useful attribute for many types of scientists. We measured the creative ability of some of our scientists and engineers and related it to their performance.

Did the scientist with high creative abilities consistently outperform the less creatively able ones? Our data said "no."

Were there certain situations which affected the "payoff" from creative ability? Our data said "yes." This chapter tells of our explorations and discusses these situations.

The chapter will first describe some of the distinctions we made in thinking about creativity, and some ideas about how the environment may affect creative payoff. Then it will describe how we measured the creative abilities of some of the scientists. And finally, it will identify some of the environmental situations which seemed to be important when one tries to help creative people do their best. Many aspects of the environment didn't seem to make much difference; these will be briefly described.

Creativity, Creative Ability, Originality, and Productivity

Before turning to the data, several concepts need to be distinguished which have often been used rather loosely.

"Creativity," as we will use the term, is a characteristic of a person's output—his work, his performance. When we say a person's work shows

[1] A previous analysis of these data formed part of a doctoral dissertation by F. M. Andrews, "Creativity and the Scientist," University of Michigan, 1962. A shortened version appeared as "Factors Affecting the Manifestation of Creativity by Scientists," *Journal of Personality*, 1965, vol. 33, no. 1, pp. 140–52. The analysis reported here was mainly supported by a grant from the National Institutes of Mental Health.

high creativity (or that the person is "creative"), it means that others have found his performance both *original* and in some way *useful*. Originality is simply the quality of being different or unusual. Many people can propose unusual solutions to problems. People in mental hospitals frequently do this. But whether the unusual solution meets the specifications of the problem, that is, whether it is useful, determines whether an original solution is creative. Thus creativity implies originality, but original products may or may not be creative.

"Creativity" is sometimes used synonomously with "productivity." These two need to be distinguished also. A scientist, or anyone else, may be highly productive in a routine way. If so, although he is productive, he is not being creative. Only if his productivity is characterized by a high degree of *unusualness* (originality) and *usefulness* will we say that the scientist shows high creativity.

People differ in their ability to be creative, and we would like to distinguish creative ability (the potential) from creativity (a characteristic of the product). It may be possible for a person to have high creative ability but not actually be creative. Some ideas as to how this might occur are the topic of the following section. Before turning to it, however, it should be noted that creative ability is only one of many abilities a person may have. Even though a scientist may have only a small amount of creative ability, he nevertheless may be a productive and highly valued colleague because he makes good use of other abilities.

Environmental Effects

The environment in which a person works may affect the likelihood of his making good use of whatever creative ability he has. Without pretending to be able to describe the entire creative process, we can perhaps identify gross aspects of it.

It seems reasonable to expect that somehow a person first must get an original idea which is useful for the problem at hand. (His capacity for getting such original and useful ideas is his creative ability.) Even after he has the idea, he must be willing and able to make this idea known to others. Now the presentation of new ideas may be risky, particularly when the new idea is contrary to accepted ways of doing or thinking about things, or where the suggestion that something could be done better implies a criticism of one's superiors or colleagues. Thus, for a variety of good reasons, people may be *unwilling* to suggest whatever potentially creative ideas occur to them.

If a person has a good idea and is willing to make it known to others, he still may be *unable* to do so. The message may not "get through" to others. Some possible reasons might be that the person lacks the status (for example, formal training, reputation, rank, etc.) for others to take his ideas seriously; or perhaps there is no direct way for him to convey an idea to a superior who is several rungs up the hierarchy. Thus the communicative aspects of a situation may prevent a person from making good use of his creative ability.

Finally, a potentially creative idea may fail to pay off even if a person is willing and able to convey it to others. For a variety of reasons (many of them good), a situation may be rather *inflexible* and therefore not open to the shifts which a creative idea would require. For example, after several years a research project may have its methods and goals clearly defined. The job is to follow the decided-upon route and achieve those goals rather than to set off in new directions. There might be little opportunity for taking advantage of potentially creative ideas on such a project. Thus another aspect of the environment, its flexibility, may determine whether creative ability gets translated into creative performance.

"Payoff" from Creative Ability

There have been numerous studies of the characteristics and backgrounds of creative people (see, for example, the studies done by Barron,[2] Taylor,[3] or Anne Roe[4]). There have, however, been few attempts to distinguish creative ability from creative performance, and to examine the relationship between them in a variety of situations. This latter, in effect, asks the question, "What conditions affect the 'payoff' from creative ability?" or, alternatively, "Under what conditions is the 'payoff' from creative ability likely to be high?" Our own explorations have sought answers to these kinds of questions.

The Remote Associates Test

Before factors which might affect the payoff of creative ability in scientists could be examined, three kinds of data were needed: a measure of the scientist's creative ability, measures of his technical performance, and knowledge about the situation in which he worked. The data provided by the scientists on their questionnaires and by the judges who evaluated their performance met the last two needs. Chapter 1 describes how these data were obtained.

When the data were originally collected, however, no measures of creative ability were obtained. Therefore, about two years after the first data collection, we went back to some of the respondents and asked them to take a creative ability test. And 355 did so.

[2]F. Barron, *Creativity and Psychological Health,* Van Nostrand, Princeton, New Jersey, 1963.

[3]C. W. Taylor, "A Tentative Description of the Creative Individual," *A Source Book for Creative Thinking,* S. J. Parnes and H. F. Harding (eds.), Scribners, New York, 1962.

[4]Anne Roe, *The Making of a Scientist,* Dodd, Mead, New York, 1953.

Of the numerous ways which have been proposed to measure creative ability, we chose a test developed by S. A. Mednick, the Remote Associates Test.[5]

In several studies based on widely different groups of people (architecture students, psychology graduate students, school children, and IBM suggestion award winners) Mednick and others have found that people earning high scores on the Remote Associates Test (RAT) tended to be rated as highly creative. (It is important to note that in all of these situations there were probably few environmental constraints.)[6]

The test itself is a timed pencil-and-paper test which requires the person taking it to find a fourth word which can be associatively linked with three other words. For example, given the three words, "rat," "blue," and "cottage," the person taking the test would have to write "cheese" to get the item correct (rat-cheese, blue-cheese, cottage-cheese). Mednick argues that the mental process required for answering this type of item is the same as the mental process used in being creative. This is an ability to think of things which are not commonly associated with the "inputs" of the situation, but which can be so associated, and which at the same time meet the requirements for a good solution. Although it is not difficult to think of unusual associations, it is harder to think of associations which are both unusual and useful. In the foregoing example, "cheese" is not the word which is most commonly associated with any of the three given words. However, only "cheese" meets the requirements of being colloquially linked to all three. Obviously, the test requires an excellent command of American English, and several foreign-born respondents had to be omitted from the analyses to be described.

Creative Ability and Technical Performance

Our first attempts to relate creative ability and technical performance were disappointing. Scores on the RAT were correlated with each of the four criteria of performance. In each of the five analysis groups, the relationships were inconsistent and close to zero. These performance measures had been adjusted for length of experience (see Chapter 1), but analyses with the unadjusted measures indicated that the lack of relationships was not due to the adjusting process.

Of course, all scientists and engineers did not have an equal opportunity or desire to perform creatively. Thus one of the first analysis steps was to separate them on the basis of their responses to a combination of

[5]Mednick has discussed the theory on which he developed the Remote Associates Test in "The Associative Basis of the Creative Process," *Psychological Review*, 1962, vol. 69, pp. 220–32.

[6]Validity coefficients ranged from .4 to .7. Details are available in S. A. Mednick and M. T. Mednick, *Manual: Remote Associates Test, Form I*, Houghton Mifflin, Boston, 1966.

four items which seemed to indicate their desire for being creative. These items were: self-estimated effectiveness at creativity, self-estimated usefulness in providing original ideas to others, a preference for mapping broad features of new areas (leaving the detailed study to others), and a belief that the respondent was not very effective at tasks requiring methodicalness.[7] Even this separation of respondents did not produce results which seemed meaningful.

Thus creative ability alone (as measured by the RAT) seemed a very poor predictor of a man's scientific performance. Some scientists who got high RAT scores performed well, but about an equal number did poorly.

It will be left to the reader to decide for himself whether he considers these results discouraging. On the one hand, they may indicate that even the scientist with low creative ability can play a useful role; on the other, they could indicate that the labs from which our scientists were drawn were not making good use of the creative abilities of their personnel. (Of course, another interpretation is that the creative ability test was not actually measuring that ability. However, data will soon be presented which indicate that creative ability, as measured by the RAT, did relate to scientific performance—in certain situations.)

Situations Which Affected the Payoff of Creative Ability

Having discovered that creative ability, as measured by the RAT, was (by itself) a poor predictor of performance, we asked a somewhat different question. Were there particular situations in which high creative ability would pay off handsomely in terms of high performance? Or were there situations (perhaps opposite to the previous situations) in which the most creatively able people would do exceptionally poorly? If situations which affected the payoff from creative ability were identified, they might explain the lack of any direct relationship between RAT scores and performance.

As previously outlined, we had a number of expectations about the type of situation which might affect the payoff. Where a person was unwilling or unable to make creative ideas known to others, or in inflexible situations, the gain was expected to be low. These ideas provided a rough guide to the type of factors to be considered. We did not limit ourselves to these factors, however, and explored freely where other features of the situation promised to be interesting.

Time in Area; Time on Project. One factor which produced consistent and fairly marked effects seemed to be a time effect. It was found that

[7] The items proved to be interrelated in each of the primary groups and were combined into a single index score which weighted the items approximately equally.

creative ability was more helpful (had higher payoff) for those who had been working in their area of specialization, or on their main project, for a *short* time. Tables 1 and 2 show the data.

Since it was only after a complex analysis that these results emerged, the tables are themselves complex. Time will be taken to describe how Table 1 was constructed, and, therefore, how the analysis was done.

Each of our five major groups was divided into two subgroups according to whether the scientist had been specializing in the area in which he was most proficient a relatively long or short time. The questions used for determining this are shown in the following box.

> *Question* 2. Within a discipline or field, an individual may develop an area of specialization—a content area about which he knows a great deal. If you have such areas of specialization, please list them below in order of proficiency. (Limit to areas in which currently active.)
>
> *Question* 3. Approximately how long ago did you begin working actively in each area of specialization just listed?
>
> [Respondent checked one of seven categories from "less than six months ago" to "ten years ago or more."]

Then, for each of these subgroups, the relationships between creative ability (RAT score) and each of the four measures of performance were examined.

For Ph.D's, the average amount of time spent in the main area of specialization was roughly ten years. Thus, compared with their peers, those who had spent 0–9 years were *relatively* new in their area, whereas those with ten or more years experience were the old-timers. (Non-Ph.D's tended to have specialized in their areas of greatest proficiency more recently than the Ph.D's. In each of the non-Ph.D groups, people who had six or more years experience constituted the old-timers.)

The relationship between RAT and output of papers for development Ph.D's who were relatively new in their area was mildly positive ($r =$ .17); high creative ability went with high output. Since this figure exceeded an arbitrary cutoff criterion of $\pm .15$, this is shown in Part *a* of Table 1 by a + sign. For the old-timers, however, the figure was mildly negative ($r = -.11$); high creative ability went with low output. Since this was less than our criterion, a 0 appears for this relationship in Part *b* of the table. Had the relationships for the newcomers and the old-timers been the same, we would have concluded that time in area had no effect on the relationships. But the relationships were not the same: the payoff from creative ability tended to be higher for the newcomers

TABLE 1 *The following shows relationships between creativity, as measured by the RAT, and four criteria of performance, adjusted for length of experience. The five major groups were divided according to the length of time the scientist had been working in his major area of specialization, and correlations (r's) computed separately for these subgroups. The predominance of positive relationships among the newcomers (Part a) indicates that creative ability tended to "pay off" for them. Among the old-timers (Part b), however, high creative ability tended to go with low performance. Part c takes the same data and shows the difference between the two correlations. The predominance of positive differences indicates that creative ability helped the newcomers more. (If the correlation or difference was at least ±.15, it is shown as a + or −; relationships statistically significant at the 5% level are indicated by double symbols.)*

		Papers or patents†	Reports	Contribution	Usefulness	Scorecard +'s	Scorecard −'s
a. Relationships for those a *short* time in area of specialization*							
Ph.D's in development labs	(*N* = 21)	+	+	+	0	3	0
Ph.D's in research labs	(*N* = 7)	0	+	+	+	3	0
Engineers	(*N* = 17)	+	+	+	++	5	0
Assistant scientists	(*N* = 19)	0	−	+	0	1	1
Non-Ph.D scientists	(*N* = 13)	+	−	+	0	2	1
						14	2
b. Relationships for those a *long* time in area of specialization*							
Ph.D's in development labs	(*N* = 31)	0	0	0	0	0	0
Ph.D's in research labs	(*N* = 13)	+	0	+	0	2	0
Engineers	(*N* = 29)	0	0	−	−	0	2
Assistant scientists	(*N* = 30)	0	0	0	−	0	1
Non-Ph.D scientists	(*N* = 20)	0	−	−−	−−	0	5
						2	8
c. Difference: *r* for *short* time minus *r* for *long* time							
Ph.D's in development labs		+	+	+	−	3	1
Ph.D's in research labs		−	+	+	+	3	1
Engineers		+	+	++	++	6	0
Assistant scientists		0	−	+	+	2	1
Non-Ph.D scientists		0	0	+	+	2	0
						16	3

Note: N's vary slightly due to missing data.

* For Ph.D's, a *short* time in area was 0–9 years; a *long* time, 10 or more years. For other groups, these figures were 0–5 and 6+ years, respectively.

† For engineers, patents were used instead of papers.

than for the old-timers. A rough indication of the size of this effect was obtained by subtracting the correlation for the old-timers from the correlation for the newcomers. The difference was +.28 and is shown in Part *c* of the Table by a + sign.

Of course, these correlations were not large, and neither was the difference between them. With the small number of cases available, none of these relationships was "statistically significant."[8] Nevertheless, when the *same* relationships for other performance measures and other groups were found, we concluded that the observed trends were probably more than just a chance fluctuation.

As shown by the scorecard in Table 1, the tendency was for creative ability to enhance performance of newcomers (note predominance of +'s in Part *a*), whereas it may even have hurt the performance of old-timers (Part *b* is mainly 0's and −'s). These findings are combined in Part *c*, where the predominance of positive differences indicates that creative ability tended to help more (or hurt less) for those who had specialized in their area for only a relatively short time.[9]

Table 2 is set up in the same way as Table 1. Instead of splitting the groups according to length of time in main area, in this table they were split on length of time on most important project (see the following box).

> *Question 6.* Please list below the major projects or assignments on which you are working, or for which you have supervisory responsibility. Often these will correspond to separate budget items. List in order of importance to you.
>
> *Question 7.* Approximately how long have you been working on each project or assignment listed?
>
> [Respondent checked one of six categories from "less than six months" to "five years or more."]

Again, the tendencies were for creative ability to enhance the performance of scientists who had been working on their projects a relatively short time, and to detract from the performance of those who had been involved longer.

Thus it seemed that the person with high creative ability was most useful when he was a relative newcomer to a project or an area. Perhaps this was the time when there was still sufficient flexibility for creative ideas to have their impact. After a person had been engaged in an area of specialization or on a project for a relatively long time, the useful work may have consisted in following out the leads developed earlier rather than developing new ones.

[8] Since many factors are known to affect scientific performance, we did not *expect* to find many relationships large enough to reach statistical significance.

[9] Some tables (including Table 1) are based on fewer than the 355 scientists for whom we had RAT scores because certain data were not available for those scientists who answered the "short form" of our questionnaire.

TABLE 2 *The following shows that creative ability paid off more for those who had been on their main project a short time than for those who had been involved longer. As in Table 1, correlations (r's) were computed between creative ability and four criteria of performance separately for subgroups determined on the basis of years spent on the main project. (If the correlation or difference was at least ±.15, it is shown as a + or −; relationships statistically significant at the 5% level are indicated by double symbols.)*

		Papers or patents*	Reports	Contribution	Usefulness	Scorecard +'s	Scorecard −'s
a. Relationships for those *0–2* years on project							
Ph.D's in development labs	($N = 21$)	+	+	0	+	3	0
Ph.D's in research labs	($N = 8$)	0	+	+	+	3	0
Engineers	($N = 36$)	0	0	0	0	0	0
Assistant scientists	($N = 30$)	0	−	+	+	2	1
Non-Ph.D scientists	($N = 22$)	+	−	−	−	1	3
						9	4
b. Relationships for those *3+* years on project							
Ph.D's in development labs	($N = 33$)	−	−	0	−	0	3
Ph.D's in research labs	($N = 11$)	+	0	+	0	2	0
Engineers	($N = 12$)	0	+	−	−	1	2
Assistant scientists	($N = 28$)	0	0	−	− −	0	3
Non-Ph.D scientists	($N = 11$)	−	−	−	−	0	4
						3	12
c. Difference: *r* for *short* time minus *r* for *long* time							
Ph.D's in development labs		+	+	+	+	4	0
Ph.D's in research labs		−	+	+	+	3	1
Engineers		0	−	+	+	2	1
Assistant scientists		0	−	+	+ +	3	1
Non-Ph.D scientists		+	0	0	0	1	0
						13	3

Note: N's vary slightly due to missing data.

*For engineers, patents were used instead of papers.

A recent study by Houston and Mednick[10] suggests that the more creative scientists may themselves prefer movement to new projects and areas. Houston and Mednick found that people with high RAT scores showed a higher "need for novelty" than those with low RAT scores. Further evidence is provided by Barron,[11] who found that his most

[10] J. P. Houston and S. A. Mednick, "Creativity and the Need for Novelty," *Journal of Abnormal and Social Psychology,* 1963, vol. 66, pp. 137–41.

[11] F. Barron, "The Needs for Order and Disorder as Motives in Creative Activity," *Scientific Creativity: Its Recognition and Development,* edited by C. W. Taylor and F. Barron, John Wiley and Sons, New York, 1963.

creative subjects preferred complex things and enjoyed bringing order out of chaos. This may be precisely the kind of activity involved in making the first attack on a new project or area of specialization.

We wondered whether these findings were an artifact due simply to younger people making better use of their creative abilities. This was checked by dividing the groups into relatively young and old subgroups and examining the relationships between creative ability and performance in each subgroup, as in Table 1. Age proved to have no consistent effect on these relationships. We also looked at differences in career level and found that creative ability was likely to be just as helpful for the senior-level scientist or research supervisor as it was for junior or intermediate-level scientists. (Data not shown.) Thus the findings shown in Tables 1 and 2 seemed not to be attributable simply to differences in the age or career level of the scientist.

Coordination in Chief's Group. One other factor was found which affected the payoff of creative ability for all five of our groups of scientists. This was the degree of coordination within the group headed by the scientist's chief. Creative ability tended to help more in groups which were relatively uncoordinated. Using the same methods as in Table 1, but splitting the groups according to the scientist's estimate of the coordination in his chief's group, the data shown in Table 3 were obtained. The actual question is shown in the following box.

> *Question 25.* To what extent do members of the group headed by your chief coordinate their efforts for some common objective?
>
> Coordination of effort is:
>
> Nil; each member's work is separate from rest ____
> Slight; for about one-quarter of the work ____
> Moderate; for about half of the work ____
> Substantial; for about three-quarters of the work ____
> Full; almost all the work within the group is coordinated ____

Not surprisingly, the degree of coordination seen by Ph.D's was somewhat less than the coordination experienced by assistant scientists and engineers. Nevertheless, each scientist was compared with his peers to determine whether coordination was *relatively* high or low.

The predominance of positive differences in Part *c* of Table 3 shows that creative ability paid off more in relatively uncoordinated groups. In uncoordinated groups it tended to enhance performance (see Part *a*), but it hurt performance in coordinated ones (see Part *b*). Although one cannot say for sure, it seems reasonable to think that this may be another manifestation of the "flexibility phenomenon": a scientist with creative ability needs room to act on his ideas and hunches. Work teams which

TABLE 3 *The following shows that creative ability was more likely to enhance performance when activities within the chief's group were only loosely coordinated. The relationships between creative ability and four measures of performance tended to be positive for scientists in loosely coordinated groups (Part a), but tended to be negative for those in tightly coordinated groups (Part b). (If the correlation or difference was at least ±.15, it is shown as a + or −; relationships statistically significant at the 5% level are indicated by double symbols.)*

		Papers or patents†	Reports	Contribution	Usefulness	Scorecard +'s	Scorecard −'s
a. Relationships for those in *loosely coordinated* groups*							
Ph.D's in development labs	(N = 23)	0	+	0	0	1	0
Ph.D's in research labs	(N = 11)	+	+	+	+	4	0
Engineers	(N = 27)	0	+	0	0	1	0
Assistant scientists	(N = 23)	0	0	0	−	0	1
Non-Ph.D scientists	(N = 19)	0	0	0	0	0	0
						6	1
b. Relationships for those in *tightly coordinated* groups*							
Ph.D's in development labs	(N = 29)	0	−	−	0	0	2
Ph.D's in research labs	(N = 7)	0	0	0	−	0	1
Engineers	(N = 18)	0	0	−	−	0	2
Assistant scientists	(N = 30)	−	−	0	0	0	2
Non-Ph.D scientists	(N = 12)	+	−	−	−	1	3
						1	10
c. Difference: *r* for *loose* minus *r* for *tight*							
Ph.D's in development labs		+	++	+	0	4	0
Ph.D's in research labs		+	+	+	+	4	0
Engineers		−	+	+	+	3	1
Assistant scientists		+	+	0	0	2	0
Non-Ph.D scientists		−	+	0	+	2	1
						15	2

Note: N's vary slightly due to missing data.

*For Ph.D's and non-Ph.D scientists, *loose* was defined as "slight" or less, for other groups this was "moderate" or less. Scientists who indicated higher levels of coordination were considered to be in relatively tightly coordinated groups.

†For engineers, patents were used instead of papers.

coordinate their activities closely to attain the team's mission may provide effective means for achieving an objective but, the data suggested, may stifle creativity.

We also looked at several measures of the scientist's influence and autonomy, and describe these now.

Influence. Chapter 2 showed that the ability of a scientist to influence those who make decisions about his goals seemed to be associated with performance (see Chapter 2 for the wording of the question). Scientists with more influence tended to perform at higher levels. Would influence affect the relationship between creative ability and performance? Our expectation was that scientists who could influence those around them would make more effective use of their creative ability. These scientists might be more able and willing to suggest new ideas, and might be better able to get others to act on these new ideas. As shown in Part *c* of Table 4, this expectation was supported for three of the groups—the Ph.D's in development or research labs, and the non-Ph.D scientists. The effects of influence were inconsistent, however, for the engineers and assistant scientists. Part *a* of Table 4 shows that high influence enhanced the payoff from creative ability for Ph.D's in development labs (but not other groups). On the other hand, creative ability appeared to hurt the performance of non-Ph.D scientists who lacked influence (note Table 4, Part *b*).

There was another item which measured a somewhat different form of influence—the ability of a scientist to influence the person with most weight in deciding his *resources.* The groups were also split according to this item, and results were roughly similar to those shown in Table 4. Although the items were positively related (people who had high influence with respect to goals also tended to have high influence with respect to funds), the relationship was not very strong. The important thing appeared to be that scientists who felt they could influence others who made important decisions about their work were likely to have higher gains from their creative ability.

Further investigation showed, however, that getting high payoff was not simply a matter of a scientist having freedom to do whatever he pleased. The groups were split on still another item—the scientist's perception of his own autonomy (his own weight in decisions about his goals, see Chapter 2 for question wording). This factor produced no consistent effects on the relationships between creative ability and performance. Thus a scientist with great freedom himself might or might not get high gains from his creative ability; the important thing seemed to be his ability to influence the *other* people who affected his work.[12]

Dedication. Another factor which we thought might affect the payoff from creative ability was the scientist's dedication to his job. Scientists who were strongly involved in their work might have been more likely to achieve creative ideas and communicate them.

The data provided some surprises. Our expectation was confirmed for Ph.D's in research labs, among whom the *general* level of involvement

[12] Unfortunately there were too few cases to examine the effect of particular *patterns* of influence sources, as described in Chapter 2.

TABLE 4 *The following shows that being able to influence the person who has most weight in determining one's technical goals enhanced the payoff from creative ability for Ph.D's in research or development, and non-Ph.D scientists; the effects of influence were mixed, however, for engineers and assistant scientists (see Part c). As in previous tables, we compared the strength of the correlations (r's) between creative ability and four criteria of performance in subgroups of the respondents. (If the correlation or difference was at least ±.15, it is shown as a + or −; relationships statistically significant at the 5% level are indicated by double symbols.)*

a. Relationships for those with *high* influence*		Papers or patents†	Reports	Contribution	Usefulness	Scorecard +'s	Scorecard −'s
Ph.D's in development labs	(N = 40)	0	0	0	0	0	0
Ph.D's in research labs	(N = 5)	+	+	+	+	4	0
Engineers	(N = 34)	0	0	0	−	0	1
Assistant scientists	(N = 15)	+	−	−	−	1	3
Non-Ph.D scientists	(N = 22)	0	0	0	0	0	0
						5	4
b. Relationships for those with *low* influence*							
Ph.D's in development labs	(N = 15)	0	0	0	0	0	0
Ph.D's in research labs	(N = 14)	0	0	0	0	0	0
Engineers	(N = 15)	−	0	0	+	1	1
Assistant scientists	(N = 41)	0	−	0	0	0	1
Non-Ph.D scientists	(N = 11)	−	− −	−	−	0	5
						1	7
c. Difference: r for *high* influence minus r for *low*							
Ph.D's in development labs		0	+	0	+	2	0
Ph.D's in research labs		+	+	+	+	4	0
Engineers		+	+	−	−	2	2
Assistant scientists		+	0	−	−	1	2
Non-Ph.D scientists		+	+	+	+	4	0
						13	4

Note: N's vary slightly due to missing data.

* For research Ph.D's, *high* influence was defined as "great" or more, for all other groups this was "considerable" or more. Scientists who indicated lower levels than these were considered to have relatively low influence.

† For engineers, patents were used instead of papers.

was higher than in any of our other four groups. Among research Ph.D's, creative ability tended to pay off more for those who were very highly involved than for those who were less involved. But for the four other groups of scientists, results were exactly opposite to expectation! Those who were *not* especially involved in their work were the ones who got high effects from their creative ability. Creative ability was associated with low performance among those most highly involved.

In Table 5, which shows the results, the predominantly negative differences for the first four groups listed in Part *c* indicate that among these types of scientists high involvement was *not* the better condition for creative payoff—it was better to be less involved.

TABLE 5 *This again shows relationships between creative ability and four criteria of performance. This time the groups were subdivided according to degree of involvement in work. Among research Ph.D's, creative ability tended to show higher payoff if the scientist was highly involved (note the positive differences in Part c). Among the other four groups, the higher payoff came from those whose involvement was lower than average (note the negative differences). (If the correlation or difference was at least ±.15, it is shown as a + or −; relationships statistically significant at the 5% level are indicated by double symbols.)*

		Papers or patents†	Reports	Contributions	Usefulness	Scorecard +'s	Scorecard −'s
a. Relationships for those *highly involved* in work*							
Ph.D's in development labs	(*N* = 24)	0	−	0	0	0	1
Engineers	(*N* = 11)	+	0	−	− −	1	3
Assistant scientists	(*N* = 42)	0	−	0	−	0	2
Non-Ph.D scientists	(*N* = 21)	−	0	−	−	0	3
		Totals (omitting research Ph.D's)				1	9
Ph.D's in research labs	(*N* = 11)	+	+	+	+	4	0
b. Relationships for those *moderately involved* in work*							
Ph.D's in development labs	(*N* = 32)	0	+	0	0	1	0
Engineers	(*N* = 39)	0	0	+	+	2	0
Assistant scientists	(*N* = 17)	+	−	+	+	3	1
Non-Ph.D scientists	(*N* = 12)	+	0	+	0	2	0
		Totals (omitting research Ph.D's)				8	1
Ph.D's in research labs	(*N* = 7)	+	+	0	−	2	1
c. Difference: *r* for *high involvement* minus *r* for *moderate*							
Ph.D's in development labs		0	−	0	−	0	2
Engineers		+	0	−	− −	1	3
Assistant scientists		−	0	−	− −	0	4
Non-Ph.D scientists		− −	0	−	−	0	4
		Totals (omitting research Ph.D's)				1	13
Ph.D's in research labs		0	+	+	+	3	0

Note: *N*'s vary slightly due to missing data.

*For Ph.D's and engineers, *highly involved* was defined as "very strongly" or more, for other groups this was "strongly" or more. Scientists who indicated lower levels of involvement were considered to be relatively uninvolved.

†For engineers, patents were used instead of papers.

We don't have a ready explanation for these results. One speculation is that scientists need a certain detachment to be creative. Perhaps the Ph.D in the research lab has it because of his status and what the world expects of him. Involvement helps him to focus on the problem. Among more applied scientists, however, involvement may heighten the pressure to follow paths known to lead to a solution. For them, the detachment needed to try unknown alternatives may come only with lower involvement.

Communication. Another of the factors we expected would be important in enhancing creative performance was communication. Before a creatively able person's ideas can become recognized, they must be communicated to other people—his superiors and colleagues. Without adequate communication potentially creative ideas may go unrecognized. Also, communication might itself stimulate the generation of new ideas.

It was a problem to know how to measure the "adequacy of communication," for communication occurs in many ways. One factor examined was the frequency with which a scientist communicated with his chief and colleagues. The question is shown in the following box.

Question 41. As a general rule, how frequently do you communicate with each of your supervisors and colleagues on work-related matters? (Whether by conversation, memos, seminars, etc.)

	Supervisor	Colleagues				
	I	II	III	IV	V	VI
Few times a year, or less	____	____	____	____	____	____
Few times a month	____	____	____	____	____	____
Few times a week	____	____	____	____	____	____
Daily	____	____	____	____	____	____

The effects of these items, as shown in Parts I and II of Table 6, were in accord with our expectation for the engineers, but showed marked inconsistencies for the other four groups. For engineers, those who were in contact with their supervisor and/or colleagues relatively frequently were likely to get greater gains from their creative ability.

Why the expected effects did not appear for the other groups is not clear. Perhaps these items inquiring about frequency of communication were not sensitive enough to tap differences in the ability of creatively able people to get their ideas across to others.

In another attempt to measure adequacy of communication, the groups were split according to the number of reports or talks prepared by the scientist (adjusted for the length of the scientist's experience, as described in Chapter 1). The question is shown in the following box.

> *Question 75.* Over the past five years, about how many of the following have you had:
>
> .
> .
> .
>
> Unpublished technical manuscripts, reports, or formal talks (either inside or outside this organization) ——

We reasoned that if a person wrote more than the average number of reports for scientists of his type and experience, then perhaps his ideas would have a better than average chance of being communicated to others. This measure of communication produced consistent effects in the expected direction for two groups: the research Ph.D's and the assistant scientists. Those who wrote an above-average number of reports tended to be helped more by their creative ability (see Table 6, Part III). Results for the other groups were mixed, but none was consistently opposite to expectation.[13]

Thus our general expectation that good communication would enhance the payoff from creative ability was substantiated for three of the five groups. Other groups showed inconclusive results which may have reflected a need for more sensitive measures of communication.

Some Factors which Did Not Affect the Relationship between Creative Ability and Performance

So far, a number of factors have been identified which seemed to affect the payoff from creative talent. A number of other factors were interesting because, surprisingly, they did *not* affect it.

For example, it is sometimes assumed that supervisors who are "supportive"—offering enthusiasm or appreciation—provide settings which are auspicious for creativity. However, the data, based on the scientist's perceptions of his supervisor, showed no consistent effects. High performance was as likely to accompany creative talent under nonsupportive supervisors as under supportive ones. The supervisors' *own* creative ability was also examined, with the thought that those having high talent might be particularly sensitive to such talent in their subordinates. Again, no consistent effects.

Certain characteristics of the scientist himself, his environment, and his work also seemed to be unimportant with respect to the usefulness of creative ability. Included among these were the scientist's independence, his orientations toward science and his organization, the way he

[13] The groups were also split according to output of published papers or patent applications. No consistent effects were observed.

TABLE 6 *The following shows the effects of three communication items on the relationship between creative ability and performance. The effects, as in previous tables of this chapter, are indicated by the magnitude and sign of the difference between correlations (shown in this table). Frequent contact with chief or colleagues enhanced the payoff from creative ability for engineers. An above-average output of reports or formal talks enhanced the payoff for research Ph.D's and assistant scientists. (If the difference between correlations was at least ±.15, it is shown as a + or −; relationships statistically significant at the 5% level are indicated by double symbols.)*

	Papers or patents[a]	Reports	Contribution	Usefulness	Scorecard +'s	Scorecard −'s
I. Contact with chief						
Difference: *r* for frequent contact minus *r* for infrequent[b]						
Ph.D's in development labs (*N*'s = 32, 33)	−	−	0	+	1	2
Ph.D's in research labs (*N*'s = 17, 24)	−	++	0	−	2	2
Engineers[c] (*N*'s = 25, 35)	+	+	+	+	4	0
Assistant scientists (*N*'s = 51, 59)	−	+	+	0	2	1
Non-Ph.D scientists (*N*'s = 32, 28)	0	+	−	0	1	1
II. Contact with colleagues						
Difference: *r* for frequent contact minus *r* for infrequent[d]						
Ph.D's in development labs (*N*'s = 23, 18)	0	0	−	−	0	2
Ph.D's in research labs (*N*'s = 11, 4)	+	0	0	+	2	0
Engineers[e] (*N*'s = 17, 28)	++	+	+	+	5	0
Assistant scientists (*N*'s = 21, 28)	0	−	+	+	2	1
Non-Ph.D scientists (*N*'s = 15, 15)	0	0	−	−	0	2

Note: N's vary slightly due to missing data.

[a] For engineers, patents were used instead of papers.

[b] For engineers, frequent contact with chief was defined as "daily," for other groups this was "weekly" or "daily." Scientists who indicated lower frequencies were considered to have relatively infrequent contact.

[c] Differences are due to positive correlations for engineers in daily contact with their chiefs, and correlations close to zero for those with less frequent contact.

[d] People scored as having "frequent" contact with colleagues indicated an average contact of at least several times a week with their 1–5 closest colleagues. Scientists indicating less frequent average contact were considered to have relatively infrequent contact.

[e] Differences are due to positive correlations for engineers in frequent contact with colleagues, and negative correlations for those in infrequent contact.

III. Output of reports, talks	Papers or patents[a]	Reports	Contri- bution	Useful- ness	Scorecard +'s	−'s
Difference: r for above average output minus r for below average[f]						
Ph.D's in development labs (N's = 35, 31)	0	−	0	−	0	2
Ph.D's in research labs[g] (N's = 28, 19)	0	+	+	++	4	0
Engineers (N's = 37, 23)	−	+	0	+	2	1
Assistant scientists[h] (N's = 59, 52)	0	+	++	+	4	0
Non-Ph.D scientists (N's = 26, 32)	−−	+	−	0	1	3

[f] In determining whether a scientist's output was above or below average, he was compared with other scientists in the group who had comparable lengths of experience.

[g] Differences are due to positive correlations for research Ph.D's with high output, and negative correlations for those with low output.

[h] Differences are due to positive correlations for assistant scientists with high output, and correlations close to zero for those with low output.

preferred to approach his work, the size of his group, and his autonomy.[14] None of these factors seemed to affect the payoff from creative potential.

SUMMARY AND IMPLICATIONS

The chapter concludes in the format of a conversation with an imaginary research director.

> You folks have cited a great many "factors" which you have looked at. What do you consider your main findings?

Okay, let's summarize. First, creative ability, as we measured it, did not relate to any measure of performance for any group of our scientists. The reason, however, was because some scientists were in situations where their creative ability "paid off" for them, but others were in situations where creative ability seemed to hurt their performance. The situations which seemed to enhance the payoff from creative ability were the following: working on a project or specializing in an area for a relatively short time; being part of a work team where coordination was not too high, and where one had the ability to influence important decision-makers; and having reasonably good facilities for communicating new ideas to others. You might say that creative ability was less likely to

[14] Chapters 2 and 6 described these measures.

pay off, and may even have hurt a man's performance, if he was in a *restrictive* situation.

> Several times you have commented that creative ability may hurt a man's performance. Do you really mean that?

In at least one sense, we do. Under some conditions the men with the high creative abilities did less well than those with less talent.

Let's consider a hypothetical example of how this might occur. You accidentally put a scientist who has high creative ability—that is, who is constantly seeing unusual ways of doing a job, or thinking about a problem—into a very restrictive situation. When he gets one of his potentially creative ideas, he feels he would be risking his job if he were to suggest it; or he does suggest it, but finds no one will listen. Our scientist can't "turn off" his flow of new ideas. What might be the result? He becomes disappointed and dejected. If he doesn't leave the job, he may slow down, feeling that his talents go unrecognized. Soon he is doing less well than his unfrustrated colleagues. In this sense, his creative ability is hurting his performance.

> I am still puzzled at why your RAT test didn't always relate to performance if it is supposed to measure creative ability. After all, that is what we want in my lab, creativity.

Our best explanation is that creative ability can hurt as well as help, depending on the situation. When the statistics for everyone in our sample were taken together, the minuses canceled out the pluses. Another hunch is that some R & D labs may not want as much creativity as they claim. They need experts who can solve problems by well-established techniques. If so, real creativity won't be rewarded. The man with a *new* approach won't have a chance to report or publish it. He won't earn higher status, nor even the respect of colleagues.

> You haven't said much about where you found the most creatively able people; did they tend to work in particular kinds of situations? I am wondering whether they are really a "different breed of cats" from those who are less creative. . . .

We looked at this and found that scientists who scored high on the RAT test answered the questionnaire pretty much like those who scored low. If creatively able people selected themselves into certain kinds of situations and avoided others, the data didn't show it.

> If I believe your findings, what would you suggest I do to make use of them in my lab?

You probably already have a rough idea which of your scientists are likely to come up with new, useful, *original* ideas. These are your creatively talented people. Some are probably doing well now, others maybe not so well, though they may have done well in the past.

Consider how flexible each man's situation is—as he sees it. Maybe encourage the man to take up a new specialty or work on a new project if he has been involved in his existing area or project for many years. Maybe shift him to a new group where he will have some ability to influence the course of events if his present group doesn't allow this. If he is extraordinarily involved in one project, perhaps a challenging second project he could "play" with would provide the detachment which seems helpful to many scientists. And what are his opportunities for trying out original ideas on others? Make sure such opportunities are easy to find.

As a final suggestion, look at the reward system in your lab. Many labs say they want creativity, but instead give the biggest rewards to those who are productive along *well-established* paths. If you want creativity, do you reward it when it occurs?

10

AGE[1]

Performance Peaked at Mid-career, then Dropped—but Less among Inner-motivated Scientists and Those in Development Labs.

Does the scientist's creative potential fall off after reaching its peak in his late 30's? If such a trend does occur, are there any clues as to why? Are there certain conditions under which scientists continue a creative career throughout their life span?

These were not questions with which our investigation had started. But we found ourselves forced to consider them. Some of the insights we gained proved valuable in our main task of studying the composition of a stimulating environment.

This chapter begins with H. C. Lehman's evidence that pioneering discoveries in various scientific fields were most likely to occur in the late 30's or early 40's, and thereafter declined in frequency. In examining our own data, we found one peak of scientific performance at roughly the same age, but also another peak 10 to 15 years later. A further breakdown showed interesting contrasts among age curves in university, industrial, and government settings. A search for factors that might account for these differences led to a set of motivational items indicating "inner motivation" or self-reliance. Among individuals who were strongly motivated by their own ideas, performance resisted the normal erosion of later years. The chapter concludes with some practical suggestions, and observes that the effects of age apparently may be modified by both external and internal "research climates."

THE "CREATIVE YEARS"

In 1953, findings by Lehman[2] aroused a controversy which still provokes debate at scientific meetings: does a scientist's creative potential

[1]This analysis was supported by grants from the National Science Foundation. A preliminary version appeared in D. C. Pelz, "The 'Creative Years' and the Research Environment," *IEEE Transactions on Engineering Management,* 1964, vol. EM-11, no. 1, pp. 23–29.

[2]H. C. Lehman, *Age and Achievement,* Princeton University Press, Princeton, New Jersey, 1953.

fall off after reaching its peak in his late 30's? Lehman has continued to probe this explosive topic with more documentation[3] and usually arrives at the same outcome: on the average, the likelihood of outstanding achievement peaks in the late 30's and early 40's, and thereafter declines. He found the peaking to appear earlier in abstract disciplines (mathematics, theoretical physics), and later in more empirically based disciplines (geology, biology). He found the peaking to be sharper for the most outstanding achievements, and flatter for minor achievements.

At first blindly attacked, these findings in some circles have come to be just as blindly accepted. (A physicist once remarked to one of the authors, "If you haven't made your mark by 35, you might as well quit the field.")

Instead of continuing to contest the reality of the phenomenon (the weight and carefulness of Lehman's data seem to us overwhelming), it would be more creative, we think, to begin examining why. Although numerous hypotheses have been advanced, the nature of the causal mechanisms has hardly been tested.

Some of the hypotheses are more flattering to the scientific ego than others. (a) The notion of a general decline in intellectual potential with age seems not to be supported with existing studies of intellectual abilities.[4] (b) A favorite alternative hypothesis is that the more able achievers are drawn off into teaching, administration, and committee work not productive of scientific output. If so, then able individuals who continue to devote the bulk of their time to research should continue to be just as productive. If we had some way of identifying "ability" other than by the man's achievements, this hypothesis could be tested. (c) A third interpretation is that after the young scientist has struggled and built his reputation, he tends to relax; the strength of his achievement motivation (whether to become famous, or to have the fun of discovering, or both) diminishes. If so, then among those who continue to be achievement-oriented, the drop in performance should not occur.

(d) A fourth hypothesis is that as the scientist becomes a specialist, he loses the fresh viewpoint needed for breakthroughs. Having made a name as an expert in one field, he cuts himself off from discoveries in new fields. It is these discoveries, however, which are the "outstanding achievements" in the histories from which Lehman draws his data. If this is the case, those scientists who resist the temptation to specialize are more likely to continue achieving. Furthermore, if the scientist

[3]"The Chemist's Most Creative Years," *Science*, May 23, 1958, vol. 127, pp. 1213–1222; "The Age Decrement in Scientific Creativity," *American Psychologist*, 1960, vol. 15, pp. 128–134.

[4]For one entry to the literature see Nancy Bayley, "On the Growth of Intelligence," *American Psychologist*, 1955, vol. 10, pp. 805–818.

changes his field periodically, or at least the problems in the area of his specialization, his performance should continue on a high level. (e) A fifth hypothesis is similar to the fourth, and concerns technical obsolescence: the scientist loses touch with recent advances and grows stale. If so, then time-off periods for study and refresher courses, intensive seminars and the like should tend to maintain achievement.

The Age Curves were Saddle-shaped

The performance scores used in this and the following chapter were different in one important respect from those used previously. Since we wanted to see how performance varied with age, it would have been inappropriate to adjust the scores for experience and seniority (Adjustment II,

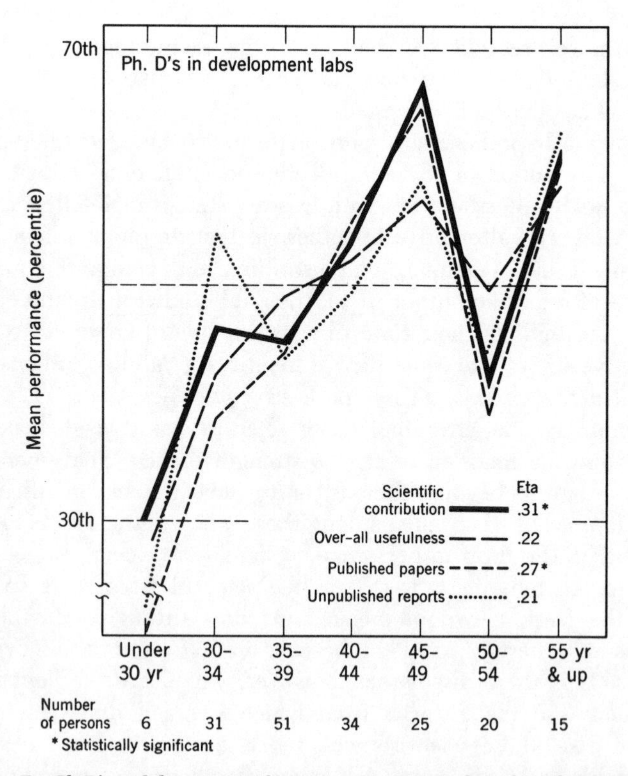

Chart 1-A. *For Ph.D's in laboratories where executives stressed product development rather than scientific publication, all four performance measures (for the previous five-year period) reached a peak when the individual was in his late 40's. All measures sagged in the early 50's and recovered in the later 50's. For papers published, a mean five-year output at the 30th percentile would correspond roughly to 4 papers per man, and a mean at the 70th percentile to 11 per man. For unpublished reports, the corresponding figures would be 5 and 16.*

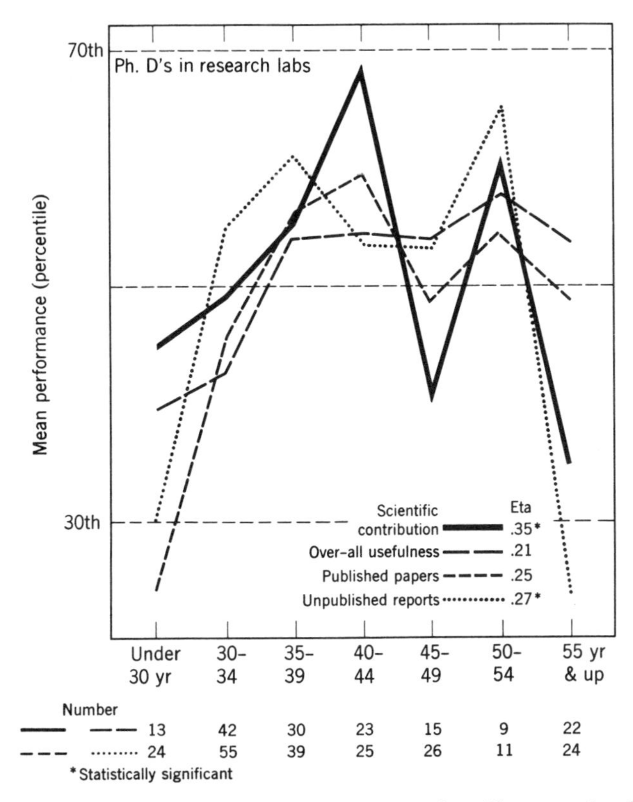

Chart 1-B. *For these Ph.D's in laboratories stressing scientific publication rather than product development, significant scientific contributions were highest in the early 40's (five years earlier than in development labs). After a dip there was another surge in the early 50's, and thereafter a sharp decline. Output of unpublished reports showed a similar pattern, although over-all usefulness—and to some extent paper publication—continued at moderate or high levels at later ages. For five-year paper publication, mean scores at the 30th or 70th percentiles would correspond roughly to 5 and 12 per man, respectively; for unpublished reports, 5 and 14.*

described in Appendix C). Accordingly, adjustments were made only for differences attributable to type of laboratory and education level (Adjustment I).

We then examined how adjusted performance related to chronological age, and beheld an intriguing pattern: not one hump per curve, but two. One peak did appear during the "creative years" between the late 30's and late 40's; the maximum varied for different subgroups. But then in all groups there was a renascence 10 to 15 years later, in the 50's (the comeback was clear in four of the five groups, and hinted at among engineers.) The data are plotted in Charts 1-A through 1-E.

Chart 1-A shows how four performance measures varied by age among

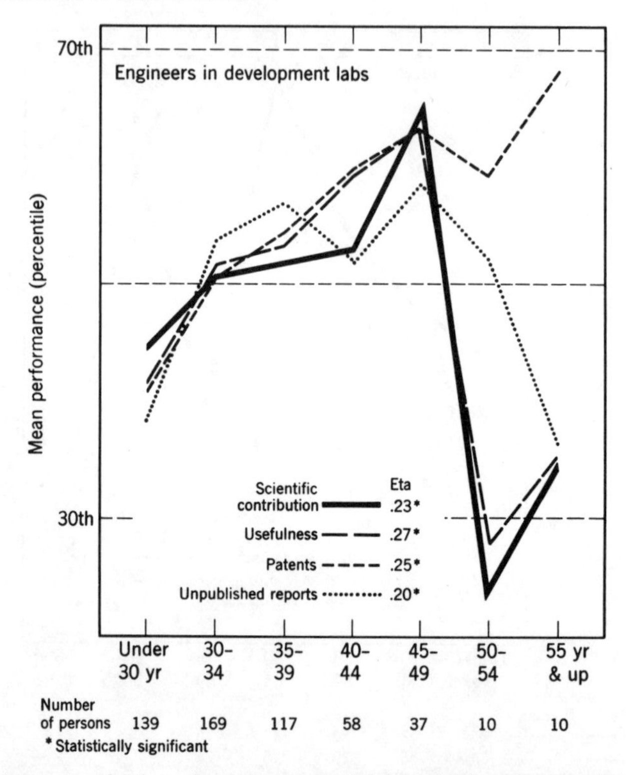

Chart 1-C. *For non-Ph.D's in development-oriented labs not dominated by Ph.D's (called "engineers" for simplicity), contribution and usefulness reached a maximum in the late 40's, and thereafter dropped sharply, with a slight recovery in the late 50's. Older individuals continued to patent, however. Possibly they were supervisors drawing on the work of their group. Mean scores for five-year patent output at the 30th and 70th percentiles were about 0 and 2 per man; and for reports, about 4 and 10.*

Ph.D's in development-oriented laboratories. Mean scores for scientific contribution (over the previous five years) reached a peak when the individual was in his late 40's, then dropped sharply in the early 50's, but recovered in the late 50's. Over-all usefulness showed a similar pattern with not so sharp a drop, and recovery just as high as before.

Were these curves affected by the "subjectivity" of colleague rankings? Was the late recovery, perhaps, simply a tribute by the senior man's colleagues to a lifetime of achievement, in spite of our instructions to the contrary? The curves for publication of papers, and writing of unpublished reports, disproved this interpretation. They showed exactly the same trends as the judgments—a clear peaking at 45 to 49 years, then a five-year trough, and a recovery at 55 and older.

The results for Ph.D's in research-oriented laboratories appear in Chart 1-B. A similar set of twin peaks appeared, with the interesting difference that the peaks and the trough occurred five years earlier in research than in development. Sharpest effects appeared for the judgment of scientific contribution, which reached a maximum in the early 40's. If we remember that the performance in question occurred over the previous five years, and so subtract two or three years from the plotted age, this peak is almost identical with that obtained by Lehman for outstanding achievements.

The renascence in the early 50's, however, did not correspond to Lehman's trends. Note that it coincided with heavy output of unpublished reports, mirrored to a lesser extent by some rise in publications. Perhaps this was more a "productive" than a "creative" period. Perhaps the

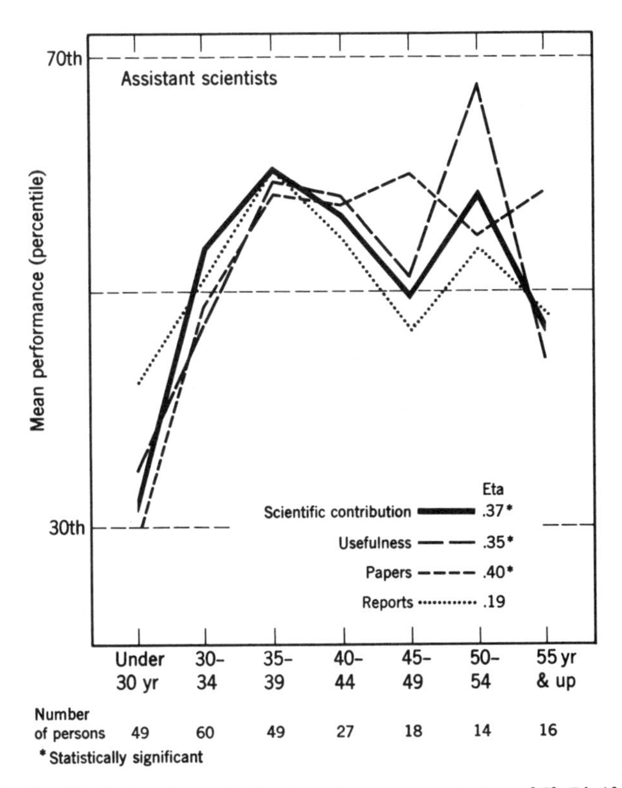

Chart 1-D. *Non-Ph.D's in laboratories having a heavy concentration of Ph.D's (for simplicity we have called them "assistant scientists") showed a peaking on all measures in the late 30's, and another peak 10 or 15 years later. These scientists reached their maximum usefulness in the early 50's. The 30th and 70th percentiles for five-year paper publications were 0 and 4 per man, respectively; for unpublished reports, 1 and 8.*

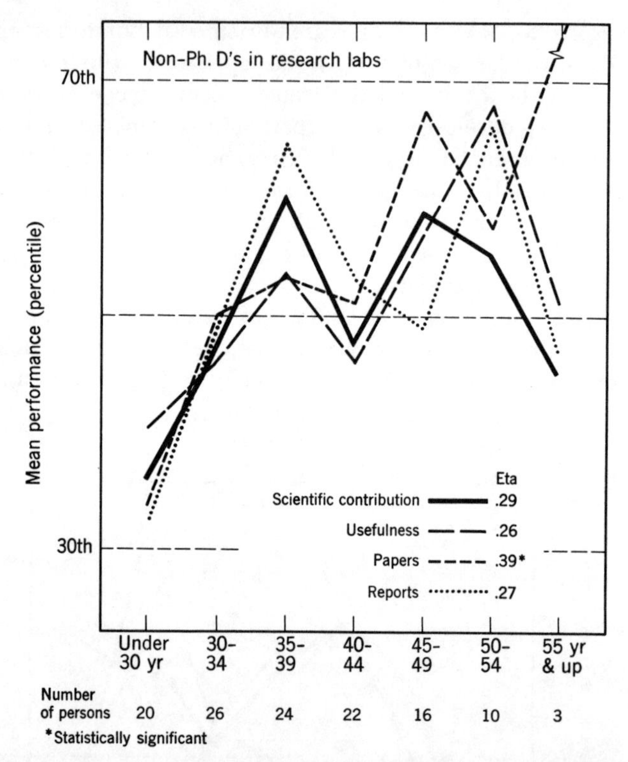

Chart 1-E. *Among non-Ph.D's in research-oriented government labs containing few Ph.D's, the same saddle-shaped trends appeared, with a preliminary peak in the late 30's, and another in the late 40's (for contribution and papers) or early 50's (for usefulness and reports). The data for age 55 and up, based on only three cases, may be unstable. The 30th and 70th percentiles for paper publication were about 0 and 4 per man; for reports, 3 and 8.*

researcher was busy with follow-up studies to confirm his earlier pioneering. Perhaps as an established professor or section head, his own efforts could now be multiplied with the help of graduate students and junior associates.

Note also that despite the sharp drop in scientific value of his work in the late 50's, publication of papers continued, and his over-all usefulness to the organization remained at a high level.

The remaining charts show data for nondoctoral scientists. Like the Ph.D's in development labs, engineers' performance rose steadily to a peak in the late 40's. Then contribution and usefulness dropped precipitously, with a slight recovery in the late 50's (not nearly so complete as that of the Ph.D's). There were very few cases over 50, however. Perhaps by now the more capable men had moved out of R & D entirely.

Another possibility was that these older men were now section heads who claimed credit for patents produced by their section (patent output continued to climb), even though their own scientific reputation fell sharply.

Chart 1-D shows data for assistant scientists. Again the saddle shape appeared clearly in three of the four curves. A preliminary peak occurred in the late 30's, followed by a 10- to 15-year sag, with a recovery in usefulness, contribution, and reports in the early 50's. Their usefulness to the organization was greatest at this age. Since few of them were supervisors, one cannot apply the interpretation (which might be advanced for Ph.D's) that the late recovery was on the work of subordinates. Some characteristic of the individual himself must have been responsible.

Chart 1-E presents the data for our relatively few non-Ph.D scientists. Again the saddle shape appeared, rather similar to that of the assistant scientists. A preliminary peak in at least three of the measures occurred in the late 30's, followed by a 10- or 15-year sag, with a recovery in scientific contribution and paper publication in the late 40's, and a peak in usefulness and report writing in the early 50's. Only three cases were available for the late 50's, so these values may be unstable; three of the measures dropped, although this oldest group published phenomenally.

Some Comments

The remarkable saddle shape in four of the five groups (echoed faintly among the engineers) stands in distinct contrast to the continuing decline in Lehman's data for major contributions. That this recovery was not simply respect for past achievement is shown by the parallel rise in published papers and unpublished reports in these age brackets. Perhaps the later contributions will not be noted in histories of the field as landmarks. But they are probably solid achievements, nevertheless, even if they simply extend the breakthroughs of the man's creative 40's.

Although the saddle-shaped curves surprised us, this was not the first time they had appeared.

Several years ago Pelz was analyzing some data obtained by Seymour Lieberman and Leo Meltzer on a nationwide sample of physiologists to study the connection between age and number of citations in recent annual reviews of discipline. A similar saddle-shaped pattern appeared for highly motivated physiologists.[5] At the time, he was not sure whether

[5]The curves are presented in D. C. Pelz, "Motivation of the Engineering and Research Specialist," *Improving Managerial Performance, AMA General Management Series*, 1957, no. 186, pp. 25–46. Available as Publication #1213 from the Survey Research Center, University of Michigan.

this was accidental or genuine. The persistence of the pattern in the present data suggests a real phenomenon.

When Oberg[6] plotted supervisory ratings of over-all value to the organization among scientists and engineers in a large company, he also found a bimodal curve. The twin humps were contributed partly, he found, by two separate populations—those in research and development peaking between 30 and 40, and those in engineering (outside of R & D) rising steadily to about age 55. But even within the research labs, there was some resurgence in the late 50's.

What does this recovery mean? When we have discussed it with groups of scientists, they have offered several suggestions. By now the scientist is a supervisor, perhaps, or a teacher; he stopped producing for a while, but now is publishing jointly with subordinates or students. Or perhaps the two peaks represent different kinds of contribution: the earlier one, creative discoveries; the later one, synthesis of a lifetime's progress. Other interpretations focus on the trough more than the peaks. In the mid-40's, the man has his heaviest financial needs and family troubles. His children are going to college and his marriage may be growing stale; he wants a safe, secure job to pay the bills. When the crisis period is past, he can return to the risks of a difficult assignment or a new job. Then someone always suggests a physiological explanation: the male change of life.

A different idea is the possibility of two distinct subpopulations, peaking 15 years apart. If so, we have thus far been unsuccessful in isolating them.

Whatever the explanation (and with our data most of the foregoing hypotheses cannot be tested), there is some reassurance here. The scientific career may be productive for considerably longer than Lehman's data would indicate. *Productive*, but not necessarily "creative" in the sense of the outstanding discoveries which Lehman found to peak most sharply between 35 and 45.[7] It is important to note that older individuals continued to be useful to the organization in many ways, despite their drop in strictly scientific achievement.

One may also note the difference between research and development laboratories. *Older individuals did better in development.* That is to say, the peaks and troughs appeared about five years later. It seems likely that the research scientists in our sample used up their scientific resources at a faster rate, so to speak, than those in development. Or to put it

[6] W. Oberg, "Age and Achievement and the Technical Man," *Personnel Psychology*, 1960, vol. 13, pp. 245–259.

[7] Careful scrutiny of Lehman's curves shows that he too sometimes obtained a second, lower hump in the late 50's—note, for example, Table 3 in *Age and Achievement* (1953).

another way, perhaps the wisdom of cumulative experience was more useful in development than in research.

This difference parallels Lehman's finding that abstract disciplines (such as mathematics) flourished at an earlier age than those requiring extensive observation (geology, astronomy).

Note one feature of the curves for unpublished reports. Among Ph.D's both in development and research, and to a lesser extent for the other groups, this measure was the quickest to rise. One has the impression that preliminary reports contained the raw material out of which later publications and patents, and later scientific reputation, were fashioned. As a practical step, the young scientist should be urged above all to set down his early work in writing, or at least in the form of notes for seminar presentations. He can then rework this in professionally acceptable form. But without this raw material, produced by age 35, he may lack a foundation for solid later achievement. (For more details on this suggestion, see the closing paragraphs of Chapter 6.)

Comparisons among University, Government, and Industrial Settings

One of the popular interpretations of Lehman's curves is that creative scientists become so eminent that they are thrust into supervision, lectureships, and chairmanships; they don't have any more time for research.

Some data presented in Chapter 4 tend to undercut this argument by showing that some time spent in nonresearch activity (provided it was not too great) did not detract from achievement.

But another variation is sometimes advanced—that good people are promoted out of research altogether, and the weak ones left behind to pull down the curves. The data in Charts 1-A through 1-E were obtained from the entire research staff at all levels up through research director. If the better people were promoted within the lab, they would appear in our data, and their high performance at later ages should pull the curves up rather than down.

How can we tell whether the good people stayed in research? One scientist suggested: "Why don't you separate the university scientists from those in government? In the university you generally get promoted within the ranks; it's in government or industry that the good man may get pulled out." And so we did. We divided the research Ph.D's into the categories of university and government. And for good measure (not because of any prior expectations, but only for the sake of a tidy consistency) we divided the other four groups as well, according to type of setting. Some fascinating patterns appear in the following set of charts.

First, let us look at the results for the groups which had started us

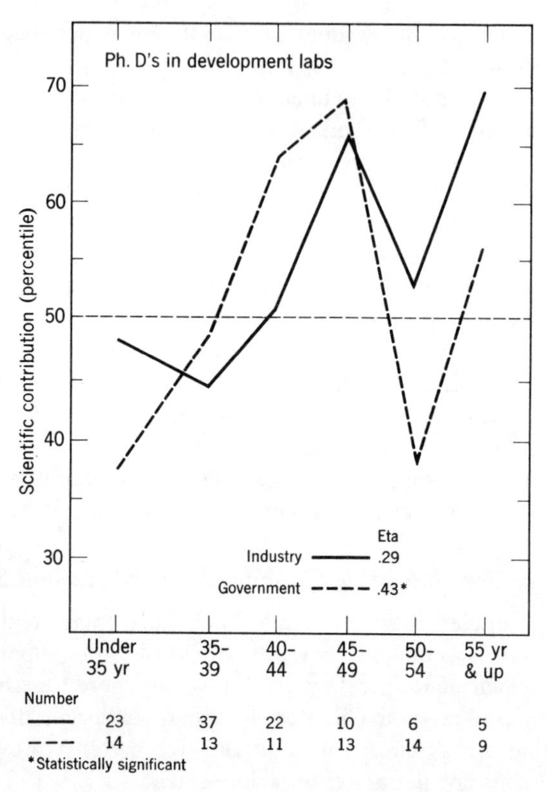

Chart 2-A. *When Ph.D's in development labs were separated into those in industry and government, the same saddle shape for scientific contribution appeared in both. But although the curve for industrial scientists recovered to the same high level in the late 50's, for the government scientists it returned only to average.*

on this area of study—university Ph.D's versus those in government research labs (Chart 2-B). In the late 50's, university men dropped just as sharply as did those in government. It is unlikely that the best scientists were promoted to deanships and presidencies (such positions are scarce); it is more plausible that the older professors were still in the research setting, but contributing less. The "up and out" theory was not supported. (As a postscript we might add that the usefulness of the older men as teachers and mentors continued high.)

But look at the interesting contrast with government in the height of the two main peaks. After the late-40's trough in the university (which was not really low), contribution recovered as high as ever in the early 50's. In government, though, the late-40's sag may have been deeper (caution: there were few cases to go on), and the 50's recovery was only up to an average level. Why? Let's withhold speculation while we look at some more data.

Chart 2-A shows Ph.D's in development, divided into industrial and government settings. Again the twin peaks appeared in both. But whereas the general trend in industry was still rising by the late 50's, scientists in government had staggered at the early-50's trough, and their comeback was only average.

Comparison of engineers in industry and government (Chart 2-C) showed that the marked achievement we saw in Chart 1-C in the late 40's occurred exclusively in industry. Government engineers reached their maximum early, and between ages 30 to 50 the curve was essentially flat. Both groups declined sharply after 50, with industrial engineers recovering slightly better (on the basis of a few cases).

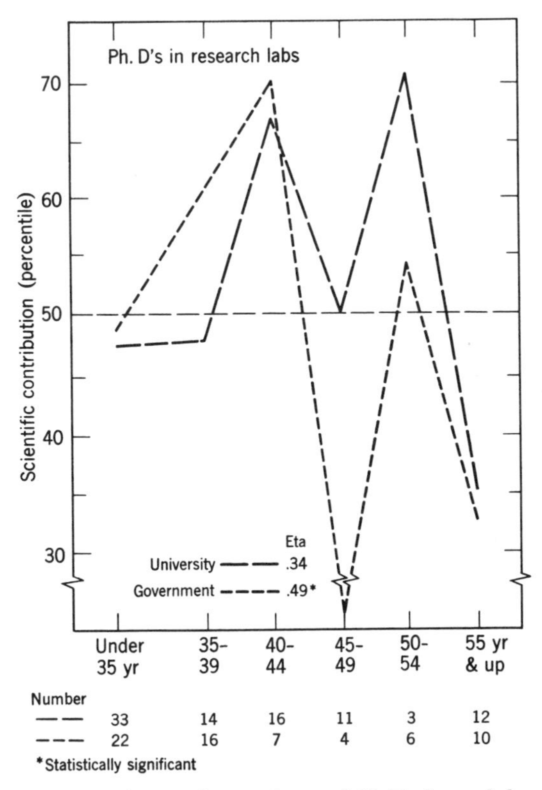

Chart 2-B. *After the mid-50's, contribution of research Ph.D's dropped sharply, both in university and government laboratories. Just before this, university scientists surged back to the same high level as in the early 40's, whereas the recovery among government scientists was much lower.*

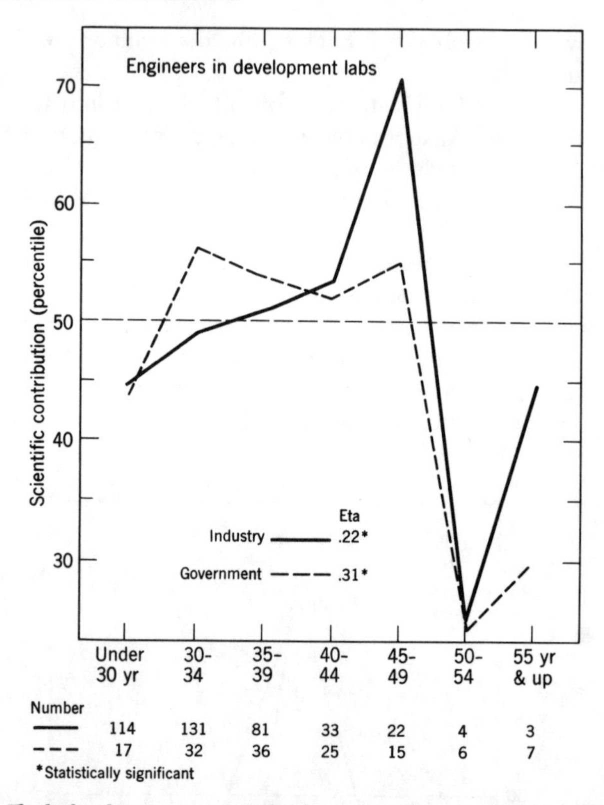

Chart 2-C. *The high achievement among development engineers in their late 40's was found only in industrial laboratories. Among government engineers, contribution was essentially flat during the 30's and 40's. Both groups dropped sharply in the 50's; the low recovery thereafter was slightly better in industry.*

The assistant scientists (Chart 2-D), however, showed a surprising and dramatic reversal. It was the assistants in *government* who climbed steadily to high achievement in the 50's, and even after that dropped only to average. In industry, however, after good achievement between 30 and 45, the assistants' contribution plunged, struggled upward in the early 50's, then relapsed. Why the difference between subordinate scientists in the two settings? And why the reversal between government and industry, compared to the previous charts? Some possible clues will soon be discussed.

Let us pause to summarize some trends.

First: despite over-all differences in slope, note the remarkable similarity among the two curves in each graph in the incidence of peaks and valleys. Ages of best achievement in research labs were similar, whether the individual worked in the university or in government. Ages of best achievement in development laboratories were also similar, whether in industry or government, whether Ph.D or not. Regardless of setting, research men achieved sooner, and dropped sooner, than was true in development.

Second: at the same time, one is struck by the downward tendency among older government Ph.D's and engineers. They tended to do well at first, and then to decline. But in industry and university, performance of Ph.D's and engineers generally improved with age (going by the tops of the peaks).

And third: assistant scientists showed exactly the opposite pattern. Was it possible that the same government atmosphere could fail to keep the Ph.D stimulated, whereas it encouraged the non-Ph.D?

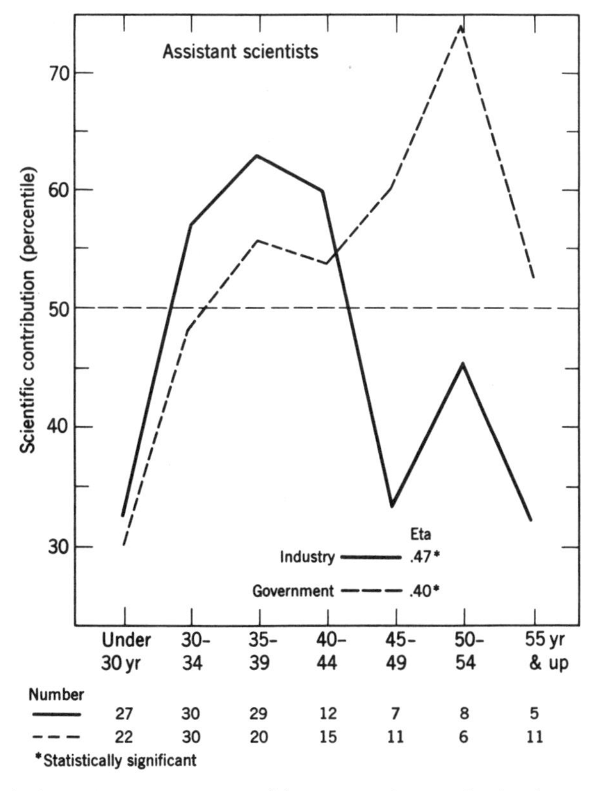

Chart 2-D. *A dramatic contrast appeared between assistant scientists in government and industry. Those in industry flourished between 30 and 45 and then withered, whereas those in government continued to gain in achievement. Later data suggested a sharp difference in level of self-direction with increasing age in the two settings.*

MOTIVATIONAL TRENDS IN THE VARIOUS SETTINGS

What could account for these institutional differences in the age curves? Four kinds of hypotheses might be advanced.

(a) *Lack of inner stimulation.* In the government laboratories which we happened to study, was the atmosphere for some reason less stimulating than in the industrial and university departments in our sample? As our scientists aged, did they fail to find as much challenge in government as in the other settings? If so, could the same explanation account for the opposite trends among assistant scientists?

(b) *Over-specialization.* Perhaps government personnel were digging themselves into narrow specialties more than scientists in industrial and university labs. Chapter 11 will show that flexibility tended to enhance the performance of older scientists. Could it be that government scientists lacked the flexibility as they aged?

(c) *Differences in retention of personnel.* Might it be that, as time passed, the better scientists in government were leaving? Were academic and industrial departments refreshed by an inflow of competent older people? (If so, such a process might not only alter the performance curves in older brackets, but the research climate in these brackets as well.) And if such a process occurred, did we have any clues as to what might induce the better scientists to change location?

(d) *Differences in selection of personnel.* Even without differences in retention, the curves might make sense if the kinds of people attracted to each institution had changed over the years. For example, did the government laboratories hold more attraction for security-seeking men two decades ago than at the present time?

These are knotty questions, and we have only a few indirect clues on hypotheses (c) and (d). We have much fuller data concerning hypotheses of type (a) and (b) on the current climate. While studying the following pages, the reader should remember that our sample was not drawn systematically from each type of institution; the results do not necessarily apply to other government or industrial or university situations.

Lack of Inner Stimulation?

One set of motivational items seemed particularly promising. We observed that the general level of these motivations in different settings, and their bumps and troughs at different age levels, roughly paralleled the performance curves. Perhaps, we thought, they could help to explain the puzzling performance curves.

Chapter 6 described an index of motivation from one's own ideas,

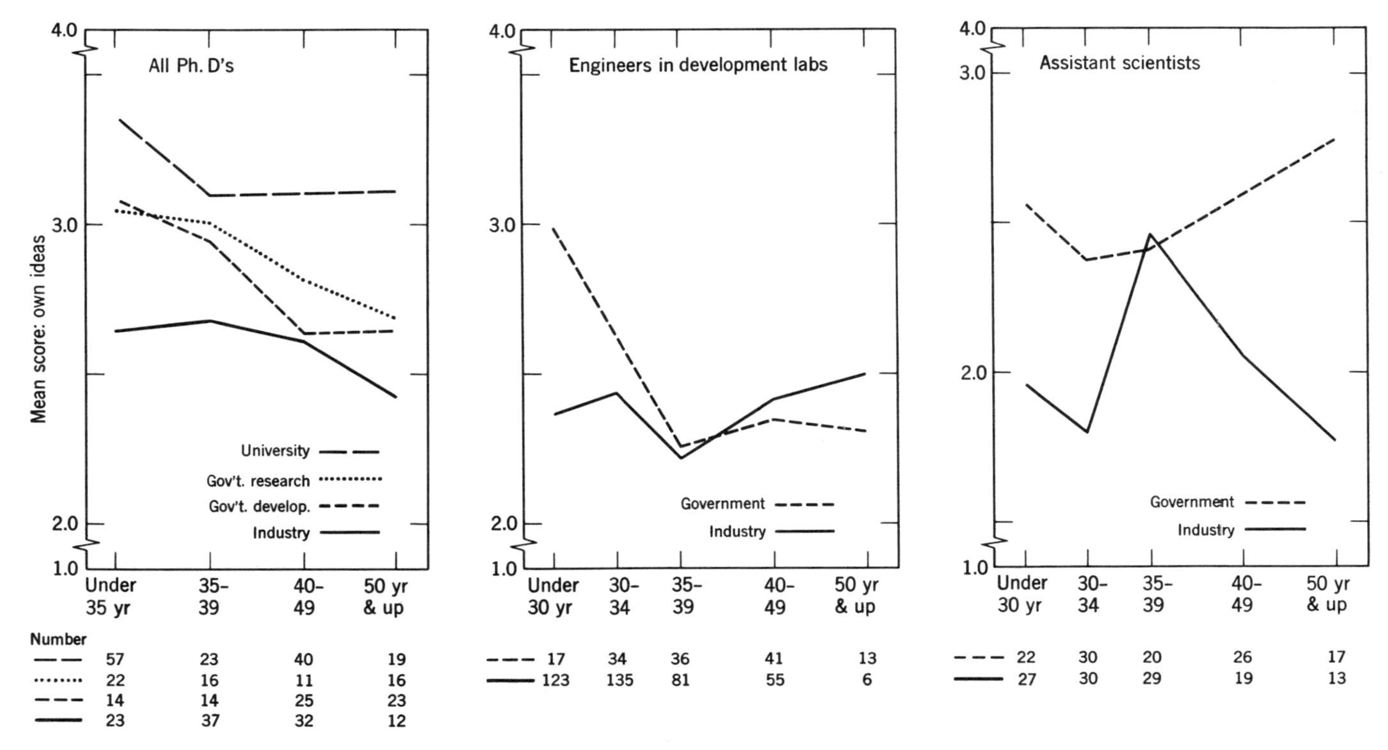

Chart 3. *Mean motivation from own ideas (self-reliance) is plotted against age within each group. Government Ph.D's and engineers dropped slightly more than other groups, but assistant scientists in government tended to rise. Total scores on this index were divided roughly into quartiles, scored 1 to 4.*

based on desire for freedom, stimulation from own previous work, and from own curiosity. It might be called "self-reliance," or "desire for self-direction." Chapter 6 showed that it fairly consistently accompanied high performance.

Chart 3 plots the way the average response on this measure varied with age within eight kinds of settings.

Among Ph.D's, desire for self-direction tended to diminish in each group with increasing age; but the drop was a little steeper in government laboratories than in university or industrial. Among engineers, the contrast was even sharper; government engineers dropped distinctly in desire for self-direction, whereas industrial engineers were largely unchanged over the age range. The sharper drop among these government scientists and engineers may have weakened their resistance to the "normal" inhibitions of advancing years.

By contrast, among assistant scientists, those in government showed an increase in motivation from own ideas after age 40, whereas those in industry dropped—paralleling the performance rise at older ages among the former, and the performance drop among the latter.

Here, then, was a major clue. In those situations where scientists continued to be effective as they aged, we found a strong emphasis upon the individual's *own ideas* as a source of motivation. They were self-reliant; active rather than passive toward their environment.

Effects of Self-direction over the Life Span. If these hunches are correct, some further tests with our data should be possible. Let us see what happened to performance at various ages among individuals who stood relatively high or low in self-direction. Performance of the first should stay rather high in later years. Performance of the latter should drop more with age, and perhaps fluctuate more widely.

Each primary analysis group was divided into those scoring high and low on the "motivation from own ideas" index, and within each half, the relationship of age to scientific contribution was plotted (see Charts 4-A through 4-D).

Our expectations were generally confirmed. Among Ph.D's in development labs (Chart 4-A), those strongly motivated by their own ideas were superior in performance in almost every age bracket. Their slump in the early 50's was only mild. Among Ph.D's in research labs also (Chart 4-B), those with strong inner motivation surpassed those with weak motivation in every age bracket except one.

Among engineers in development labs (Chart 4-C), similarly, those with strong inner motivation surpassed those with weak inner sources in every age bracket except one. Among assistant scientists finally (Chart 4-D), we see that strongly self-reliant scientists jumped ahead of the

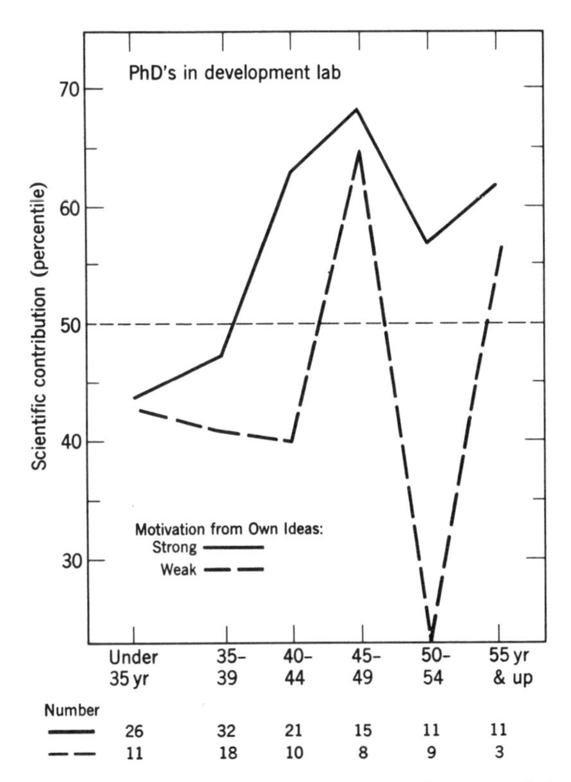

Chart 4-A. *Development Ph.D's who were strongly motivated to pursue their own ideas continued to achieve throughout their life span, and resisted the early 50's sag which appeared for the weakly motivated.*

weakly motivated in the early 30's, held this lead with one exception in the late 40's, and came back to their highest achievement in the early 50's. Those with weak inner motivation, however, declined steadily in scientific contribution from the mid-30's on.

Thus the development of strong inner motivation in scientists may be an important factor in prolonging their achievement over a broad span of their scientific career.

From these results, the reader should not jump to the conclusion, however, that scientists will continue to achieve over a lifetime if they are simply left alone and protected from distractions. Previous chapters have indicated that strong outer stimulation also plays a part.

From other data not reported here, it is evident that strong motivation of several varieties may help in prolonging achievement throughout the life span. In a previous publication (see footnote 1) we presented data on how performance varied with age when our sample was divided

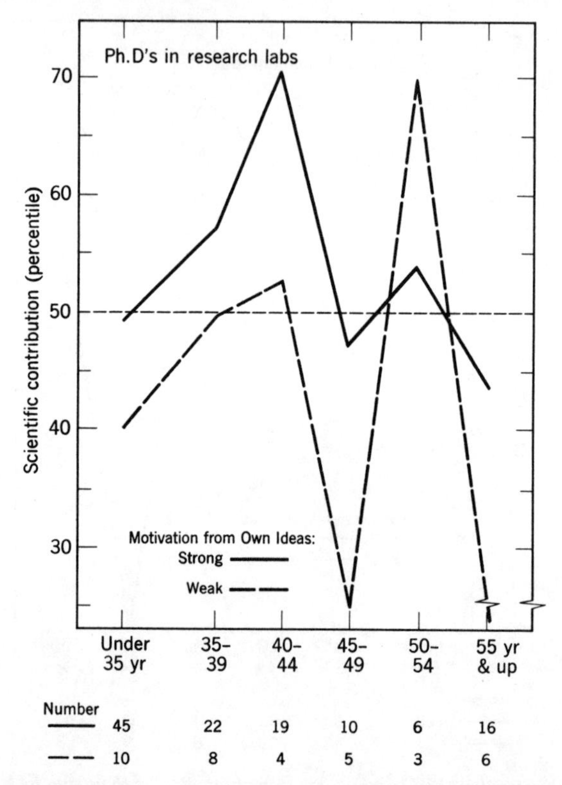

Chart 4-B. *Ph.D's in research also contributed more at each age if strongly motivated from within, with a brief exception.*

among those with relatively strong or weak "involvement" in their work (details on this measure were given in Chapter 5). Strongly involved people maintained a higher level of performance at various ages than did the weakly involved, although at the peaks of performance there was no difference in involvement. (At the height of creative power, it is perhaps not the total intensity of motivation which separates the excellent performers from the good, but rather the quality or source of motivation.)

Overspecialization?

Chapter 11 will show that best performance occurred for most groups under conditions which encouraged the development of both wisdom and flexibility. Would these factors help account for the different trends observed in government and industry? Was it possible that members of government laboratories were overemphasizing the accumulation of

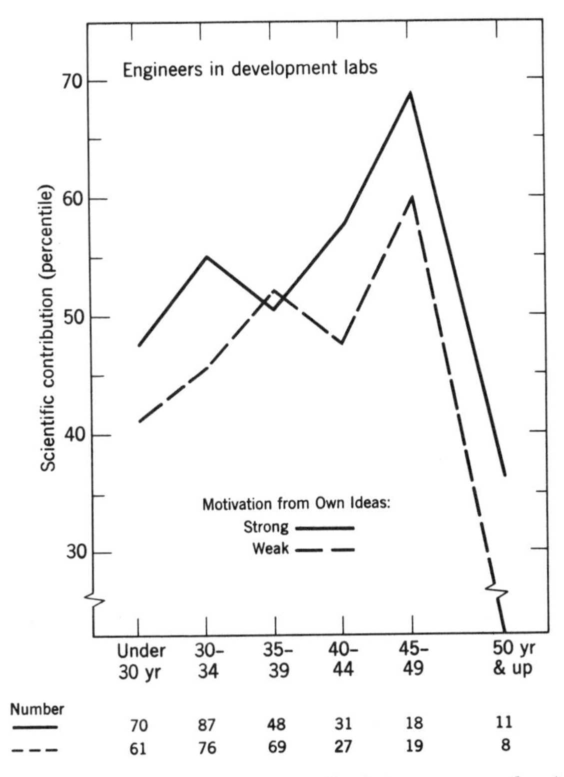

Chart 4-C. *Engineers showed the same trend as Ph.D's (previous two charts): the self-reliant among them rose to higher achievement and dropped less than did those with weak inner motivation.*

wisdom (by digging into narrow specialties, for example) and losing needed flexibility?

We cannot take space here to plot all of the data. (Chapter 11 will show the measures we used to assess wisdom and flexibility.) Few of the curves we examined supported the hunch that our government engineers were becoming too specialized with age—if anything, the reverse was true. Relative to engineers in industry, they had ample diversity and breadth of perspective. Thus this hypothesis failed to account for the trends we observed.

Differential Turnover?

To test hypothesis (c) directly—that able older scientists were leaving government—it would have been necessary to compare scientists who had left with those who stayed; and we had almost no information on the former.

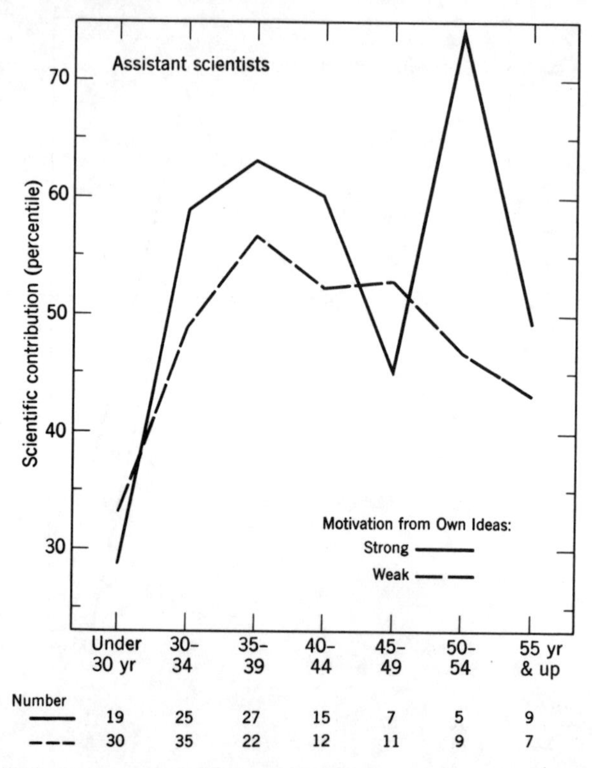

Chart 4-D. *With one exception, assistant scientists who relied on their own ideas remained high in scientific contribution through the early 50's, whereas those with weak inner motivation declined after the 30's.*

An indirect test was possible, however. If the hypothesis were true, we should observe that older scientists remaining in government had been in their organization for a longer period, on the average, than older industrial or academic scientists had been in *their* organizations. We therefore examined how average length of service with the laboratory varied as individual age increased.

For Ph.D's in development, Ph.D's in research, and assistant scientists, there was no systematic difference (data not shown). In some age brackets, scientists in government had longer tenure, but in other brackets those in the industrial or university situations had longer tenure.

For our group of engineers, this interesting difference appeared: before age 35, those in government on the average had longer tenure; but after 40, those in industry had increasingly longer service. In other words, the government labs were keeping their younger people longer than industry, but recruiting more older employees—directly opposite to our hypothesis.

Our indirect evidence, then, did not support the hypothesis that government labs were losing their better older scientists.

Differential Selection?

According to hypothesis (d), the kind of persons attracted to, or selected by, government service two decades ago may differ systematically from those attracted or selected recently.

Our questionnaire told us very little, of course, about the man at the time of hiring. We did find out his age at receiving the bachelor's degree, and the length of time between B.S. and Ph.D (if he earned one). Earlier studies had suggested that men who took a long time to earn their doctorate were likely to publish less frequently. Our own data confirmed a somewhat lower level of performance among slow earners of the Ph.D (see Appendix C), and we eliminated such effects in the adjustments of performance scores used in Chapters 2 to 9 and 12.

Time between B.S. and Ph.D. Did government organizations select scientists who differed in this respect? Chart 5 revealed some clear differences.

Among Ph.D's in research labs, those aged 40 or over in government

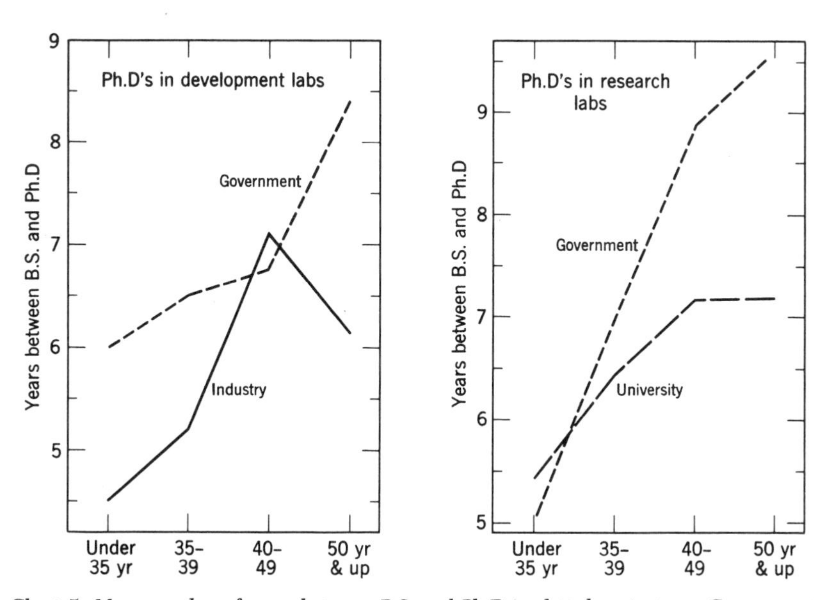

Chart 5. *Mean number of years between B.S. and Ph.D is plotted against age. Government Ph.D's in development labs took one to two years longer to earn their degree (in three out of four age brackets) than did those in industry; so did government Ph.D's age 40 and over in research labs, compared to those in the university. Note a general trend: over the past two decades, the time taken to earn a Ph.D has declined. Number of persons is the same as in the left portion of Chart 3.*

took 2 to $2\frac{1}{2}$ years longer to complete the doctorate following their B.S. than did our university scientists. Among Ph.D's in development, there was a difference of one to two years in each age bracket except 40 to 49. (Note, incidentally, that younger scientists in general earned their Ph.D's more quickly than older scientists—possibly an indication of better financial support for Ph.D candidates nowadays.)

Age at Baccalaureate. If government labs were attracting those who had delayed their Ph.D, did they also attract late earners of the bachelor's degree? We won't take space to show the data, but the answer, for engineers and assistant scientists, was clearly "no." In all age brackets, our government engineers earned their B.S. *sooner* than did those in industry (about a year's difference, on the average). Among assistant scientists there was no consistent difference. Our small group of government non-Ph.D's in research labs, however, earned their B.S. sooner than either group of assistant scientists in every bracket except one. (Government Ph.D's received their B.S. at either the same age as the other groups, or somewhat later.)

In short: although the hypothesis of differential selection did apply to government labs in the hiring of Ph.D's, it did not apply (or applied in a reverse fashion) to the hiring of non-Ph.D's.

SUMMARY AND IMPLICATIONS

> Your saddle-shaped curves provide a distinct contrast to Lehman's single-peaked achievement curves. Is there a contradiction here?

We do not think so. Lehman, you will recall, also found that when he examined a larger number of lesser contributions—good work, but not the earth-shaking discoveries that highlight the history of science—these continued at a steady rate over a wider span of the scientific career.

Guilford and his associates speak of a distinction between "divergent" thinking (which generates new elements) and "convergent" thinking (which pulls together common aspects of given elements).[8] Both are essential in science. Our current hunch is that the earlier peak represented work of a more "divergent" or innovative type, whereas the later peak represented work more "convergent" or integrative in character.

Chapter 4 contains a hint, for example, that investigators midway in their career did best if they had broad rather than narrow interests, whereas those at later supervisory levels could be eminent with specialized interests. We shall carry this line of inquiry further in Chapter 11.

[8] J. P. Guilford, "Three Faces of Intellect," *American Psychologist*, 1959, vol. 14, pp. 469–479.

What is your thinking about the Lehman curves as such? You listed five kinds of hypotheses that have been offered to account for them. Do your data support or disprove any of these?

As we mentioned, other studies don't lend much support to hypothesis (a) that the decrease in productivity is due to a decline in intellectual powers. However, the notion (c) that performance may decline because individuals relax their zeal or motivation after having achieved receives considerable support from our data. We saw in Charts 4-A through 4-D that among strongly motivated individuals, especially those with strong self-reliance, achievement resisted the normal erosion of age.

On the other hand, hypothesis (b) that capable men become so burdened with administrative responsibilities that they are drawn away from creative achievement seems to us a weak explanation. Chapter 4 gave some evidence that a small amount of administrative distraction is not inconsistent with high achievement. Perhaps those who allow themselves to be drawn mainly into nontechnical activities do so because their scientific motivation has already declined, and they cease to find as much challenge or gratification in technical work as they do in exercising administrative power.

Then you would prefer the hypothesis that performance declines because of reduced motivation among older scientists?

That is important. Some clues in the next chapter also support notion (d) that Lehman's curves for outstanding achievement (not so much our own curves which reflect productivity as well as creativity) are attributable mainly to the scientist losing the fresh viewpoint needed to pioneer. Chen Ning Yang (who, with Tsung Dao Lee, won the Nobel prize in physics for their overthrow of the principle of conservation of parity) remarked in an interview that, "As you get older, you get less daring. You have seen so much—therefore, for every new thought you have, you immediately marshal a large number of counter-arguments."[9]

This process *may* be irreversible. A certain quality of naiveté may be indispensable to great discoveries; and this naiveté may be possible only in a *young mind.*

But perhaps the process is not irreversible. We are still far from having determined the limits of the mind. What would happen if a mature and able scientist were assured enough financial support to start again as a student? Even if he could afford it, a reputable investigator would find it hard to humble himself to the status of a novice. But who can tell what neglected capacities for curiosity and daring might be released?

Hypothesis (e) on technical obsolescence is a variation on this theme.

[9] *Newsweek*, Jan. 22, 1962, p. 49.

Systematic attention to renewing and broadening one's technical skills seems promising as a way of prolonging the creative, or at least the productive, years.

Is there anything a research manager can do about the late-40's sag?

For one thing, be on the lookout for it. Major attention should be given to a midcareer review with each scientist. If a sag has set in, counteractive steps should be taken before morale is damaged. Perhaps this is the point for a planned program of renewal through courses, sabbatical service at a university, rotation to a different department which is breaking new ground, and the like.[10] It may help both the scientist and his director if both recognize that the sag is typical, and especially that it need not signal the end of a fruitful career.

In addition, perhaps this is the time for a deliberate decision on whether the man should stay in research as such, or move to a more applied, or perhaps administrative, position. Remember that the curves peaked later in development than in research. Perhaps it would be fruitful for him to transfer to a more applied branch of R & D, or work closely with an engineering department. (We have heard of one industrial organization which systematically encourages transfer from the research labs to development or engineering at this stage.) Perhaps the man has accumulated skills and wisdom which will make him a valued partner in the latter for many more years.

What if the man doesn't want to transfer into practical applications?

He might continue in the role of consultant rather than investigator; in general, the curves for *usefulness* remained high.

If his decision is to stay in technical work, some data to be shown in Chapter 12 may be instructive. Don't turn him loose in his own ivory tower with a minimum of distraction and a maximum of autonomy. He may simply vegetate. See that he is exposed to the stimuli of consultation and collaboration, of challenging new problems outside his main specialties.

The curves indicate, at the very least, that a decline in the later years is not inexorable—not due to some irreversible process of decay. The fact that the curves peaked later in development than in research, the fact that they differed somewhat in government compared to university or industrial labs, the fact that motivation could change the shape of the curves—all these suggest that the effects of age may be modified by both external and internal "research climates."

[10]For a description of several university programs see D. Allison, "Engineer Renewal," *International Science and Technology*, June 1964, pp. 48–54, and 109–110.

What about the differences you found between government settings and those in industry or universities? Do you have any special suggestions for a government research manager?

We're not sure. According to our data, the manager of a government *development* lab may face an odd problem: persuading his older staff members to be *less* flexible and cooperative. We found these senior scientists and engineers broad in their outlook; they were willing to change. Maybe too much so. They would do better to insist more on their *own* ideas and their personal growth.

How to encourage these motives? A number of suggestions are given at the end of Chapter 6. See that each man produces distinctive work he can call his own, and then see that a professional audience knows about it through publication in journals, through his service as a consultant, or as a speaker at conferences.

11

AGE AND CLIMATE[1]

As Age Increased, Performance Was Sustained with Periodic Change in Project, Self-reliance, and Interest Both in Breadth and Depth.

In Chapter 10 we saw how different measures of performance rose or fell at successive age levels, within various settings. Previous chapters explored a variety of factors making up effective climates for achievement—both "internal" as well as "external" climates.

We now ask: do the climates needed for high performance show any systematic change with age? What conditions of internal motivation or external stimulation are most helpful for maintaining high achievement at various stages in the scientist's lifetime?

In trying to answer these questions we shall be seeking clues as to why scientific performance may vary over the life span, and what (if anything) can be done to boost performance at the start of the scientist's career and prolong it in his later years.

The reader will recall the report in Chapter 10 of some interesting trends in performance by scientists in successive age brackets. In the present chapter we return to these age data to seek further understanding of conditions that might be responsible for the peaks and valleys.

The chapter first sketches some theoretical ideas about the importance of both cumulative wisdom or tradition on the one hand, and of flexibility or innovative capacity on the other. It then examines correlations of various wisdom and flexibility factors with performance, within four primary analysis groups. In a very tentative way, it appeared that Ph.D's after 40 should maintain diversity, whereas non-Ph.D. engineers could afford to specialize somewhat as they aged. Both gained after 50 by an interest in pioneering. Periodic change of project helped, and in all groups achievement after 40 required self-reliance and willingness to risk the uncertain. The chapter closes with some practical suggestions.

[1]This analysis was supported mainly by a grant from the National Institutes of Health, U. S. Public Health Service.

Some Theoretical Expectations

Our thinking was stimulated by some remarks of Thomas S. Kuhn, a historian of science, on the role of both *tradition* and *innovation* in scientific breakthroughs.[2] Kuhn suggested that major advances in science have not come from a deliberate intention to innovate, but rather from an "essential tension" between tradition (systematized knowledge) and innovation (restructuring of knowledge). The former, he suggested, resembles Guilford's concept of "convergent" thinking, and the latter his concept of "divergent thinking."[3] Correspondingly, it seemed to us that work of the highest scientific value might occur when two disparate conditions were simultaneously present: one condition providing "wisdom" (cumulative experience, traditional knowledge), and another condition providing "flexibility" (capacity to innovate, freshness of approach).

It seemed reasonable to suppose that as an individual grows older his supply of "wisdom" is likely to increase, whereas his "flexibility" may decrease. Under this premise, younger individuals may be handicapped by lack of sufficient wisdom, and older individuals by their loss of flexibility, whereas scientists in the middle years have the advantage of both wisdom and flexibility. This premise might account for the peaking of achievement in the middle years.

A hint of such a process is apparent in the data on the age of *groups* in Chapter 13. Interest in "breadth" decreased as group age increased, whereas interest in "depth" increased; group performance was maximum close to where the curves crossed.

If these assumptions were valid, certain predictions should follow. Younger scientists should perform better if their particular climate enhanced wisdom, and older scientists should do better if *their* climate encouraged flexibility. As the reader will see, the actual situation was not quite so simple.

The Analysis Procedure

To test these ideas, the following method of analysis was used. We divided our primary analysis groups by age bracket. Among the members of each bracket, we obtained the *correlation* (product-moment coefficient, r) between numerous measures of motivation or stimulation, and our measures of performance. (The performance measures were the same as those in Chapter 10, adjusted to remove systematic effects of educational level and type of setting, but not effects of time since degree.)

[2] T. S. Kuhn, "The Essential Tension: Tradition and Innovation in Scientific Research," Chapter 28 in C. W. Taylor and F. Barron (eds.), *Scientific Creativity: Its Recognition and Development*, John Wiley and Sons, New York, 1963.

[3] J. P. Guilford, "Three Faces of Intellect," *American Psychologist*, 1959, vol. 14, pp. 469–79.

If a certain motivational factor was, in fact, conducive to high performance for persons in a certain age bracket, this effect would appear as a strong positive correlation between that motivation and performance. If the motivation, on the other hand, tended to inhibit performance at that age, the result would be a negative correlation. The following charts, then, plot the value of *correlations* between various climate factors and scientific contribution at successive ages.

Correlation coefficients, of course, reveal only the linear component in the relationship, and are insensitive to curvilinear trends—for example, the possibility that at a certain age a moderate-strength motivation may enhance performance more than either weak or strong motivation.

In this analysis we were limited by the numbers of cases available, and have therefore combined some age brackets. In every group, the 40 to 49 bracket contained either the earlier of two achievement peaks or the later. Because of limited numbers, the data on non-Ph.D's in research-oriented labs had to be omitted.

In addition to looking at effects within age brackets, we also looked at effects within each "career level"—apprentice, junior, senior, supervisor (these are described in Appendix E). Age and career level were highly correlated, of course. However, results using the two approaches sometimes disagreed. Since results by age brackets seemed a little clearer than those by career level, we arbitrarily show only the former. The reader should not assume that findings for younger scientists (aged 30 to 34, for instance) are typical of junior investigators, nor that results for older scientists (for example, 50 years and up) are true of supervisors.

Some Indicators of Flexibility

According to the foregoing theoretical ideas, achievement of older scientists should be prolonged if their climate contains factors promoting flexibility of approach. We felt that the measures of diversity described in Chapter 4 might perform this function. A scientist who possessed several areas of specialized knowledge, or who performed several different kinds of R & D activities, would be in a better position to bring a fresh viewpoint to bear on a given problem than a scientist who confined himself to a few specialized skills or research functions.

Results with number of areas of specialization are given in Chart 1. For Ph.D's, the trends were about as expected. The work of younger scientists (under 35 years) was unaffected by the individual's number of specialties (as indicated by correlations close to zero between number of areas and scientific contribution). But among older scientists, particularly research-oriented Ph.D's in their 40's, number of specialties showed a definite positive connection with contribution.

The vertical axis in the following charts shows the size of the *correlation* (product-moment coefficient, *r*) between some motivational factor and scientific contribution as judged by colleagues. The four lines in each chart show the size of these correlations within four of our primary analysis groups at successive age levels.

The data used in these correlations were obtained from the 500 long-form respondents. The measure of scientific contribution is the same as that used in Chapter 10. It has been adjusted to rule out mean differences between Ph.D's and non-Ph.D's, but effects due to length of working experience since the degree have not been removed. In computing the correlations, sampling weights were used; the actual (unweighted) numbers of cases were as follows:

<table>
<tr><th>Age</th><th>Under 30</th><th>30–34</th><th>35–39</th><th>40–49</th><th>50 & up</th></tr>
<tr><td>Ph.D's in development</td><td colspan="2">15</td><td>26</td><td>33</td><td>26</td></tr>
<tr><td>Ph.D's in research</td><td colspan="2">21</td><td>19</td><td>23</td><td>15</td></tr>
<tr><td>Engineers</td><td>43</td><td>67</td><td>50</td><td>52</td><td>7</td></tr>
<tr><td>Assistant scientists</td><td>18</td><td>25</td><td>23</td><td>14</td><td>9</td></tr>
</table>

Because of the very few persons aged 50 and up among engineers or assistant scientists, correlations in these subgroups will be unstable.

Nondoctorals, however, showed a different trend. Among these it was the younger individuals (30 to 34 years) who mainly seemed to benefit from possession of several specialties. With increasing age, the advantage of breadth diminished.

In a rough way, several of the following charts will present a similar picture. With advancing years, Ph.D's were high performers if they maintained breadth and flexibility; whereas among non-Ph.D's those with breadth of skill performed well before 40, but after this point high achievers often specialized. In Chapter 4, such a trend appeared for engineers in terms of career levels rather than age.

A hint of the same generalization appeared for the number of research and development functions (Chart 2). Ph.D's in development excelled after age 40 if they retained an interest in different kinds of R & D functions; Ph.D's in research were aided by this source of diversity in their late 30's. Among assistant scientists, though, diversity in R & D functions was a slight handicap after 40.

Another interesting pattern was generated by the individual's self-reported interest in "mapping broad features of important new areas, leaving detailed study to others." Chart 3 plots the correlations of this

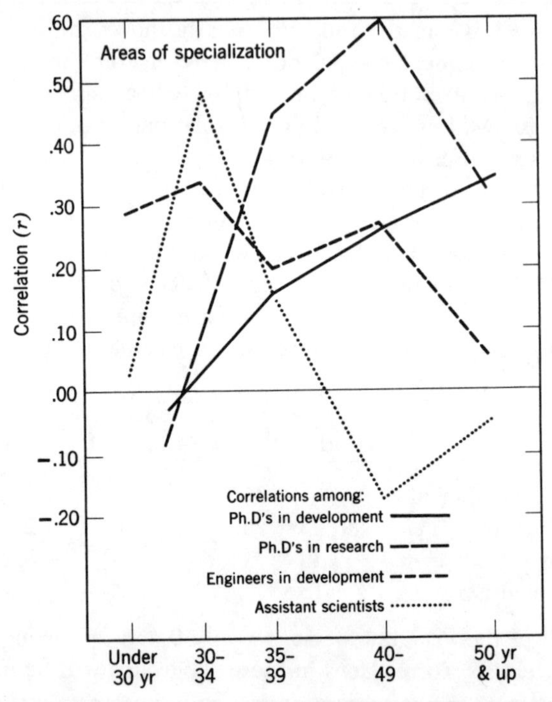

Chart 1. *Correlation between number of areas of specialization and scientific contribution. Knowledge in several areas accompanied high performance among older Ph.D's and among younger non-Ph.D's.*

measure not with scientific contribution, but with scores of overall *usefulness* to the organization, which yielded a clearer picture.

In all four groups, scientists aged 50 and over were more useful when they retained a lively interest in broad pioneering. (Other ups and downs in this chart are puzzling.) Perhaps this can be said: even though non-doctorals might gain by specializing after 40 (according to Charts 1 and 2), they would do well after 50 to stay receptive to new developments in their field.

The interest in broad pioneering is well illustrated by a passage from a recent book by Hans Selye on his lifetime of experience as a scientist.[4] In the preface he writes: "The specialist loses perspective and by now I am sure that there will always be a need for integrators, for naturalists who keep trying to survey the broad fields. I am no longer worried about missing some of the details. There must remain a few of us who train men

[4] H. Selye, *From Dream to Discovery: On Being a Scientist,* McGraw-Hill Book Co., New York, 1964, preface.

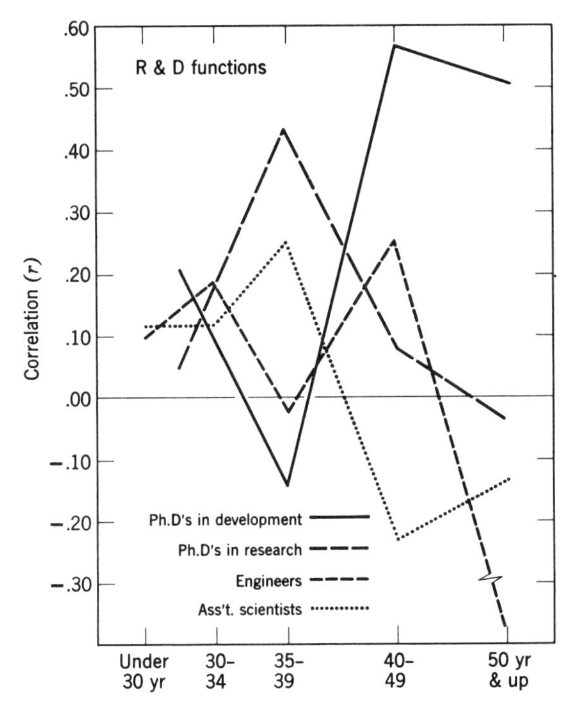

Chart 2. *Correlation of number of R & D functions with scientific contribution. After 40, Ph.D's in development achieved well when they performed several functions, and so did research Ph.D's in the late 30's.*

and perfect tools to scan the horizons rather than to look ever closer at the infinitely small."

Some Indicators of Wisdom

According to the foregoing theoretical ideas, younger scientists might be aided if given the opportunity rather early to develop a cumulative body of knowledge about some area. Such a source of "wisdom," combined with the natural flexibility of youth, should permit early achievement. Two possible measures of wisdom are illustrated in the following charts.

Time in Area. When we asked the individual to list his various projects, we also asked how long he had been working actively in each. The longer one has worked on a given project, the greater should be his systematic knowledge or wisdom about it. When we divided our groups according to age brackets, the interesting pattern in Chart 4 appeared.

Chart 4 plots correlations between length of time in one's main project, and scientific contribution. Strong fluctuations appeared, which suggested cyclical effects.

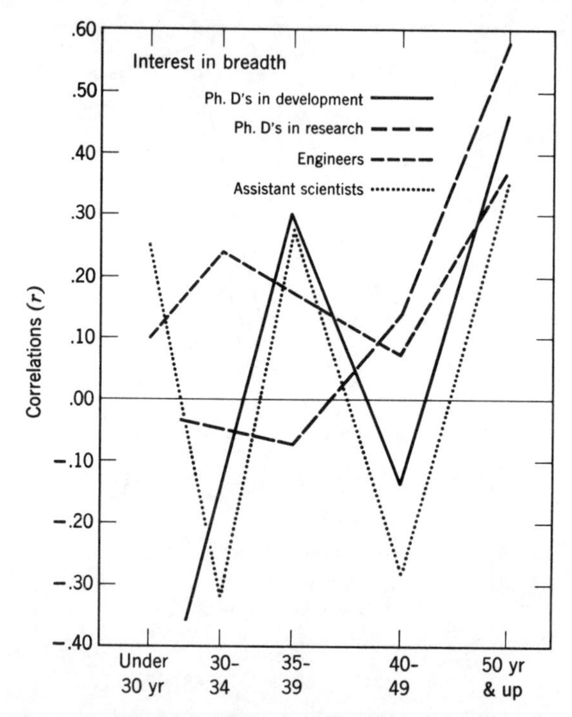

Chart 3. *According to these correlations with over-all usefulness, scientists in all groups were more useful in their 50's when they retained an interest in breadth: broad mapping of new areas.*

Within all four groups, those in the youngest age bracket performed distinctly better if they had spent a relatively long time in their main project compared to others of the same age. (Among the youngest Ph.D's and assistant scientists, the average time in main project was about two years; among engineers, one year.) Five to ten years later, however, this advantage disappeared. In the late 30's, for example, both groups of Ph.D's, as well as the engineers, showed no benefit, and perhaps were inhibited, by long experience in their main project. One has the impression, for these three groups, that by the late 30's the individual should have shifted to a new main project.

Then another build-up occurred (during the 40's for these three groups, earlier for assistant scientists) in which cumulative experience in a given project was once more an advantage; but for most groups this advantage again dissipated within the next ten years.

Note that the *general* slope of all curves was downward; younger individuals gained more by prolonged exposure to a given project than did older persons. The strong fluctuations superimposed on this over-all

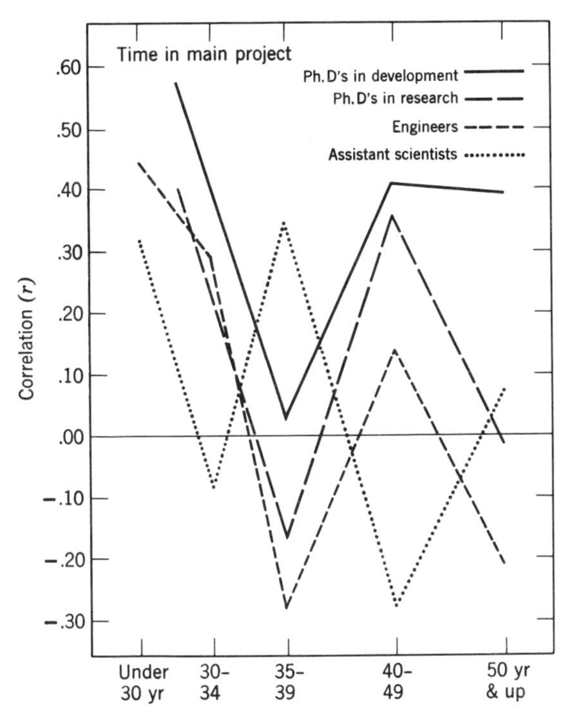

Chart 4. *According to these correlations, the youngest scientists in all groups contributed more if they had spent a relatively long time in main project. This advantage disappeared in five to ten years, but reappeared after another interval.*

trend, however, suggested the need of periodically terminating old projects and starting afresh on new ones.

Breadth. Another measure that might be an indicator of wisdom was self-estimated interest in "probing deeply and thoroughly in selected areas, even though narrow." If narrow specialization permits the building of cumulative knowledge about a topic, then according to the foregoing theory it should benefit the performance of younger scientists, and inhibit that of older men. The data shown in Chart 5 seemed to say precisely the opposite—and we show the chart partly to illustrate the complexity of these phenomena.

Younger Ph.D's, instead of being helped by "probing deeply," seemed to be hindered; and instead of being hindered after 40 by this orientation, development Ph.D's and assistant scientists seemed to gain, as did research Ph.D's and engineers after 50.

Perhaps the lesson to be gained is that one should not consider length of time in a given area of knowledge or a given project as equivalent

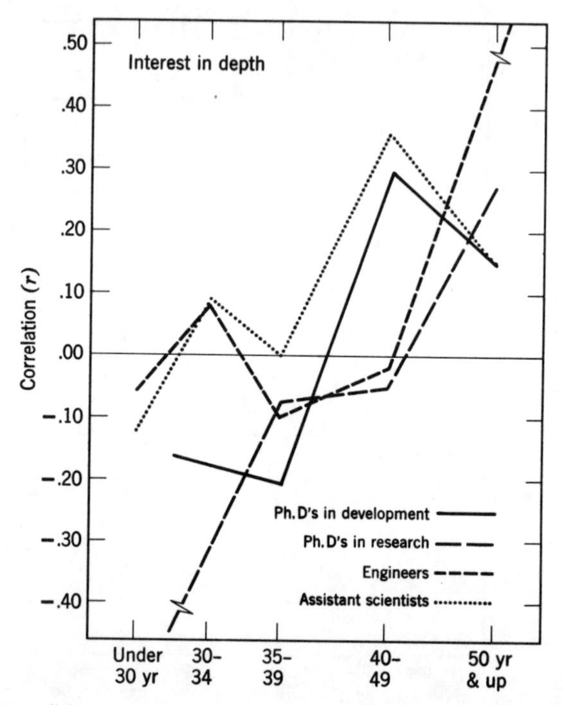

Chart 5. *Paradoxically, an interest in depth—probing deeply into narrow areas—was also an advantage in later years.*

to "narrow specialization." Perhaps the younger Ph.D should remain in one problem or research area, but learn as much as he can about many aspects of this area; he should not limit himself to a narrow facet. As he grows older, the Ph.D scientist should then broaden his outlook by moving into new problems or additional areas of knowledge, and perhaps at the same time he can afford to probe more deeply into a narrow facet of one of these areas.

Perhaps the key for mature scientists lies not in the dominance of breadth *over* depth, but rather in the presence of both breadth *and* depth.

Security and Self-confidence

We now leave the wisdom-flexibility framework, and consider another dimension. At an early stage in the project, we suspected that one of the conditions for high achievement might be "security" or lack of "anxiety." Security or confidence might be provided either from the situation (a supportive chief, for example), or from inner resources. The scientist lacking in either source of security would tend to stay on well-trodden paths where he might be productive but not creative.[5]

This concept seemed to appear in a factor analysis of responses from university and industrial scientists separately, where Andrews obtained in both groups a parallel factor which he labeled "motive to avoid risks."[6] It included items such as preference to obtain acceptable if not spectacular results, desire to probe deeply in narrow areas, and preference for the role of "right-hand man" for a more experienced advisor.

Chapter 6 reported on two measures which seem to indicate inner sources of security, that is, *self*-confidence. One was "motivation from own ideas," or desire for self-direction; in general, it correlated positively with performance. The other—negatively related to performance—was the preference for acceptable results (one of the items in Andrews' factor). For convenience we labeled it "caution."

Did the correlation of these measures with performance change with advancing age? Results are given in Charts 6 and 7.

Self-reliance. In Chapter 10 we saw that scientists with a strong desire for self-direction resisted the erosion of advancing years. Chart 6 examines the same data, treated by using correlations rather than means.

In three of the four groups the lines of correlation generally sloped upward: strong desire for self-direction was especially beneficial to performance after 40. (But among Ph.D's in research, this factor was even more strongly correlated among younger men under 35—a puzzling discrepancy from the simple generalization.)

Attitude of Caution. In Chart 7 we note that negative correlations were found among research-oriented Ph.D's at all age levels, but among the other three groups only after age 40. Caution, then, was a handicap, especially for older scientists.

In short: after age 40, continued achievement depends on high self-confidence, as evidenced in the individual's willingness to take risks and to rely on his own judgment.

Does this mean that security is unimportant before 40? Not neces-

[5]D. C. Pelz, "Uncertainty and Anxiety in Scientific Performance," working paper on analysis plans for study of scientific personnel, January 1960. Available as Publication #1588 from the Survey Research Center, University of Michigan. The term "uncertainty" referred to lack of predictability in the scientist's intellectual environment. Scientists would be more creative (we thought) if their work contained ingredients of uncertainty, although at the same time they felt secure or confident. One source of uncertainty may be *diversity* in the work.

[6]F. M. Andrews, "An Exploration of Scientists' Motives," Analysis Memo #8, March 1961, available in Publication #1825 from the Survey Research Center, University of Michigan.

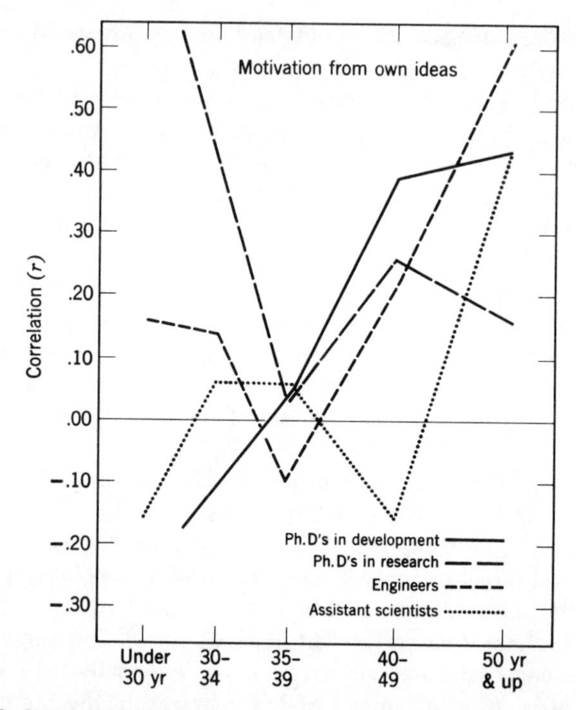

Chart 6. *Self-reliance or motivation from own ideas was more strongly correlated with contribution in the 40's or 50's than in the late 30's.*

sarily. But younger scientists, perhaps, can gain security from the *situation*: from their group, their colleagues, their chief. With advancing years, security must lodge increasingly in the man *himself*.

Hans Selye comments aptly: "Perhaps the most important thing I have learned is self-confidence; nowadays I no longer waste so much time in justifying my ways to others and to myself. It is difficult for an objective young man to have self-confidence when he still lacks the evidence to prove that he is on the right track."[7]

Self-confidence and self-reliance, of course, are not qualities that suddenly appear at age 40. Experiences throughout the 30's are critical in building them. How can they be developed? The concluding paragraphs of Chapter 6 contain a number of practical suggestions.

As a summary to date: in general it appeared that Ph.D's after 40 (whether in research-oriented or development-oriented labs) should remain flexible by means of diversity (knowledge of several specialties, and periodically entering new areas); whereas non-Ph.D's—especially engineers

[7] Selye, *loc. cit.*

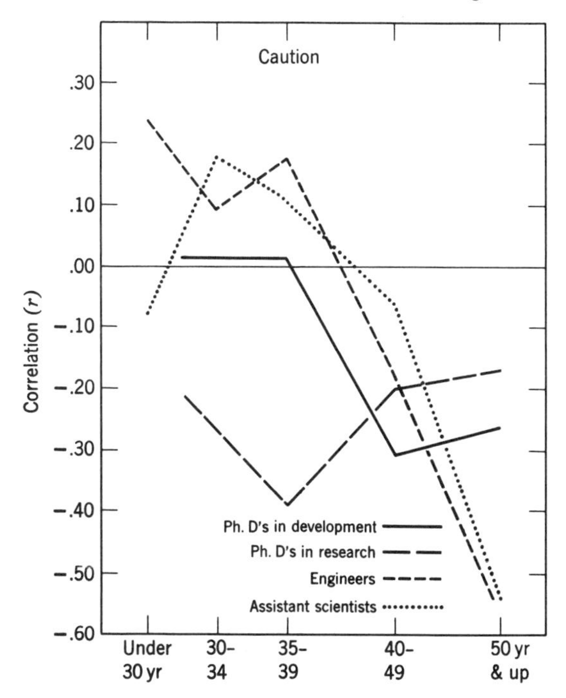

Chart 7. *According to these correlations, cautious individuals (who preferred to be sure of acceptable results) contributed less after 40.*

in development-oriented labs—could afford to specialize as they aged. Another useful concept was that of self-confidence; in all groups, achievement after 40 required self-reliance and willingness to risk the uncertain.

SUMMARY AND IMPLICATIONS

In the following conversation with an imaginary reader, we shall attempt to translate these very tentative findings into practical action.

> This chapter contains a lot of detail. What would you consider your single most important point?

Perhaps simply to emphasize again—as we did at the conclusion of Chapter 10—that the rise or fall in scientific performance over the life span of an individual is not fixed, but can vary with the kind of research climate.

More specifically, younger Ph.D scientists achieved sooner if the natural "flexibility" of their youth could be complemented with a substantial body of knowledge about some area; older Ph.D's continued to achieve if their cumulative "wisdom" was balanced with breadth of perspective. Engineers had a different pattern, though.

> What does this mean in practical terms? As a research manager, how should I handle my younger scientists?

Let's take a young Ph.D first (in either a research- or development-oriented lab). See that he is attached to a single project for three or four years, where he can get a broad exposure to all facets of one problem. *Don't* assign him to a narrow piece of the problem, but see that he reads and talks widely about it from many angles. It might be helpful to encourage an older, experienced man to serve as his mentor.

During the first five to ten years, as he begins to produce, be sure that he is recognized for his achievements. Pay him more if you can; or at least see that he has a chance to explain his work to key executives. See that he gets a chance to undertake more challenging assignments, as he is ready for them, and that he has continued opportunity to learn. This opportunity for self-actualization may be critical during the first ten years.

Then, in his mid-30's, make sure he gains a chance to stand on his own feet, independent from his supervisor or mentor. It might be wise to shift him to a new project—let him start afresh with a new chief and several new colleagues. Don't let him rest on achievements in his first area, but push him into other areas. Now he should start broadening out.

> To what extent do these suggestions apply to the young nondoctoral man?

Our data suggest that the young engineer (for example, under 35) in a development-oriented lab, as well as the young assistant scientist in a Ph.D-dominated lab, should be pressed rather soon to develop several specialized skills. This doesn't mean a rapid rotation among projects; all groups of young scientists did better if allowed to stay for several years on their main project. But the young nondoctoral man can be encouraged to dig into several topics related to this project.

The other comments concerning early recognition, rotation to a new project after several years, and increasing independence from the immediate supervisor should apply just as much to non-Ph.D's as to Ph.D's.

> What about the post-40's? Is there anything I can do as a research manager to help my older scientists sustain achievement?

Helpful for all groups after 40 is a well-developed desire for self-direction, and self-confidence about venturing into risky areas. If the man lacks these qualities by 40, it may be too late to build them. Nourishing these attributes is your job in the 30's, rather than later.

The pattern of laboratory policies and procedures will affect the climate of self-confidence. Is a senior scientist liable to having his funds chopped if some individual or group at the management level doesn't like his annual progress report? Such arbitrary decision-making, with little control by the scientist himself, is likely to breed cautious dependence rather than independent self-confidence.

The post-30's engineer, in addition, should have at least one area about which he knows a great deal.

Keep an eye on the relationships of your scientists in their 40's with their chief. If you find that some of these men are good lieutenants under a chief they admire—working closely with him rather than acting on their own initiative—then it is time for a change. Encourage them to join a new group or project where they will be leaned upon rather than leaning. But it's advisable to do this *before* 40 rather than after.

> What about the period just before 40? Are there any special features of the 35 to 39 year bracket that I should be watching for?

This is a very interesting period, and may be a critical one for later achievement. It seems to be a period of transition—a kind of midcareer adolescence.

We have the distinct impression that this is a period of creative change. The effective individual exposes himself to external stimuli. He is shedding his previous identities, searching for new ones. Help him. This is the time for self-renewal. Give him a sabbatical, or send him to the university for retraining. Assign him for six months to another department. Have him participate in long-range planning seminars.

Out of such external bombardment he will (you hope) set new directions for himself—the foundation for sustained achievement through the 40's.

12

COORDINATION[1]

In Loosely Coordinated Settings, the Most Autonomous Individuals Did Poorly—Perhaps Because They Were Isolated from Stimulation.

In previous chapters we have been concerned with characteristics of the individual scientist: his age; his autonomy and influence over decision-makers; the content of his work; his orientations and motivations; his communication with colleagues; his satisfaction, creativity, and similarity to others. In this chapter we undertake to remap some of the same territory from a different perspective, that of the organization in which the scientist functions.

Specifically, we shall abstract from our variety of settings the extent to which they are tightly or loosely coordinated. Then we shall ask: what kinds of individual characteristics are most effective in organizations which differ in degree of coordination?

One major question we wished to examine was: how much autonomy for the individual is conducive to high scientific performance as the situation varies from tight to loose?

This chapter recalls previous results which showed that autonomy correlated with performance of nondoctorals, but not with performance of Ph.D's. The chapter then describes the classification of various work situations along a scale from "tightly" to "loosely" coordinated, and successively presents data to answer these questions: Did our measures of motivation and stimulation (discussed in Chapters 3 through 7) correlate differently with performance as coordination of the situation varied (Part A)? Did measures of autonomy and of influence (Chapter 2) correlate differently with performance at different levels of coordination (Part B)? How did autonomy correlate with various motivating factors at each level (Part C)?

In general, we found that (A) the looser the situation, the more strongly high levels of motivation (both internal and external in source) accompanied

[1]This analysis was supported by grants from the National Science Foundation. These findings were presented in D. C. Pelz and F. M. Andrews, "Autonomy, Coordination, and Stimulation in Relation to Scientific Achievement," *Behavioral Science*, 1966, vol. 11, pp. 89–97.

high performance. But (B) the individual's autonomy and influence were most effective in situations of only moderate looseness. Part (C) gives some clue as to why; there were numerous hints that maximum autonomy in a very loose setting may isolate the individual from stimulation.

Some Points of Departure

Since science is a search for generalized principles, the search is prodded by inconsistencies. An important and puzzling inconsistency was described in Chapter 2 for the measure of autonomy, defined as the proportion of weight which a scientist himself exerts in deciding his technical goals. Autonomous nondoctorals (that is, those who exerted 30% or more of the weight themselves, with no other person or group exerting this much) were substantially above-average in performance. But autonomous Ph.D's were only average.

Other inconsistencies appeared in our data, less striking than the foregoing result. Although intergroup competition correlated well with performance of Ph.D's in research, it correlated not at all for assistant scientists (data not shown). Provision of opportunity for self-actualization and for status advancement (Chapter 7) correlated with performance in most groups, but again less so for assistant scientists. Patterns of effective communication (Chapter 3) also seemed to differ among assistant scientists compared with other groups.

What might account for these differences? As one step in our search, we felt it was necessary to know more about the kind of *situation* in which the scientist was functioning. In preliminary interviews we had been impressed with the variation of different laboratories along a dimension that might be described as "tightness" to "looseness" of structure. Some settings resembled a typical bureaucracy; that is, jobs were assigned with little leeway to deviate, and all activities were closely coordinated by a supervisory structure. At the other extreme, situations such as university departments were highly decentralized; individuals largely set their own goals, with a minimum of coordination.

Measuring the Coordination of the Situation

As one of the preparatory steps in our exploration, we had already prepared a measure of departmental "coordination versus autonomy." (See Appendix H.) Within our 11 organizations were 53 departments, including 26 industrial and 20 government branches consisting of several sections reporting to one department head, and 7 university departments. They ranged in size from 10 to 45 professional members, with a median of 22. Mean responses were computed for supervisors and nonsupervisors separately within each department on 16 questionnaire items relevant to coordination. Intercorrelations among the scores indicated a clustering or consistency of responses among several of the items, which enabled us to place each department along a scale from "highly coordinated" to "highly autonomous." All of the university departments fell at the autonomous end of the scale; over half of the departments in government labs were moderately autonomous; and industrial departments split evenly between moderately autonomous and coordinated.

In addition, we felt it advisable to consider the factors that defined our five primary analysis groups. It was likely that Ph.D's would be allowed more discretion than non-Ph.D's; that scientists in research-oriented laboratories would be more loosely coordinated than those in develop-

TABLE 1 *A scale of situational coordination was constructed by dividing the five primary analysis groups according to departmental score as coordinated or autonomous. The total response in each subgroup on individual autonomy, or own weight in setting goals, was used to assign the group to one of five levels.*

			Own weight in goals*		
Level	Type of department	Number	None (0–9%)	Considerable (30–100%)	Difference
I. Very tight					
Assistant scientists	Coordinated	50	58%	14%	−44
II. Moderately tight					
Non-Ph.D's in res. labs	Coordinated	85	6%	13%	7
Engineers in devel. labs	Coordinated	7	31%	20%	−11
		92			
III. Mixed					
Engineers in devel. labs	Autonomous	133	17%	40%	23
Ph.D's in devel. labs	Coordinated	59	20%	45%	25
Assistant scientists	Autonomous	40	14%	39%	25
		232			
IV. Loose					
Non-Ph.D's in res. labs	Autonomous	38	16%	55%	39
Ph.D's in devel. labs	Autonomous	37	7%	57%	50
		75			
V. Very loose					
Ph.D's in res. labs	Autonomous	79	5%	81%	76

* Shown is the percentage in each subgroup responding at the low and high ends of the autonomy scale. The question itself is shown in Chapter 2.

ment; and that assistant scientists under a corps of Ph.D's would have less freedom than those in departments not dominated by Ph.D's. We subdivided the five primary analysis groups according to the department's score as coordinated or autonomous. This process yielded nine subgroups (all the Ph.D's in research labs were in autonomous departments).

How could these nine subgroups be placed along a scale of coordination? The method we used was to examine again the total responses within each subgroup on individual autonomy (own weight in deciding technical goals); and on this basis we arranged the subgroups into five distinct categories of "tightness to looseness" of coordination, as shown in Table 1.

It is important to keep in mind that entire *groups* were assigned to each level of coordination, not detached individuals; the scale described the situation, not the individual. Within each level, of course, individual autonomy could vary considerably.

Situations at level I, which we have labeled "very tight," were perhaps no more tightly coordinated than a typical industrial operation employing skilled craftsmen. Over half of this group said they had little or no autonomy.

At level II, "moderately tight," the large majority said they had *mild* autonomy (10 to 29% of weight exerted by self in setting goals).

Level III, "mixed," consisted of three subgroups marked by some factors making for tightness (for example, two-thirds were nondoctorals), and by other factors making for looseness (these non-Ph.D's worked in autonomous departments). Two persons in five reported considerable autonomy.

At level IV, three persons out of five reported considerable autonomy; and at level V, "very loose," four out of five. The latter consisted entirely of Ph.D's in research-oriented, autonomous laboratories.

A. MOTIVATIONAL FACTORS AND COORDINATION

A General Expectation

In examining how motivation or stimulation related to performance at each level of coordination, what might we expect to find? Chapter 7 presented some ideas of Chris Argyris on the conflict between a mature individual's desire for independence and self-actualization, and the demands of a bureaucracy for passive conformity. Our tightest situations (level I) resemble the latter. Here the individual who is strongly self-reliant may find himself frustrated; he may try to circumvent the system, and succeed only in arousing ire. His performance (as rated by senior colleagues) is likely to suffer. He might do better if his motivation were moderate rather than strong.

As the situation becomes looser, however, these conflicts should diminish.

The self-reliant individual should be able to achieve more. In fact, precisely because the situation does not specify what he is to do, or how, he *must* be self-reliant to achieve. Otherwise he may stagnate.

This general picture was suggested by some previous evidence. E. E. Ghiselli, for example, used a self-descriptive forced-choice inventory (in which the respondent describes himself by selecting statements out of sets of equally favorable items) to develop several measures including supervisory ability, initiative, and self-assurance.[2] He found an interesting pattern. When he compared data from workers, foremen, middle managers, and upper managers, he found that at the top levels the measures showed distinct positive correlations with performance ratings. But at lower levels, the correlations became smaller. The measure of "initiative," for example, was largely unrelated to performance of line workers, and may have hindered the performance of foremen as rated by superiors (it correlated negatively with such ratings).

"Initiative" resembles our measure of motivation from own ideas, and perhaps desire for self-actualization. The situation of a nonsupervisory employee or a foreman is more tightly controlled than that of a manager. Strong initiative in a tight situation may harm the employee rather than help.

Another hint came from an unpublished analysis by Pelz of data from an electronics manufacturing company.[3] An index of "self-motivation" was formed, based on the identical question of involvement used in our present study, plus two other questions reflecting internal motivation. Employees in engineering and research departments were considered to be in more flexible (looser) situations than those in manufacturing or central staff. Strength of self-motivation was found to correlate more positively with performance ratings in the looser departments than in the tighter ones. When the situation had little structure, stronger internal motivation was needed for achievement.

The next several charts will present data to test this expectation. We shall examine the degree of relationship between various motivating and stimulating factors on the one hand, and our four measures of performance on the other. For simplicity, relationships were measured with the correlation coefficient (r). This, of course, reflects only linear trends, not curvilinear ones.

[2] E. E. Ghiselli, "The Validity of Management Traits in Relation to Occupational Level," *Personnel Psychology*, 1963, vol. 16, pp. 109–113.

[3] D. C. Pelz, "Self-Determination and Self-Motivation in Relation to Performance" (prepublication draft, mimeo), January 1962, Survey Research Center, University of Michigan, 21 pp.

Some Results with Internal Motivations

We saw in Chapter 5 that one of the items consistently related to performance was "involvement in one's work." It is perhaps our single best indicator of total commitment or dedication.

According to the conception just sketched, the looser the situation, the more dedicated an individual should be if he is to achieve. In tight situations, involvement should correlate only mildly with performance, whereas in loose ones it should correlate strongly. The data are given in Chart 1-A.

The following charts resemble those in Chapter 11. The vertical axis does not show mean performance, but rather the size of *correlation* (the product-moment coefficient, r) between two measures. Along the horizontal axis, scientists are arranged according to the level of tightness or looseness of their situation, as previously described in Table 1.

Chart 1-A, for example, shows that in the tightest situations (level I), the strength of individual involvement correlated mildly (+ .14) with scientific contribution—the highly involved scientists performed slightly above average; whereas in the loosest situations (level V), the correlation was stronger (+ .36)—highly involved scientists performed significantly better than average.

The data used here were obtained from the 500 long-form respondents. The performance measures have been adjusted to rule out effects of length of working experience and type of setting, as described in Chapter 1.

By "papers/patents" is meant publication of papers by members of all groups except engineers, for whom patents were substituted.

In general, the expected results appeared for three of the four performance measures. As situations became looser, involvement showed an increasingly positive correlation with scientific contribution, publication of papers, and (slightly) writing of reports. Usefulness, oddly, correlated with involvement in most situations, but not in the loosest. Perhaps in the latter, highly involved scientists were following their own interests and ignoring the needs of the organization. (As we shall see throughout, usefulness often went with high motivation or stimulation at various levels of coordination.)

Chapter 6 indicated that effective scientists in several settings were self-reliant and motivated by their own ideas and previous work. We correlated the index of "own ideas as a source of motivation" with the four performance measures, at each of the coordination levels, with results as shown in

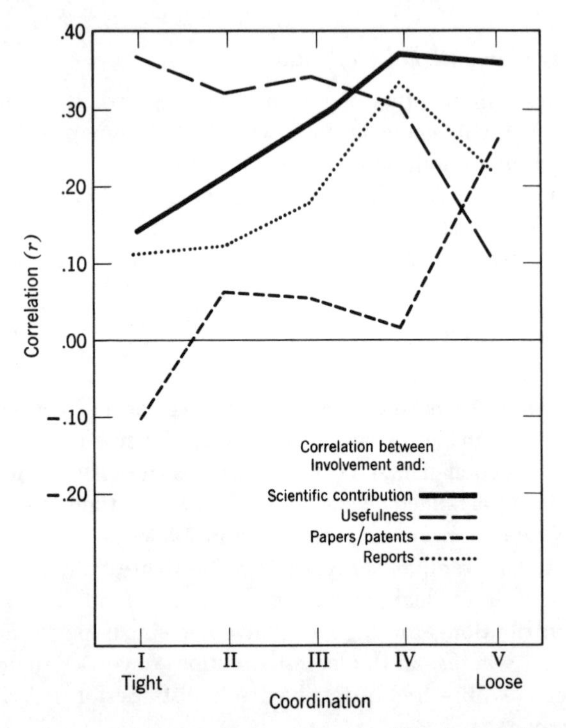

Chart 1-A *plots the correlations between involvement and the four performance measures in situations varying in level of coordination. The looser the situation, the more highly involved scientists tended to perform better. See text for interpretation.*

Chart 1-B. Again the correlations for scientific contribution showed the same result: the looser the situation, the more strongly inner motivation accompanied high contribution. The other measures showed no consistent trend.

Note in Chart 1-B an odd result in loose situations (level IV): inner-motivated scientists seemed to publish *fewer*-than-average papers. A similar dip for publication will appear in several charts. Why? One clue (not plotted) was that paper publishers at this level felt they had little influence regarding resources. Another clue (not plotted) is that they scored low on "provision of status advancement." Our hunch is that these scientists, either Ph.D's in development labs or non-Ph.D's in research labs, occupied low status positions where they felt ignored. They could not afford the luxury of following their own interests. Rather, they spent their energy turning out papers in order to build their prestige.

Chapter 7 discussed conflict between desires of the individual for self-actualization and independence, and demands of the organization for

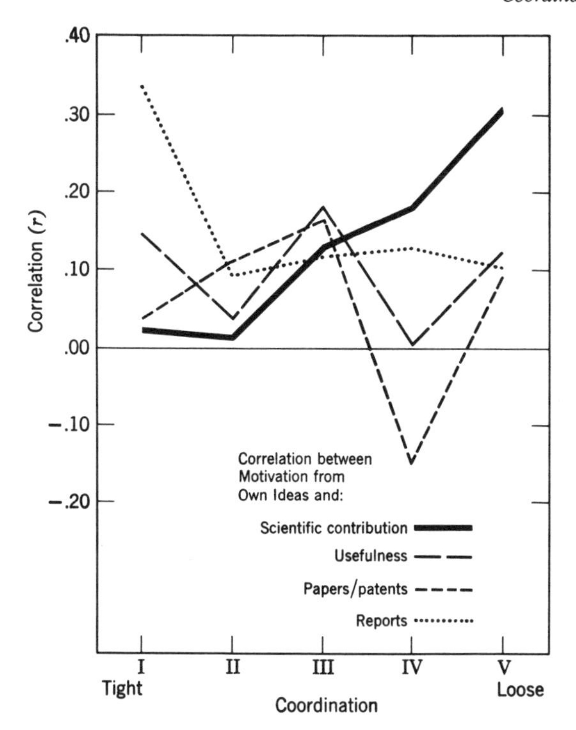

Chart 1-B. *Motivation from one's own ideas did not affect the scientific value of one's work in tight situations (levels I and II), but increasingly accompanied high contribution in looser situations. In a number of charts, the odd fact will appear that motivated individuals at level IV published fewer papers than average.*

dependence and conformity. Such conflict should be evident in tightly controlled situations, such as those at level I. Here, a strong desire for self-actualization might not only fail to aid performance, it might tempt the individual to tackle problems that were not relevant, or to use discretion that was not permitted. Chart 1-C shows what happened when we correlated the measure of desire for self-actualization with the four performance measures. (The index was based on importance attached to using present knowledge, learning new knowledge, working on challenging problems, and having freedom to carry out one's own ideas. The last item was also one of three in the previous index of inner motivation; therefore the two measures are not independent.)

In tight situations (levels I and II), a strong desire for self-actualization actually seemed to inhibit contribution—at least as judged by the senior staff members. But as looseness increased, desire for self-actualization became increasingly helpful for achievement.

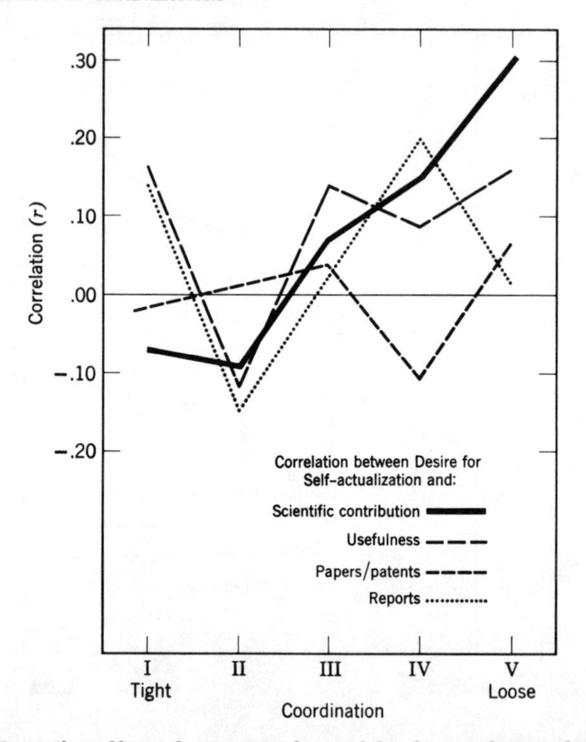

Chart 1-C. *Desire for self-actualization may have inhibited scientific contribution in tight situations (levels I and II) by arousing conflict with the organization, but it was increasingly helpful as the situation became looser. People with this motivation were useful, however, at most of the levels.*

Note, though, that in the tightest situations (level I), the self-actualizing tendency enabled the person to turn out useful reports. It is plausible that when the situation is rigidly defined, high energy from any source may enable the person to be *productive,* even if not creative. (In Chart 1-B note the high correlation with reports at this level.)

Stimulation from Other People

Next let us consider some external sources of stimulation. Chapter 3 reported that performance generally rose as the individual communicated more often with his colleagues, and exchanged information with a wider number of colleagues. If interaction with colleagues is motivating, then according to the previous hypothesis such interaction should relate more strongly to performance in loose than in tight situations. Results are given in Charts 2-A and 2-B.

Except for level II, frequency of communicating with colleagues

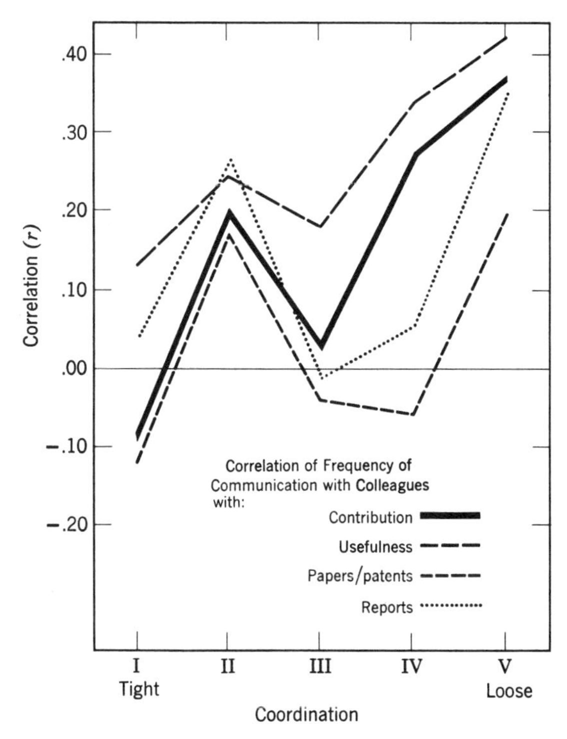

Chart 2-A. *Scientists who communicated often with colleagues did not produce work of scientific value in very tight situations (I), but did so in loose ones (IV and V). As the situation became less rigid, frequent communication increasingly accompanied usefulness. See text for discussion of exceptional picture at level II.*

(Chart 2-A) showed the predicted pattern. In loose situations (level IV and especially V), frequent communication correlated more strongly with scientific contribution, usefulness, and publication than it did at levels I or III.

The *number* of close colleagues (those with whom the scientist exchanged detailed information, Chart 2-B) correlated positively with contribution and usefulness at level V, whereas at level I the correlations were zero or negative.

Why the exception for level II? We can only speculate. Note that these people all worked in coordinated departments; the work was interdependent. At the same time, they had at least mild weight in deciding their goals. Cooperation was necessary, but it was willing cooperation among equals, not passive obedience by subordinates (as at level I). Voluntary teamwork was the key to success; and those individuals who kept their communication lines open were best able to achieve.

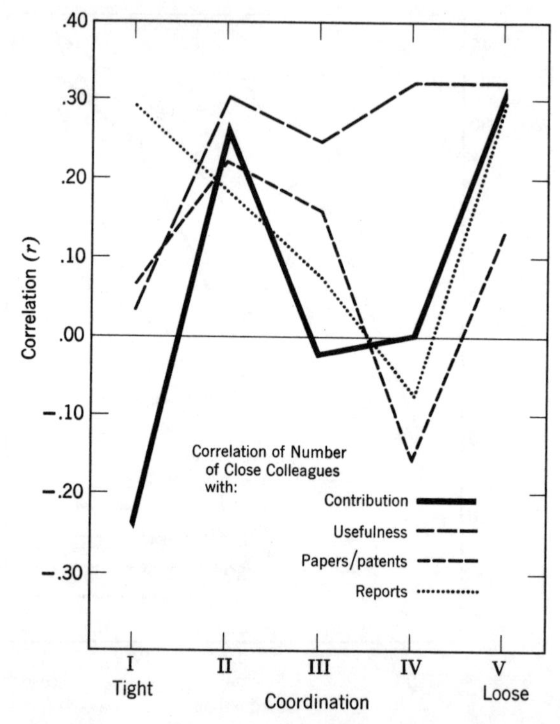

Chart 2-B. *Scientists who worked closely with many colleagues did work of low scientific value in tight situations (I), but of high value in loose ones (V). However, communicative individuals wrote many reports at both extremes, and were useful at several levels.*

What about competition? We knew (data not presented) that there was some tendency for high performance to accompany between-group competition, and (to a mild degree) competition among individuals. The correlations for these factors are plotted in Charts 3-A and 3-B.

Competition between *groups* (Chart 3-A) generally showed a rising trend in the correlation coefficients as the situation became looser. Competition seemed to inhibit performance in very tight situations, and to stimulate it in very loose ones. But competition among *individuals* (Chart 3-B) showed a mildly opposite trend. Scientists in the teamwork atmosphere at level II did better work in a climate of rivalry (friendly rivalry, according to Charts 2-A and 2-B), whereas those at level V did better without it. Possibly the Ph.D's at level V felt they had earned prestige, and did not need to impress each other.

Stimulation from Diversity in the Work Content

Chapter 4 presented evidence that diversity in the content of the scientist's or engineer's work might be an ingredient in high achievement. If diversity is a source of stimulation, then according to our general

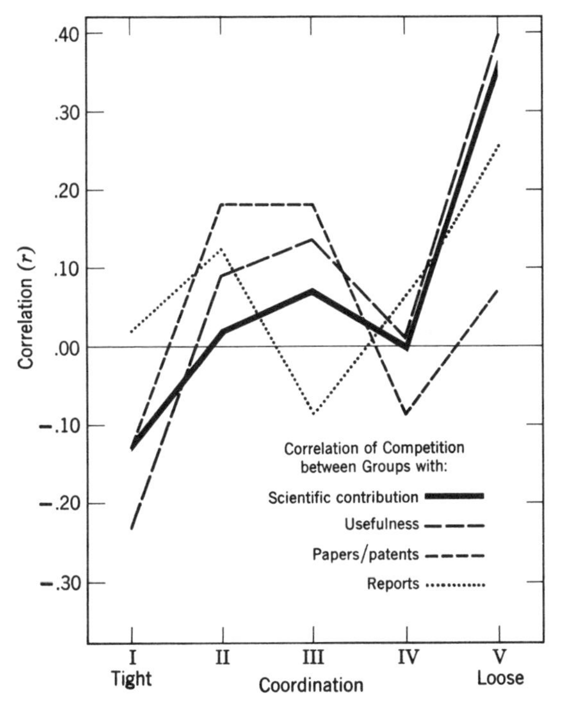

Chart 3-A. *Competition between groups may have inhibited performance in very tight situations (I), but mildly aided it in mixed ones (III). In the loosest situations (V), scientists who reported strong competition performed better on three of the four measures.*

hypothesis the correlations of diversity factors with performance should rise as the situation becomes looser. Charts 4-A and 4-B provide data to test this expectation.

With some exceptions, the expectation was fulfilled. The scientist's number of areas of specialization (Chart 4-A, p. 227) correlated more strongly with scientific contribution in very loose situations (V) than in very tight ones (I), although it also correlated well with several measures at levels II and III. Oddly, at level IV it was of no value to three of the measures.

Another measure of diversity was number of different R & D functions (basic research, applied research, invention, improvement of existing products, technical services) which the scientist performed, at least to a slight extent. This measure (Chart 4-B, p. 228) also accompanied achievement most strongly in loose situations (level IV), and was least helpful (or even inhibiting) in moderately tight ones (level II).

To sum what we have seen thus far: in loose situations compared to tight ones, several measures of motivation and stimulation correlated more strongly with performance. Our general expectation was borne out.

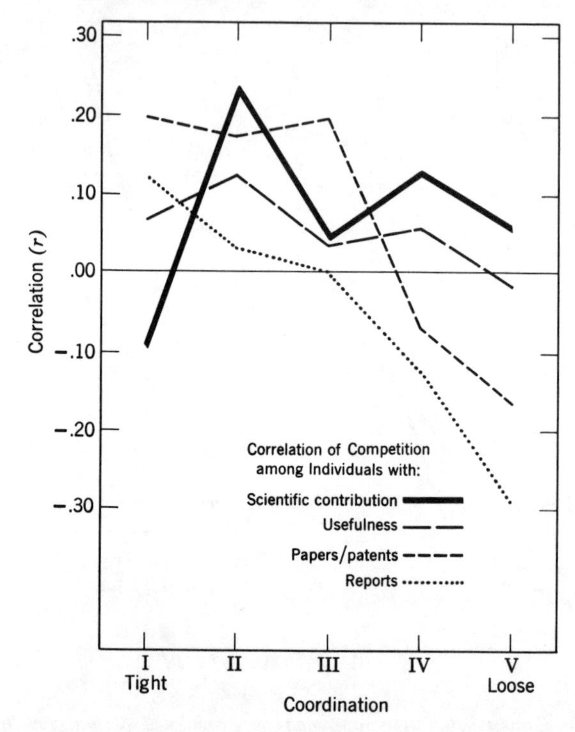

Chart 3-B. *Competition among individuals showed an opposite pattern. It characterized paper publishers and valuable contributors in moderately tight situations; but as situations became looser, individual competition aided performance less, and may have inhibited writing and publishing in the loosest situations* (V).

B. INDIVIDUAL CONTROL (AUTONOMY OR INFLUENCE)

A General Expectation

When an individual is actively seeking some goal, he is sure to encounter obstacles which must be overcome. If the individual has autonomy, he can use his own discretion to find ways around them. If he has influence with decision-makers, he can get help in removing the barriers. For these reasons, autonomy or influence should enhance performance—providing the individual is motivated to achieve.[4]

[4]Some evidence on this point appeared in an unpublished study in a manufacturing company, see footnote 3. Among individuals with high autonomy and influence (three items of such factors were combined into an index of "self-determination"), strength of inner motivation correlated more strongly with performance ratings, especially in research and engineering departments (rather than manufacturing). Where the situation is moderately loose and the individual has high control, strong motivation or stimulation may be particularly essential for achievement.

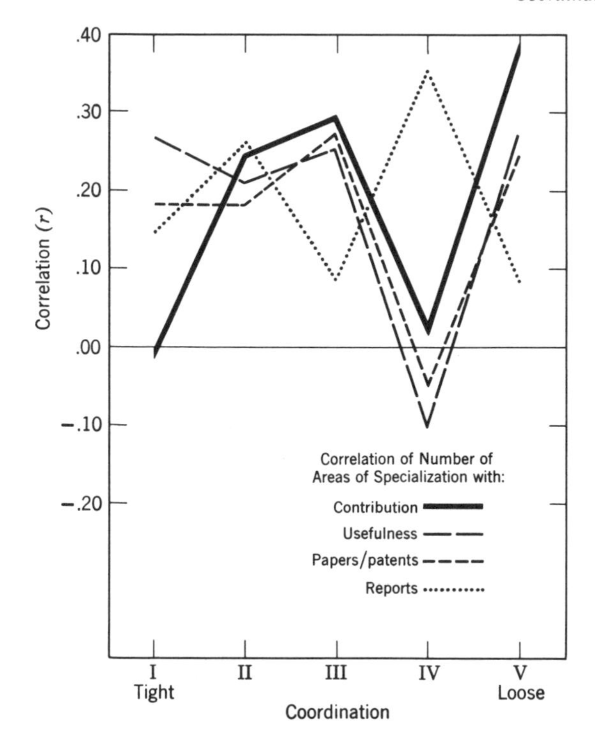

Chart 4-A. *Scientists having several specializations did work of only average scientific value in the tightest situations (I), but of high value in the loosest ones (V). Such diversity, however, accompanied high performance on other measures in intermediate situations. The text discusses the puzzling exceptions at level IV.*

We are dealing here with a topic that has been much investigated in industrial organizations under headings such as "participation in decision-making," "closeness or delegation in supervision," and "control."

Previous studies have shown that such factors often tend to go with high productivity.[5] But there have been frequent exceptions. Can we find a general framework that will incorporate both the general tendency and the exceptions?

Autonomy and influence (for simplicity we may use the term "control" to refer loosely to either) are not, strictly speaking, *motivational* in character. They do not in themselves impel the person to action. Rather, they

[5]Numerous studies at the Survey Research Center and elsewhere on influence in decision-making are summarized by Rensis Likert in *New Patterns of Management*, McGraw-Hill Book Co., New York, 1961. For an introduction to a series of studies on control, see A. S. Tannenbaum, "Control in Organizations: Individual Adjustment and Organizational Performance," *Administrative Science Quarterly*, 1962, vol. 7, pp. 236–257.

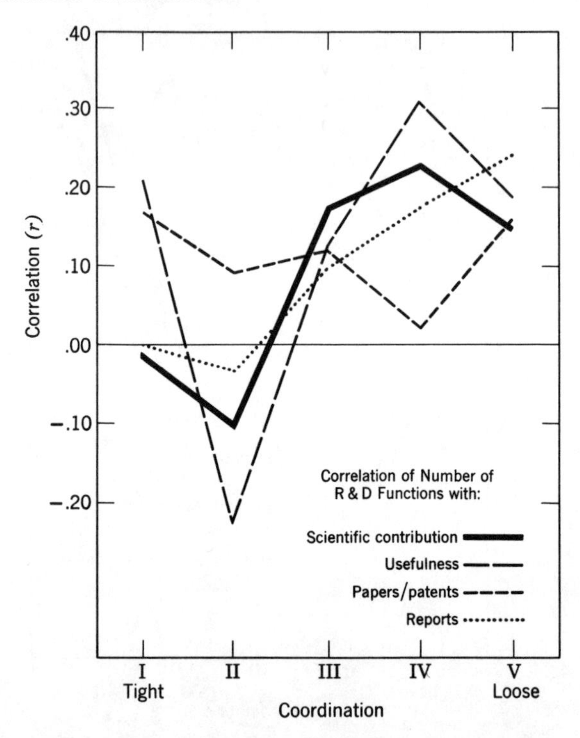

Chart 4-B. *Scientists whose work embraced several kinds of R & D activities were not helped scientifically in tight situations (I and II), but they performed better than average under looser coordination. Diversity of functions aided* usefulness *at several levels.*

serve to *facilitate.* They serve to expedite reaching goals which the individual is already seeking. (It is true that individuals who are given autonomy or influence may thereby become more interested in the work. Control may serve to arouse motivation. But on a conceptual level the two notions of control and motivation are distinct.)

How will the individual's control (autonomy and/or influence) affect his scientific performance as the situation varies from tight to loose?

If the situation is tightly coordinated, the average level of autonomy or influence will, of course, be low. What of those persons who have more control than average? Our hunch is that it will not do them much good. The situation is too rigid; only minor modifications or detours are possible. Individual control can facilitate only mildly.

In terms of correlations, we should observe either no correlation or at most a low positive one between performance and autonomy or influence in tight situations.

As the situation becomes less tight, larger modifications in procedure are possible. The person with above-average autonomy or influence can take major short cuts, or remove major roadblocks. High control can facilitate his work markedly. Assuming strong motivation, we should observe a strongly positive correlation between autonomy/influence and performance.

When the situation becomes extremely loose, impediments are fewer; many alternative paths to the goal are possible. Individual control then diminishes in utility. The person with maximum autonomy/influence enjoys only a slight advantage over the person with moderate control. The correlation between performance and autonomy or influence should drop to a low positive.

Another factor may affect our prediction. High control (autonomy/influence) may have the effect of heightening the individual's dedication. If so, the increased motivation of the autonomous individual should assist his performance in a loose situation, by our earlier hypothesis.

The net effect of these processes is that the correlation of performance with control should remain definitely positive in loose situations, though perhaps not so strong as in situations with a medium degree of coordination.

Results with Mean Performance

These expectations may be examined not only by the method of correlations, but also by examining *mean* performance as control varies. Such results are shown in Chart 5 for mean scientific contribution in relation to autonomy; results with correlations will be given in the subsequent chart.

In very tight situations (level I), an increase in autonomy was accompanied by only a mild rise in contribution. In moderately tight situations (level II), differences in autonomy made a sharper difference in contribution, and a curvilinear effect appeared: above 50% autonomy, performance dropped. In mixed situations (III), the performance rose still more steeply; and note that the optimal amount of autonomy increased from 50% to about 80%. So far, the picture fitted our expectation.

In loose or very loose situations (IV and V), however, optimal autonomy returned to moderate. Above the 50% mark, further autonomy was accompanied by a drop in contribution.

This much of a drop was surprising. From the previous theorizing, we might have expected performance in loose situations to level off as autonomy increased. But our speculation did not account for its dropping *below* average.

(This pattern is, of course, consistent with other views of the same data which were presented in Chapter 2.)

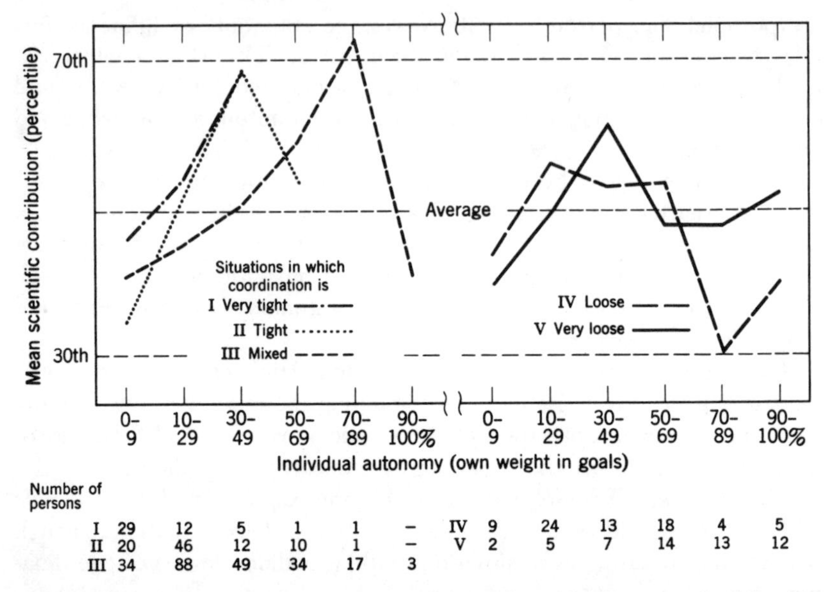

Number of persons

	0–9	10–29	30–49	50–69	70–89	90–100%
I	29	12	5	1	1	–
II	20	46	12	10	1	–
III	34	88	49	34	17	3
IV	9	24	13	18	4	5
V	2	5	7	14	13	12

Chart 5 *differs in format; it plots* mean *scientific contribution by scientists who differed in individual autonomy* (*percent of own weight in setting goals*). *As situations changed from very tight* (*I*) *to mixed* (*III*), *the optimum amount of autonomy increased; but in loose situations* (*IV and V*), *performance of the most autonomous individuals dropped.*

Results with Correlations

From the same data correlations can also be computed between autonomy and performance; these are plotted in Chart 6.

In the tightest situations (level I), autonomy showed only a mild positive correlation with contribution. As the situation became less rigid (levels II and III), the correlation increased; autonomous individuals made significantly better contribution. But in still looser situations (IV and V), the correlation disappeared, or became slightly negative. Note that autonomous scientists here published significantly *fewer* papers—or perhaps we should say those with little autonomy published more.

A measure of *influence* appears in Chart 7, derived from Question 31 as described in Chapter 2: the respondent's estimate of his influence on the person or group (other than himself) who has "most weight in choice of your work goals." Influence regarding goals was uncorrelated with contribution or usefulness in the tightest situations (I), but significantly correlated at levels II and III. Thereafter the correlations were still positive, but lower at levels IV and V.

In short: the measure of *influence regarding goals* fitted our prediction perfectly in terms of correlations with scientific contribution and useful-

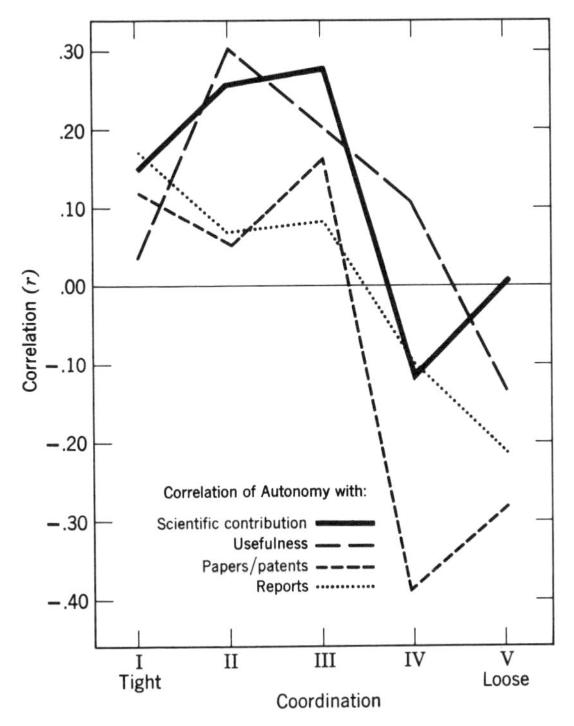

Chart 6. *In terms of correlations, autonomous individuals performed well in scientific contribution and usefulness in moderately coordinated situations (II and III). But in loose situations (IV and V), autonomous individuals were average or below in performance, especially in publication of papers.*

ness: zero correlations in the tightest situations (level I), significantly positive correlations in moderately tight situations (II and III), mildly positive in loose ones (IV and V). *Autonomy* fitted this expected pattern fairly well, but the zero or negative correlations in loose situations were puzzling.

C. SEARCH FOR AN EXPLANATION: SIDE EFFECTS OF AUTONOMY

We come now to the third step in our chain of investigation. We have seen (Part A) that as situations became looser, a high level of either internal motivation or external stimulation was increasingly important for high achievement. We also found (Part B) that autonomy was most useful in situations of moderate tightness, but in loose situations it actually seemed to inhibit performance.

Why should this be? Was it possible that complete freedom in these

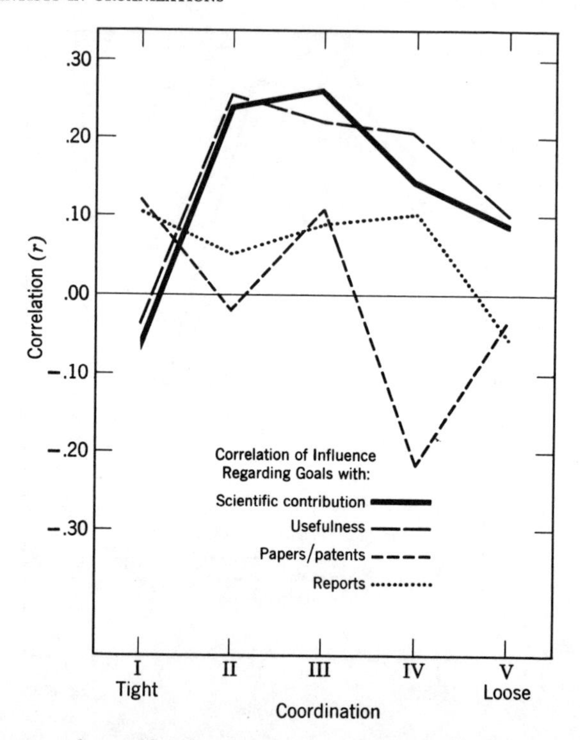

Chart 7. *Scientists who could influence other people in regard to their goals were only average performers in very tight situations (I), but high in scientific contribution and usefulness in moderate situations (II and III). After that, the connection of influence with performance dropped.*

already-free situations had certain side effects on motivation which lowered performance?

To follow this lead, we studied the correlation between autonomy and each of the motivational factors examined before, within the five levels of coordination.

Autonomy and Motivation

Chart 8 plots correlations within each of the five coordination levels, between autonomy and three measures of individual motivation.

In very tight situations (I), autonomous individuals felt strongly involved in their work. But as we saw in Chart 1-A, under tight coordination, involvement enhanced only mildly the scientific value of the man's work. Therefore the beneficial effect of autonomy on motivation was largely nullified by the rigid situation.

At level II, autonomy was not linked with motivation. But this absence

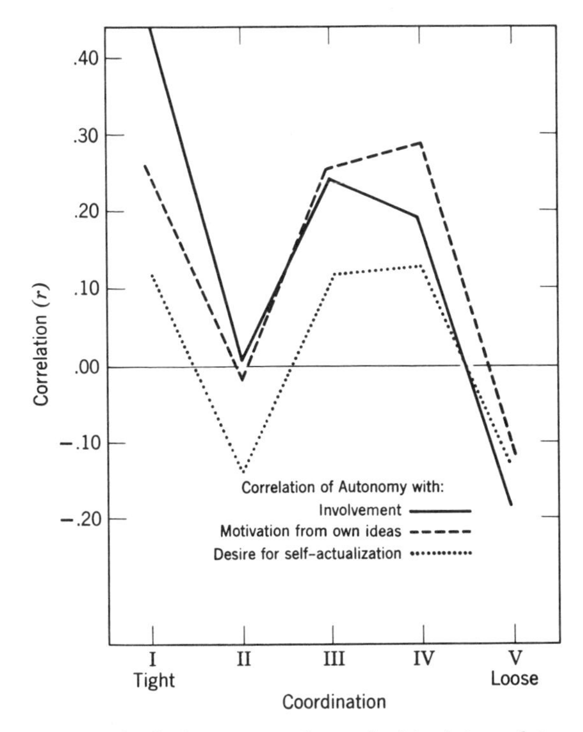

Chart 8. *Autonomous individuals were strongly involved in their work in very tight situations (I), but not in very loose ones (V). In the loosest situations, autonomy failed to correlate with motivation from own ideas, or desire for self-actualization.*

did not weaken the connection of autonomy with performance; apparently (as we shall see in a moment) autonomy brought compensating stimulations which kept performance high.

In mixed or moderately loose situations (levels III and IV), the more autonomous scientists were above average in motivation. In these situations, involvement enhanced performance (Charts 1-A through 1-C); it should follow that autonomy would also have a favorable effect, and this we noticed at level III. (Why not at level IV? We shall see in Chart 10 that autonomy at this level accompanied fewer R & D functions and fewer decision-making sources—and these losses in stimulation may have nullified the effect of internal motivation.)

Finally, in the very loose situations (V), we were surprised to note that extreme autonomy permitted (or perhaps encouraged) a *lack* of involvement in the work. It also accompanied a slight weakening of desire for self-actualization and, paradoxically, some diminishing of self-direction from one's own ideas. Why? For the completely autonomous man, no one else helps him to chart his course. Does this mean that no one else is interested? And if no one else is interested, does he lose interest himself?

Whatever the explanation, the resulting loss of motivation may have withdrawn important factors (as we saw in Charts 1-A through 1-C) in maintaining high performance.

Autonomy, Colleague Interaction, and Competition

Next let us examine various sources of external stimulation. Chart 9 shows data on interaction with colleagues, and competition between groups.

In general, correlations were very slight, but starting with level II they became progressively smaller as looseness increased, and became negative at level V. But we saw before (Charts 2-A and 2-B) that at level V such contacts and competition helped to keep achievement high.

Complete autonomy, then, appeared to isolate the individual from colleague stimulation; and this isolation may have hindered achievement.

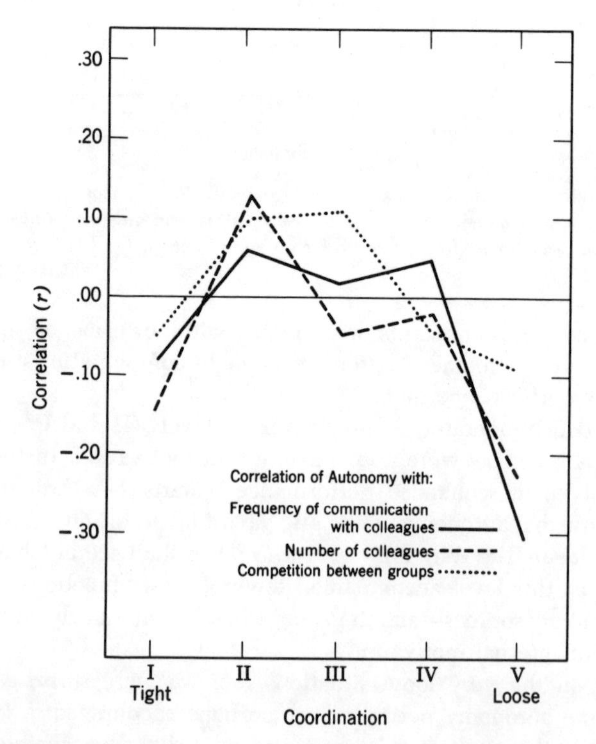

Chart 9. *In moderately tight situations (II), autonomous individuals had slightly more interaction than average, but in very loose situations (V), their interaction with colleagues was below average.*

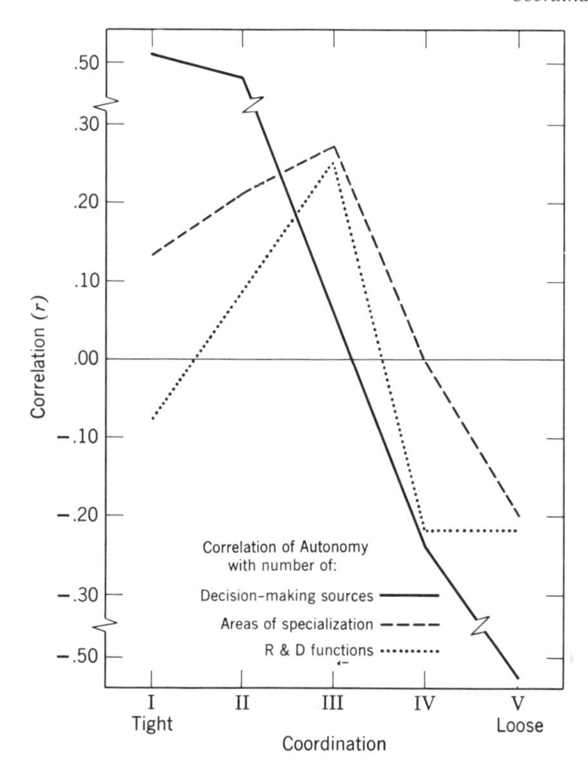

Chart 10. *Since autonomy and number of decision-making sources were derived from the same question, the negative correlations at levels IV and V were to be expected, but the positive correlations at levels I and II were surprising. Autonomous individuals were likely to have several specialties or R & D functions in moderate situations (II and III), but fewer of these in loose situations (IV and V).*

Autonomy and Decision-making Sources

In Chapter 2 we observed that for most groups, when several "decision-making sources" shared in deciding the individual's technical goals, he was likely to perform better. When we examined this measure in relation to performance at the various levels of coordination, the correlations (not shown) were generally positive at all levels except V.

The number of decision-making sources and the measure of autonomy (own weight in setting goals) were derived from the same question. It is to be expected, therefore, that a person who claimed much weight for himself could not assign much weight to other people. The correlation between autonomy and number of decision-making sources should be negative; and this turned out be the case at loose levels IV and V, as shown in Chart 10.

But note the remarkable result in tight situations (I and II). Actual amount of autonomy was low, of course. And the correlations between own weight and number of other persons having weight were strongly positive rather than negative. Individuals who had more autonomy in making decisions at the same time involved a larger number of other people. Clearly, these individuals must be involving people in addition to their supervisor—either colleagues or high echelons or both. And it was probably the involvement of higher echelons by scientists at level II that stimulated autonomous individuals here to moderately high achievement, even though they were not internally motivated.

Level I provides an instructive contrast to II. In these tightly controlled settings, the autonomous individual felt strongly involved; and he was also stimulated by several decision-making sources. But despite both inner and outer excitation, autonomy brought only a modest rise in performance (Charts 5 and 6). Presumably the rigidities of the setting prevented the autonomous and eager individual from performing much better than his directed and passive co-worker.

Autonomy and Diversity

Chart 10 also shows correlations between the two diversity factors and autonomy. At level II, autonomous individuals had more specialties, and at level III, both more specialties and more R & D functions. (Whether self-determining individuals developed more interests, or whether those with broad interests were given more freedom, cannot be determined. Perhaps some of both.) In any event, the corresponding diversity accompanied high performance at one or both of these levels (Charts 4-A and 4-B). Therefore diversity reinforced the correlation of autonomy with performance at these levels.

Under looser coordination, however, high autonomy was not accompanied by greater diversity, but the reverse. Self-determining individuals at levels IV and V were more restricted than average in the number of different R & D functions they performed, and (at level V) in the number of their specialized areas. Autonomous individuals, in short, were narrowing their interests, and this orientation (Charts 4-A and 4-B) was detrimental to scientific achievement.

SUMMARY AND IMPLICATIONS

In this chapter we have attempted to pull together findings from a number of previous chapters, and to build from them some relatively abstract generalizations about the performance of scientists in organizations. We have tried to find some way of incorporating in one framework some apparently inconsistent or even contradictory findings.

We found that the more loosely coordinated a situation was, the more necessary it was that the individual remain strongly *motivated or stimulated* if he were to continue achieving.

We also found, however, that a relatively high level of individual *autonomy* was effective mainly in the *middle range* of situations—those which were neither very tightly coordinated nor loose. In the latter, where members already enjoyed considerable freedom, the most autonomous scientists were below average in performance.

One explanation for these results may be that in loose or extremely loose settings the most autonomous scientists tended to withdraw from outer stimulation (or to reduce inner motivation), that is, to weaken stimuli which might have enhanced their performance. In very tightly coordinated situations, at the other extreme, autonomous individuals were both motivated and stimulated; but the rigidities of the setting apparently prevented these factors from enhancing creativity. Thus only in the middle-range situations were two essential conditions present: (a) high autonomy was accompanied by a number of strong motivations and stimulations, and (b) the setting was flexible enough to allow these factors to improve performance.

In the loosest settings, full autonomy may encourage complacency rather than zest, narrow specialization rather than breadth. In the strongholds of research, the isolated rooms in the ivory tower may not be the best habitat for achievement.

Suggestions for Managers

For further implications of these results, we conclude with the following dialogue.

> The picture you have drawn makes some sense. But does it have any practical advice to offer me as a research manager?

It suggests, we think, that you emphasize different techniques when dealing with different segments of your research organization. For the sake of discussion we'll stretch a point and assume that all five levels of coordination are represented in different parts of your organization. Consider first the research engineers in your development laboratories: moderately autonomous non-Ph.D's, developing new products or processes. Let's assume they correspond to level III. For these men, the kind of philosophy represented in Rensis Likert's *New Patterns of Management* should be highly effective. These men should rise to the challenge of more participation in decision-making. Stimulate them with a wide variety of problems. Make sure each man has three or four specialized skills, and a fair degree of leeway to follow up his own ideas.

> What about my design engineers—the ones who are putting together the hardware or the manufacturing processes based on protypes developed by the research men?

These perhaps correspond to level II where the job requires coordinated teamwork. Here the thing to emphasize is considerable interaction. See that each man gets to know many others throughout the organization. Schedule frequent but brief meetings of a half-dozen people each to exchange progress reports. Each individual might work on a limited number of activities, but make sure he keeps in touch with people doing related work. Maybe a series of overlapping committee assignments would serve well. See to it, also, that each man has a chance now and then to discuss his work with key executives.

> Some of my knottiest problems concern my assistant scientists in the development labs; they are professionals, all right, but essentially extra pairs of hands for the Ph.D's. And when the Ph.D's have to mesh with other teams, they can't give their assistants much chance for initiative or participation in decisions.

If these correspond to level I, you do have a difficult job here. Perhaps it will be easier, though, if you draw a distinction between creativity and productivity. Our data don't show many ways to increase creativity at this level, but there are several hints for increasing usefulness, which you might call productivity. For example, these people were useful when they had several specialized skills and carried on different kinds of R & D functions. A label for this in personnel circles is "task diversification." Also, usefulness in this group was moderately high if the members were strongly involved in the work, if they talked often with colleagues, and if several other people or groups were involved in setting their goals. And perhaps these factors all tie together. Give the man a chance to participate in meetings where the work of the section is discussed, and his interest in it will increase. His job may not allow him to be especially creative, but he will work hard within the limits of his assignment.

> What about my topnotch Ph.D's in the development labs—the ones who largely set their own pace? These are the ones I rely on for pioneering work in new products.

Assuming that these correspond to level IV, make sure that they don't limit themselves to one or two kinds of R & D functions—basic research to the exclusion of applied research, or concentration only on product improvement. Toss them a problem from time to time outside their immediate line of investigation. Also, find out whom they are talking to, and about what. If they are spending their coffee hours in hot shop-talk, fine!

If not, encourage them to collaborate with two or three others. Invite them to present their ideas several times a year to small seminars of colleagues.

> Now we come to the group for which the term "management" may be an illusion: basic research Ph.D's. They tell me just to leave them alone and not bother them. Is there anything else a manager can do, aside from seeing that they have the resources they want?

We're not sure. There are some pitfalls here. Complete autonomy for each individual can debilitate. Apparently it has this effect by isolating him from potential stimulation. Here the role of colleagues can be crucial. If he talks often with his co-workers, and has a fairly wide circle of them, you can be reassured.

How to stimulate communication? Maybe that's where, as a research manager, you can be creative. By keeping in close touch with each man's interests, you can suggest other people who could be useful to him. You can invite individuals to present their ideas at seminars, or send them off to conferences.

For these basic research Ph.D's, also, competition between groups might be stimulating. A race to establish priorities in a new scientific field—a race between institutions or groups, that is, not between members of the same group—may help to build involvement and maintain challenge.

Try to challenge each man from time to time with a problem that stretches him. (It can't be assigned, of course; you have to sell him on the problem.) Such a challenge will nourish his desire to learn and grow—and as long as this is alive, his creativity will be high. *Don't* let him rest on his reputation as the leading expert on X.

13

GROUPS[1]

Groups Declined in Performance After Several Years, but Less If the Members Became Cohesive and Intellectually Competitive

Wallace P. Wells and Donald C. Pelz

In this chapter we focus on group age—how long the members of a research or development team have been together.

This factor raises important questions for the research director. On the one hand he could argue that several years of experience will forge a set of individuals into a smooth problem-solving team. On the other hand, he might expect a group to grow stale as the members exhausted each other's stock of ideas. How long should the members of a group stay together for best performance?

When we asked technical audiences to make a prediction about the effects of group age based on their own experience, the guesses varied widely; the most typical view, perhaps, was that performance rises during the first three or four years of group life, and then drops.

The only prior evidence we know of from research labs was a study done by Herbert Shepard.[2] In 21 industrial laboratories he asked research directors to rank teams or sections on several criteria such as "creativity." The high- and low-ranking teams in each lab were asked for further information, including length of time in the group. Highest-ranking groups, he found, were those less than 16 months in group age; thereafter all rankings by management dropped consistently with increasing age. (Rankings of the group by its own members also dropped, but showed a comeback when the groups were two to five years in age.)

Was this drop in performance genuine, or did it simply reflect the manager's biases? (Since the manager probably had a hand in putting

[1] This chapter is based on a doctoral dissertation by Wallace P. Wells, "Group Age and Scientific Performance," University of Michigan, 1962; available through University Microfilms, Ann Arbor, Michigan.

[2] H. A. Shepard, "Creativity in R/D Teams," *Research and Engineering*, October 1956, pp. 10–13.

together a new group, maybe it had to look good to him at first.) Or perhaps it is in the nature of R & D that the first broad steps seem exciting, whereas the later hard work of running tests and perfecting the design seems unproductive. But, of course, it is possible that groups really do become stale over time. What would our date show?

This chapter starts with some preliminary ideas on how group age might affect the intellectual "uncertainty" and emotional security that may be required for creativity. It describes how 83 groups were located, and how averages on scientific contribution and usefulness were obtained and adjusted to eliminate effects of extraneous conditions. The adjusted scores showed a general decline in scientific contribution as group age increased, and a curvilinear effect for usefulness, peaking at 4 to 5 years of group age. In a search for factors that might account for these trends, the chapter next examines how a number of social factors changed with group age. Older groups were more relaxed than younger ones in several ways: less communicative, less competitive, and less secretive. Older groups also were more specialized in their interests. Finally, the chapter considers how these factors related to performance of younger and older groups, and reports that older groups retained their vitality if they maintained vigorous interaction and "intellectual tension."

Some Theoretical Ideas

The analysis of data was guided by some preliminary notions of the importance of both "uncertainty" and "security" for scientific achievement.[3] One condition for scientific achievement, we surmised, was some degree of intellectual uncertainty or unpredictability. The solution to a problem should not be self-evident; otherwise no search for better solutions would take place. The members of a problem-solving group should not think alike, nor should they be sure how the other members would approach a new problem.

At the same time, we felt, scientists must have a certain level of personal security or self-confidence; an insecure or anxious scientist would stick to "safe and sure" solutions. Too little anxiety, on the other hand, might mean complacency in which there is no search for new solutions.

Some *middle* level of both factors, we thought, might be the optimal atmosphere for scientific achievement.[4] Further, we suspected that the

[3]D. C. Pelz, "Uncertainty and Anxiety in Scientific Performance," working paper, 1960, 53 pp.; available as Publication #1588 from the Survey Research Center, University of Michigan.

[4]For example, E. D. Longenecker, "Perceptual Recognition as a Function of Anxiety, Motivation, and the Testing Situation," *Journal of Abnormal and Social Psychology,* 1962, vol. 64, pp. 215–21, has summarized a number of studies which indicate that an intermediate level of anxiety is associated with optimal performance on experimental tasks.

effects of each factor would be interdependent on the other. Given an atmosphere of emotional security, the scientist could benefit from a considerable amount of intellectual uncertainty. Given an atmosphere of anxiety, though, relatively little uncertainty could be tolerated.

How were these intuitive notions relevant to the phenomenon of group age? Our hunch was that new groups would experience both anxiety and uncertainty; and that as groups aged and became adjusted both to each other and to the task, they would become more secure and less uncertain.

If so, it seemed likely that newer groups would benefit from conditions lending security or reassurance, whereas older groups would benefit from conditions which prevented certainty, by facing the group with unexplored problem areas, diversity of viewpoint among members, intellectual disagreement, and the like.[5]

Data to Test These Ideas

In order to test these ideas, 83 groups or teams were identified whose members were included in our study. Forty-nine of these were in industry and 34 in government; the university contained few "groups" in the same sense. Groups were identified by examining formal organization charts to locate nonsupervisory scientists or engineers reporting to one administrative chief. We then consulted questionnaires to be sure that each group member named as his "administrative chief" or his "technical advisor" the organization chart supervisor or another group member. In a few instances where the members named the supervisor's boss as their chief, he also was included. (Supervisors were considered to be group members.)

About half the groups contained from two to five members, and half from six to 25. On the average, industrial groups were slightly smaller than government (three out of five in industry contained two to five members, whereas one out of three in government was this size).

For each of the 83 groups, the group's score (usually the mean) on a

[5] A relevant study is that by E. P. Torrance, "Some Consequences of Power Differences on Decision Making in Permanent and Temporary Three-Man Groups," in A. P. Hare, E. F. Borgatta and R. F. Bales (eds.), *Small Groups*, Alfred Knopf, New York, 1955, pp. 482–92. He compared temporary and permanent three-man bomber crews and found that the temporary crews produced a higher percentage of correct solutions to an arithmetical problem than did the permanent crews. Among other factors which might be responsible, he found that in stories written about a hypothetical conference, the low-status members of permanent crews made fewer references to disagreement than did the high-status members of these crews; but in temporary crews the low-status members referred to disagreement just as much as the high-status members. Torrance felt that the low-status members in permanent crews did not feel free to disagree and therefore withheld their ideas. The free expression of disagreement appears necessary to maintaining a climate of intellectual uncertainty.

large number of variables was determined, and the analysis was then done using these *group scores* (rather than individual scores).

"Group age" was defined simply as the average number of years that each member had belonged. Conceivably a group could have been in existence for a long time and still score as "young," if it had recruited new members.

In our data collection we did not attempt to assess the performance of groups as such (as did Shepard), only the performance of individuals. (To do so we would have had to use the judgment of a few people at the top; we wondered whether they were close enough to technical details. We wanted performance to be evaluated by a number of colleagues in close touch, nonsupervisory as well as supervisory.)

Therefore we started with the colleague judgments of the "scientific contribution" and "over-all usefulness to the laboratory" of each scientist or engineer, as described in Chapter 1. Taking the "raw" percentile scores from 0 to 99 (*not* corrected for length of individual experience, as described in Chapter 1), we obtained the arithmetical average or mean for each group. This provided two measures of the groups' performance which were then adjusted as described next.

Adjustment of Group Performance Measures. A preliminary search for the group age effect, using raw means of group performance, was disappointing. We soon realized that as group age increased among our 83 sets, various other characteristics changed too, and these were also related to performance.

In loose or autonomous departments, groups tended to be larger than in centralized or coordinated departments (managers of the latter had a freer hand to reorganize and start small new groups). But in autonomous departments the raw group performance scores were higher than those in coordinated ones by 20 percentile points on the average for scientific contribution and 14 points on usefulness.[6]

Furthermore older groups tended to be larger, to consist of individuals who were themselves older, and to contain more Ph.D's; and these tendencies were especially strong in autonomous departments (see Table 1). It appeared that research directors, especially in autonomous departments, were reluctant to reorganize teams containing Ph.D's, which tended to persist and grow larger.

Therefore when group age was allowed to vary, several other characteristics associated with performance did also, obscuring the basic effect of group age as such.

By techniques described in Appendix C, the effects of other variables were removed. The performance scores were adjusted by adding or sub-

[6]Appendix H describes how departments were classified.

TABLE 1 *As shown by these correlation coefficients, older groups tended to contain more members, older members, and more Ph.D's. The effects were stronger in autonomous than in coordinated departments.*

	Correlation with group age for groups in:		
	Autonomous departments ($N = 51$)	Coordinated departments ($N = 32$)	All departments ($N = 83$)
Group size (number of members)	.40*	.10	.33*
Average age of members	.62*	.48*	.58*
Educational status (proportion of Ph.D's)	.51*	.19	.37*

* An asterisk indicates that the correlation is "statistically significant at the .01 level." That is, if group measures such as the above were randomly paired (with an underlying zero relationship), correlations as large as those starred would arise by chance only one time out of 100.

tracting appropriate constants in such a way that the adjusted scores were no longer related to the "extraneous" factors of individual age, proportion of Ph.D's, departmental autonomy or coordination, etc.—extraneous, that is, for the purpose of studying effects of group age alone.

We originally suspected that groups might "age" at a different rate in autonomous and coordinated departments. But after making the adjustments described in the appendix, we found that the effects of age as such were much the same in both kinds of departments; therefore the results have been combined.

How Group Age Related to Group Performance

When we then examined group age in relation to the adjusted group performance scores, the curves shown in Chart 1 appeared.

The general decline observed by Shepard appeared rather clearly for the measure of scientific contribution; the measure of over-all usefulness showed a gain at 4 to 5 years of group age, and thereafter declined.

One interpretation of the dual trend may be this: for *strictly scientific or technical* advances, new members of a group do indeed provide fresh viewpoints and stimulate each other to think in new ways. But in order to be *useful* to the organization—which often means solving problems efficiently by methods which have worked well in the past—a group takes several years to solidify as an effective team.

For some later analyses we found it useful to split our population into 39 "younger" groups (0 to 3 years), and 44 "older" groups (4 to 12 years).

Table 2 summarizes the age trends in terms of correlation coefficients.

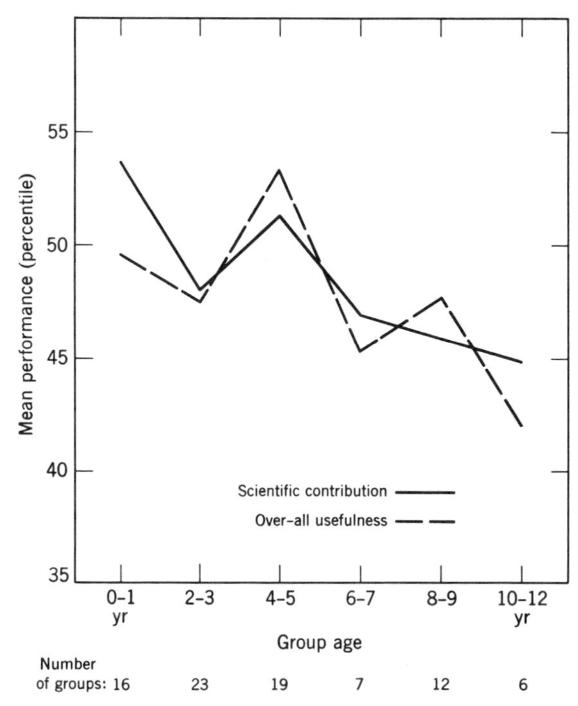

Chart 1. *Average scientific contribution was highest for newest groups—those under 24 months of group age, with some resurgence at 4–5 years. Over-all usefulness, on the other hand, did not reach its peak until 4–5 years of group age. (Performance scores have been adjusted to rule out effects due to average age of individual members, percent of Ph.D s, kind of institution, and departmental coordination.)*

TABLE 2 *The following correlations illustrate that as group age increased, mean scientific contribution (adjusted) declined among the total set of groups; mean usefulness declined most sharply with age among the 4–12-year-old groups.*

	Correlation of group age with two performance measures among:		
	Younger groups, 0–3 yrs. ($N = 39$)	Older groups, 4–12 yrs. ($N = 44$)	All groups ($N = 83$)
Scientific contribution	−.35*	−.29	−.25*
Over-all usefulness	−.11	−.37*	−.13

*Correlations this large would not arise by chance more than one time in 20.

Within the younger and older halves of the group age distribution, as well as in the total distribution, scientific contribution had a significant negative correlation with group age. Usefulness showed no strong over-all trend because of the peak in the middle, but after four years of group age it declined significantly as age increased.

Changes in Group Characteristics with Age

Our next task was to see whether we could account for these relationships. One branch of this search examined the social and psychological characteristics of young groups compared with old. Keep in mind that we did not actually observe the same groups over time; we only had data on existing teams of different ages. Keep in mind, too, that older groups differed in many ways, such as size and educational level, which seem extraneous to the aging process as such.

Communication and Cohesiveness. Two measures were used to assess communication: (a) frequency of contacting supervisors and colleagues, and (b) total time spent contacting them. These were Questions 41 and 42 shown in Chapter 3. "Group cohesiveness" was measured by the proportion of "most significant colleagues" chosen from among members of

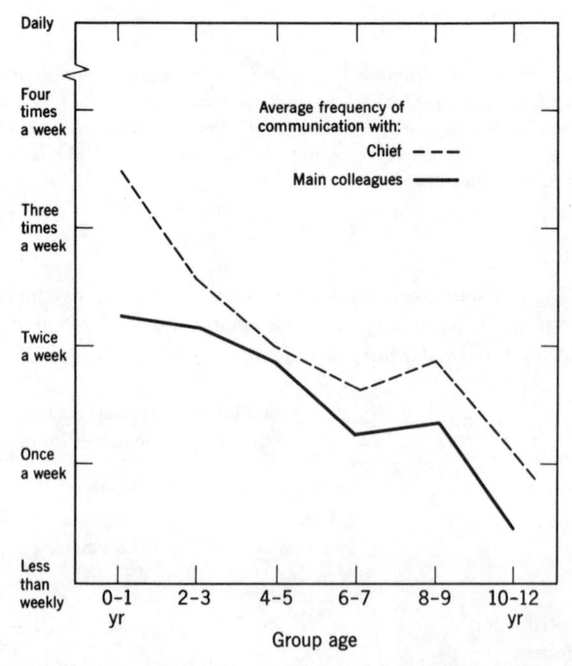

Chart 2. *As group age increased, members became more isolated. Mean frequency of contact with the chief dropped from several times a week to weekly; communication among members dropped even lower.*

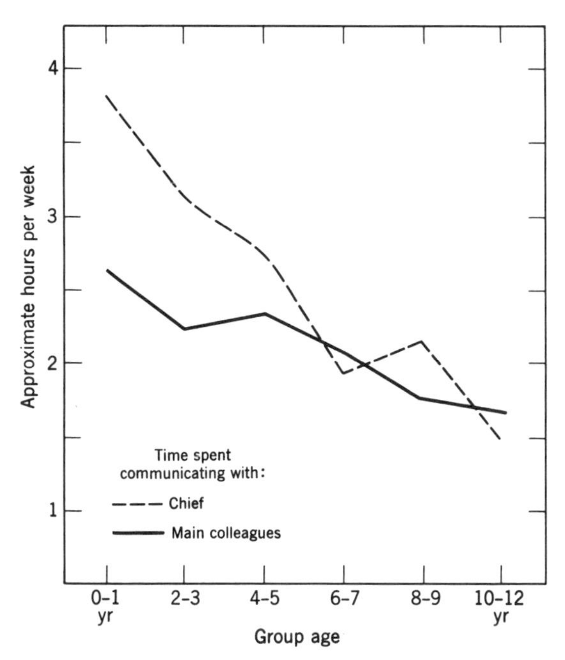

Chart 3 *shows a similar drop in contact when measured by amount of time spent per week. In new groups, members spent four hours weekly with their chief, but in old groups, only one or two hours. Members spent less time with each other than with the chief, and this also dropped with increasing group age.*

the work group (ratio of total within-group choices to total possible within-group choices). The resulting ratio proved to be independent of group size.

Charts 2, 3, and 4 show how these measures changed with increasing age. All the measures of communication declined steadily among groups of increasing age. At the same time, cohesiveness began to rise as members got better acquainted. That is, they tended to select one another more as "significant colleagues." But after 4 to 5 years together, a reverse trend set in; cohesiveness dropped to its initial level, and even lower among the oldest groups who by this time were seeing their chief only a few times a month, and their colleagues even less than that. It is even hard to imagine the latter as constituting a "group," the connections seem so tenuous.

The cohensiveness curve was remarkably parallel to the usefulness curve in Chart 1—even to the slight rise in cohesiveness at 8 to 9 years which paralleled a mild spurt in usefulness at this point. Note a similar rise in frequency of contact at this age.[7]

[7]The four communication measures all related to each other positively, that is, groups high on one tended also to be high on the others.

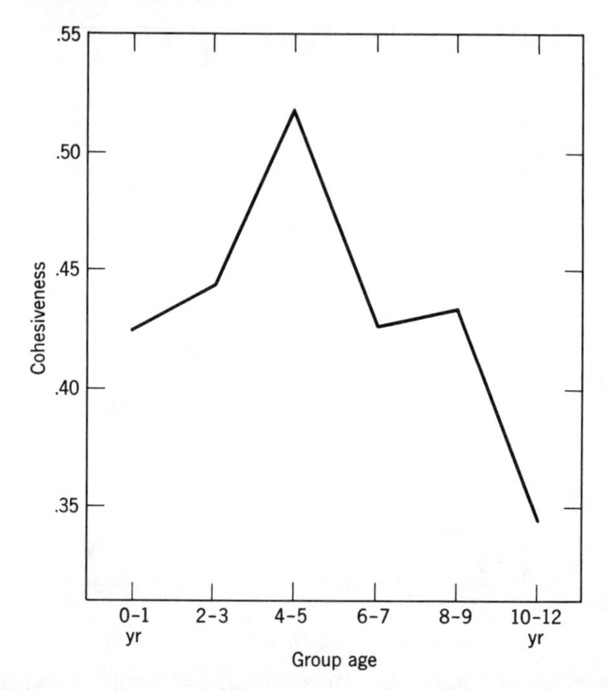

Chart 4. *"Cohesiveness"—measured by proportion of significant colleagues chosen from within the group—first rose (as expected) to a maximum at 4–5 years, but thereafter declined.*

Competition and Secretiveness. Preliminary interviews had often shown scientists aware of friendly but stiff intellectual competition with colleagues. As one confessed privately, "I want to be known as a smart cookie." This intellectual rivalry undoubtedly stimulates, but it may also inhibit. Some scientists may hold back their best ideas until they can pin down their claim to authorship or substantiate their views.

Questions 34 and 35 shown in the following boxes were designed to tap these feelings of intellectual rivalry or inhibition. For "hesitence to share ideas" we often use the shorthand term "secretiveness," though the reader should realize that secretiveness was at most mild in these situations.

> *Question 34.* To what extent are you (or your colleagues) aware of competing technically with other professionals—striving to be first or best in solving key problems?
>
> [Responses ranged from "none" to "intense" concerning:]
>
> Between myself and individual colleagues
>
> Between my immediate group (team, section, project, etc.) and other groups

> *Question 35.* Scientists and engineers are sometimes secretive about their technical ideas for fear of meeting skepticism or disapproval, failure to give proper credit, jockeying for rewards, etc. To what extent have you observed, among professionals like yourself, any hesitance to *share technical ideas freely* in the following situations (technical ideas *other* than those limited by security regulations).
>
> [Respondent rated "hesitance to share" from "none" to "severe" concerning:]
>
> My immediate groups (sections, projects, teams, etc.)
>
> Other technical groups within this organization

How did competition and secretiveness change with group age? The actual trends are plotted in Charts 5 and 6. Competition between colleagues declined up to 6 to 7 years, but after that rose again. Did members become so isolated that they no longer felt a part of the same group, and therefore began to compete again as individuals?

Competition between groups showed a slight increase at first, and there-

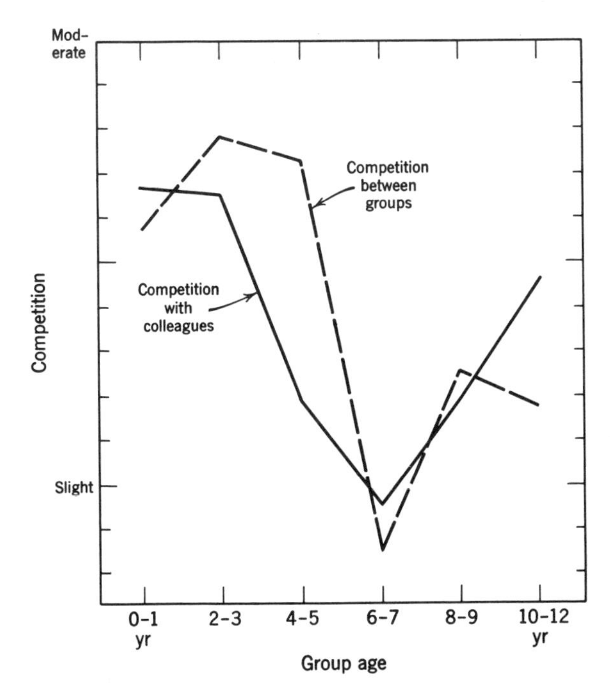

Chart 5. *Perception of competition with individuals declined as group age increased, but rose again among the oldest groups. Perceived competition between groups rose at first, and then dropped. Groups at 4–5 years felt more external than internal competition—paralleling the peak in cohesiveness and usefulness at this age.*

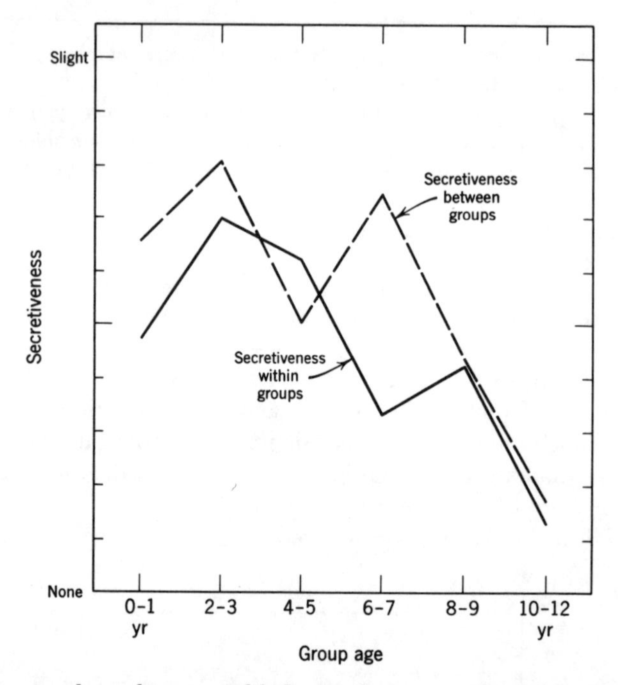

Chart 6. *In general, members reported little or no hesitance to share ideas. Even so, there was some variation with group age; perceived secretiveness within immediate groups and between groups rose slightly at first, then generally declined. Older groups were more relaxed.*

after a drop (with the same puzzling upsweep among older groups). Secretiveness within groups and between groups rose slightly at first, thereafter showed an irregular decline.[8]

It is possible, of course, that an individual group will change over time in a different pattern from that shown here. All we can say is that for our set of 83 groups, as a whole, the teams became more relaxed over time—less competitive and less secretive. This trend was consistent with our general theoretical hunch that older groups will gain in a feeling of security and self-assurance.

Similarity and Specialization. According to our initial notions, older groups should not only gain in security and assurance, but should also lose uncertainty. Their intellectual world—their approaches to problems, their knowledge of each other's abilities and ideas—should become more stable and predictable.

[8] The various measures of competition and secretiveness were themselves positively interrelated.

We explored several ways of measuring these factors. Three subsequently proved useful. Similarity was measured by Question 45 (shown in Chapter 8) which asked scientists how similar they were to their colleagues with respect to strategies used in tackling technical problems.

Specialization was assessed by two items from Question 19 (see Chapter 6). In item 19E, scientists indicated their preference for mapping "broad features of important new areas, leaving detailed study to others." Item 19F asked about their preference for probing "deeply and thoroughly in selected areas, even though narrow."

Scientists and engineers with whom we have discussed these results generally agreed with our expectation that over time, group members would become similar to one another in their approach, and would move from a "broad" perspective to a "deep" or specialized one.

The actual trends are plotted in Charts 7 and 8. As expected, mean perceptions of similarity in approach did generally increase, with a maximum at 6 to 7 years. (This was the point, we saw in Chart 5, where individual and group competition was lowest.) After that, oddly, similarity

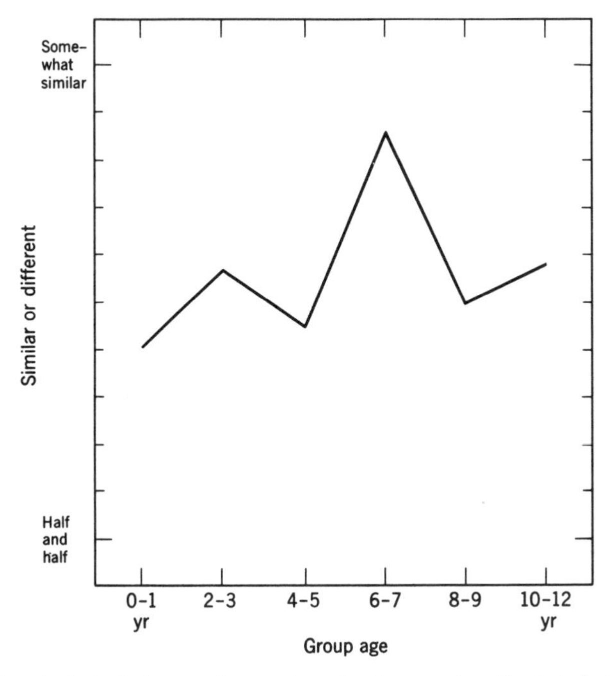

Chart 7. *Perceived similarity to colleagues in style of approach or "technical strategy" rose slightly (as expected) to 6–7 years, but then dropped.*

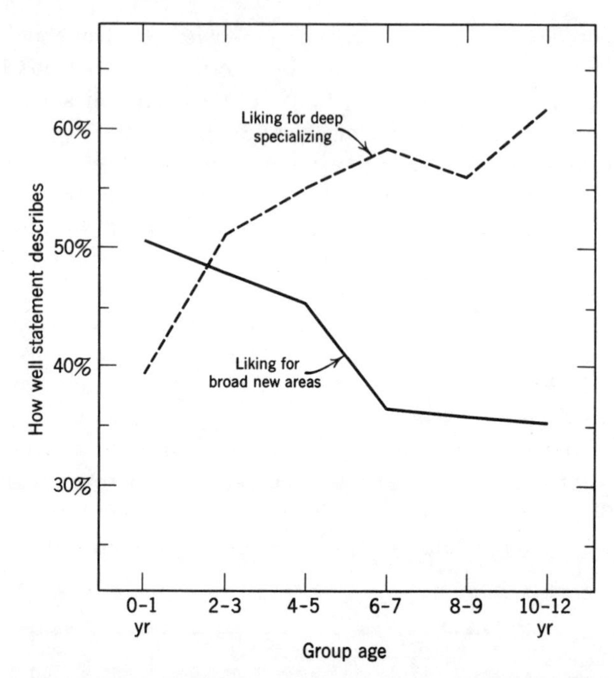

Chart 8. *Average preference for "broad mapping of new areas" dropped steadily with group age, whereas interest in "deep probing of narrow areas" rose.*

dropped again (whereas competition rose). Perhaps the members in these oldest groups were so separate that they were no longer sharing one another's viewpoints.

Chart 8 exactly confirms the expected trends. As the age of groups increased, they became less and less interested in "broad mapping of new areas," more and more engrossed in "probing deeply in narrow areas." Note the position of the two curves at group age 4 to 5 years, the point of maximum usefulness and moderately high scientific contribution. Groups which had become specialized without losing breadth—and had reached internal cohesiveness with external competition—were functioning at the peak of their power.

As groups aged further, perhaps the trend toward specialization was so strong that each member created his own niche and no longer resembled any other member. This might account for the drop both in similarity and in cohesiveness.

Effects of Communication on Performance by Younger and Older Groups

We have seen that communication declined in older groups. And from Chapter 3 we know that communication was related to individual performance. We now wanted to see whether this decline affected group performance. Specifically, we were curious about how communication related to performance in younger and in older groups.

Initially we suspected that in new groups the members might have to talk a lot in order to achieve. They would have to find out each other's ideas in order to function smoothly; therefore frequent communication should go with high performance. And we suspected that in older groups, where the members knew each other well, less talking would be needed; frequency of communication should make little difference to performance. Exactly the *opposite* proved to be the case, as shown in Table 3.

Communication with colleagues went with high performance in older groups only—mildly so for scientific contribution, more strongly for usefulness. The measure of group cohesiveness (proportion of group members chosen as "significant colleagues") showed the same pattern. However, the amount of *time* spent per week in talking did not matter. Many brief conversations rather than few long ones, among colleagues who valued each other, was the pattern in effective older groups.

TABLE 3 *These correlations indicate that frequent contact among members and strong cohesiveness went with high performance in older groups only, but frequency of contacts with chief did not matter. See text for discussion of chief's role.*

	Correlations with scientific contribution		Correlations with over-all usefulness	
	Younger groups ($N = 39$)	Older groups ($N = 44$)	Younger groups ($N = 39$)	Older groups ($N = 44$)
Frequency of communication with colleagues	−.19	.17	.04	.35*
Time communicating with colleagues	−.02	.21	−.07	.09
Group cohesiveness (mutual choice)	−.05	.36*	.12	.39*
Frequency of communication with chief[a]	.14	−.04	.05	.08
Usefulness of chief for original ideas	.31*	.11	.25	−.12
Usefulness of chief as neutral sounding board	.15	.25	−.10	.36*

* Statistically significant at .05 level.

[a] Based on responses of members other than chief.

What about the chief? We saw before (Charts 2 and 3) that communication with him also dropped in older groups. But unlike colleague communication, frequency of contacting the chief showed no over-all relationship to performance, either in younger or older groups. (The table shows only frequency; amount of time was also unrelated.)

As we pondered these results, it seemed possible that not the quantity but the *quality* of the chief's interactions might make a difference. Question 46 (see the following box) sought to tap qualitatively distinct ways in which a chief might be useful to his group.

> *Question 46.* Scientists and engineers can be useful to each other in different ways. [Nine functions were listed, including:]
>
> A. *Technical know-how,* sharing with others his wide knowledge, skill, or experience
> B. *Original ideas,* imaginative or unusual ways of looking at problems
> D. *Critical evaluation,* searching questions as to what is sound or realistic
> F. *Neutral sounding-board,* open-minded listener for others to try out their ideas
> H. *Appreciation and encouragement* for the work being done by others
>
> As far as you are concerned, which of these things does your *administrative chief* do best? That is, in which of these ways is he most useful to professionals like yourself (people who report to him or work closely with him)?
>
> [Respondent ranked the nine functions provided by his chief, from most useful to least useful (or not performed at all).]

Two of these functions related differently to performance in younger and older groups, as shown in Table 3. Younger groups benefited (especially on scientific contribution) if their chief was a source of *original ideas.* Older groups benefited (especially on usefulness) if their chief was available as a *neutral sounding board* for them to try out their own ideas. In both instances, the chief was available, but performing a different role.

Effects of Competition and Secretiveness in Younger and Older Groups

We saw before that older groups became more relaxed, with less intellectual rivalry among individuals or among groups, less hesitation about sharing ideas freely. These trends were in line with our initial hunches on greater security in older groups.

Now the question was: would relaxation harm performance? For competition, this seemed a reasonable prediction. The findings in Table 4 partially confirmed the expectation. Competition between colleagues was mildly (not significantly) associated with higher performance—and mainly

TABLE 4 *Competition with colleagues was mildly correlated with performance in younger groups, whereas intergroup competition benefited usefulness in older groups. Secretiveness hindered young groups but seemed to help older ones.*

	Correlations with scientific contribution			Correlations with over-all usefulness		
	Younger groups		Older groups	Younger groups		Older groups
Competition with colleagues	.16		.02	.26		.15
Competition between groups	.00		.21	.12		.45*
Secretiveness within groups	−.34*	<	.18	−.31	<	.35*
Secretiveness between groups	−.21		−.05	−.04		.05

* Statistically significant at .05 level.

< = Difference is statistically significant at .05 level.

in younger groups. Between-group competition went with higher performance in older groups only—particularly for usefulness.

On the other hand, scientists usually assume that secretiveness will hurt rather than stimulate performance.[9] But for the group scores we examined here, a curious result appeared. A group atmosphere of some secretiveness (remember that these scores were mild at most) did indeed inhibit performance of younger groups. But older groups seemed to benefit from mild secretiveness—particularly on the measure of usefulness.

Let us pause for some speculation. Consider first, competition. Why did competition between colleagues seem mildly effective for young groups, whereas intergroup rivalry was more effective for older groups? Perhaps newly formed groups have not had time to become internally organized; members do not know what to expect from each other. Competition against an outside team can be destructive to a group which lacks internal stability. If you have watched an under-practiced volley ball team starting to lose a match, you will have a vivid picture of disorganization produced among an unprepared group by competition. Competition between individuals, on the other hand, is a reasonable challenge to members of young groups. The resources needed to compete are in the individual's control, not in the unknown capabilities of his colleagues.

Among older groups—assuming that coordination has been established—intergroup competition should increase solidarity and cohesiveness which, in turn, enables the team to function effectively. This speculation is con-

[9] Among individuals this was so; perception of secretiveness correlated negatively with individual performance (data not shown).

sistent with the fact that older groups benefited from high cohesiveness as well as intergroup competition. A subsequent analysis showed, in fact, that cohesiveness and intergroup competition correlated .33 in older groups (statistically significant), but −.29 in younger groups. In the latter, competition disrupted cohesiveness rather than cementing it.

Second, what about the puzzling results with secretiveness in older groups? If, in fact, older groups are more relaxed, less anxious, less suspicious of each other, then perhaps "hesitance to share ideas" takes on a different meaning: more nearly the flavor of intellectual rivalry. In an older group, caution may not mean repression, but rather a careful marshalling of one's evidence in persuasive form before laying it on the table.

This interpretation reminds one of a series of results by Fred E. Fiedler, using a measure of "assumed similarity of opposites" (ASo). The respondent sorted personality descriptions so as to describe his most preferred and his least preferred co-worker, and a measure of similarity between the two profiles was obtained.

In several kinds of groups, such as basketball teams and student surveying teams, Fiedler found that leaders of effective groups showed low ASo scores—they perceived a clear difference between persons they preferred and those they rejected as co-workers. "A person with high ASo," Fiedler suggests, "tends to be concerned about his interpersonal relations, and he feels the need for the approval and support of his associates. In contrast, the low ASo person is relatively independent of others, less concerned with their feelings, and willing to reject a person with whom he cannot accomplish an assigned task."[10] Thus low Aso may reflect a quality of psychological distance or emotional reserve.

Possibly high secretiveness in older groups reflects such a quality of emotional reserve. (Note that Fiedler's result occurred for the *leader* who possessed low ASo, whereas our data are based on all group members.)

Members of effective older groups did not avoid each other, but interacted vigorously. Basically they respected one another (the measure of cohesiveness). Yet this respect, we suggest, was not based on personal liking but rather on intellectual competence. The existence of caution (perhaps the label secretiveness is inappropriate) did not stem from hostility, but rather from respect. The individual, we speculate, prepared a sound case before doing intellectual battle.

Effects of Similarity and of Specialization in Younger and Older Groups

If a matching of wits does indeed account for effective older groups, it should occur in groups whose members disagree with one another on

[10] F. E. Fiedler, *Leader Attitudes and Group Effectiveness,* University of Illinois, Urbana, 1958, p. 22.

technical matters. It should occur in groups whose members have not settled into narrow specialties, but welcome broad new problems. The findings in Table 5 support these expectations.

Diversity among group members was helpful to scientific contribution in young and old groups alike, and to usefulness of older groups. Also as expected, willingness to keep a broad interest in new problem areas was a distinct asset to the usefulness of older groups, whereas a preference for narrow specializing was just as much a handicap. (When the two measures were combined as a broad-versus-deep index, the effect was heightened.)

These results are consistent with a finding by William M. Evan.[11] Ratings were obtained on performance of several dozen R & D teams whose members had indicated for selected colleagues (a) degree of agreement or disagreement with the individual on technical matters, and (b) degree of personal liking. The more effective groups had a high degree of "technical conflict" (disagreement), but a high degree of "personal harmony" (mutual liking). In our data on older groups, "technical conflict" appears as dissimilarity (and perhaps secretiveness). "Personal harmony" appears as interaction and cohesion, but more on the basis of respect than of liking.

"Intellectual Tension"

The measures of secretiveness within groups and perceived dissimilarity were subsequently combined into a single index. A maximum score (hesi-

[11] W. M. Evan, "Conflict and Performance in R & D Organizations: Some Preliminary Findings," *Industrial Management Review*, 1965, vol. 7, no. 1, pp. 37–46.

TABLE 5 *According to these correlations, older groups benefited from dissimilarity in approach and an appetite for broad new problems, and suffered from an interest in specializing.*

	Correlations with scientific contribution			Correlations with over-all usefulness		
	Younger groups		Older groups	Younger groups		Older groups
Perceived dissimilarity in approach (similarity scale reversed)	.24		.42*	−.09	<	.36*
Preference for broad approach	.10		.25	.12	<	.55*
Preference for deep approach	−.23		.00	−.20		−.48*
Broad-versus-deep approach	.21		.13	.20	<	.58*
Intellectual tension: secretiveness within groups plus dissimilarity	−.15	<	.40*	−.35*	<	.55*

* Statistically significant at .05 level.

< = Difference is statistically significant at .05 level.

tance to share ideas plus diversity) marked a condition of both "low security" and "high uncertainty." We might label this a condition of "intellectual tension." In contrast, a minimum score indicated little tension: much security and little uncertainty.

Generally we expected that younger groups would benefit from low tension, and older groups would need progressively stronger tension. The secretiveness-plus-dissimilarity data in Table 5 indicate this trend nicely. Young groups were hindered by tension, whereas older groups were aided by it.

The same data are plotted visually in Chart 9. Among groups low on the tension index (low secretiveness plus similarity), usefulness was high during the first three years of group life and thereafter fell sharply. Among groups with a high tension score (high secretiveness plus dissimilarity), the initial performance was low. Presumably the group was struggling to

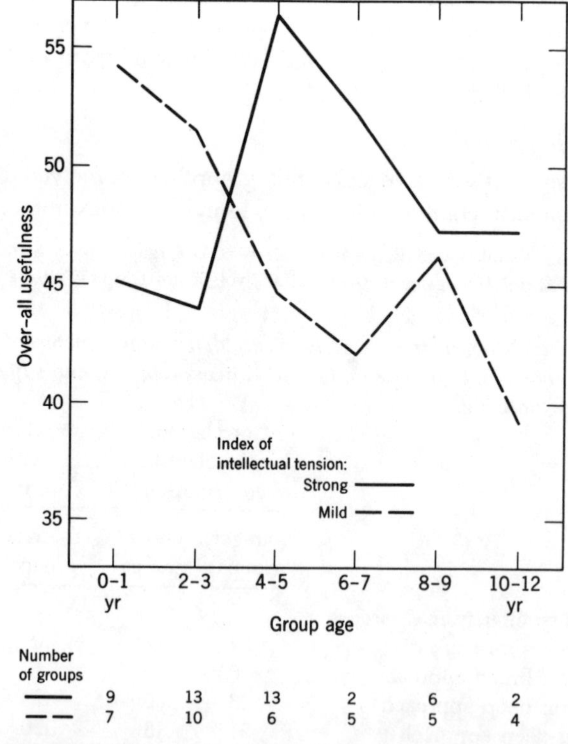

Chart 9. *An index of "intellectual tension" was combined from scores on hesitance to share ideas freely, and perceived dissimilarity. Among 37 groups with mild tension scores, usefulness started high and dropped steadily with group age; among 45 groups with strong tension, usefulness started low and peaked at 4–7 years.*

work through its internal conflicts. Once this resolution had been achieved, they surged to high usefulness between four and seven years, and thereafter dropped only moderately.

SUMMARY AND IMPLICATIONS

We examined various social properties of 83 groups in industrial and government laboratories to see if we could find clues for the general decline in scientific contribution of these groups with increasing group age, and a decline in usefulness after 4 to 5 years. We caution the reader again that with the small number of groups available it was not feasible to isolate the effects of group age as such from other factors with which it was associated. By statistical adjustment we did disentangle performance from some "extraneous" factors, but they may still affect group properties whose relationship to performance we have examined.

> Could you tell me in a few words why some older groups were able to keep their vitality?

It is partly true, although a little glib, to summarize the results by saying that "old groups which behaved like young groups continued to achieve." We did observe that old groups were more relaxed than young groups—less talkative with each other, less competitive and less secretive, and more inclined to specialize. And we found that among the older groups which resisted these tendencies—which continued to be competitive and a little secretive, communicated often, and retained a zest for broad pioneering—performance tended to remain high.

> You mean that effective older teams were simply reliving their group youth?

No, not exactly. The quality of their interaction was different from that of young groups. There was an atmosphere of intellectual struggle or tension, and at the same time a strong mutual respect.

In short: effective older groups were those which maintained the energy of young groups (interaction and competition), but replaced an atmosphere of friendly warmth with one of intellectual rivalry—sometimes toward each other, and often toward outsiders.

> What does this mean for the supervisor of an older group? Does he have a different role than in a young team?

We think so. According to these data, the effective supervisor of an older group is not technically better than they, not a source of original ideas. Rather he is a neutral sounding board; he draws out *their* ideas. And he invites the various members to challenge each other, regardless of rank.

> Won't that breed dissension? You said that group cohesion was also essential.

There is some risk. But the supervisor can build cohesion by taking a back seat. He doesn't claim credit for his subordinates' achievements, but gives the credit to the *group*. He can build mutual respect by pointing out to executives the contribution of each member. Finally, he can pit his section in rivalry with other groups, inside the organization or preferably outside.

> What about the supervisor of a *new* group? How does his role differ?

In two ways. Unlike the other he can be the technical leader, supplying original ideas himself. But he must also encourage ideas from the members *without* criticism and challenge. This atmosphere of "withholding judgment" is central to the process of brainstorming. Younger teams must be free to consider many approaches to a problem, without fear of criticism.

> What do your results mean for the laboratory manager? What can he do to maintain vitality of older groups?

The findings on "depth" and "breadth" suggest one key function. The manager must not let himself or the group assume that they have become the laboratory's "experts" in a specialized area. From time to time he challenges them with problems outside their field of competence. Occasionally he will give *another* team a problem within their specialty. He may invite several teams to sketch an attack on a pioneering new area. In this way he can maintain breadth as well as depth. Both are needed for vitality.

Appendix A

EVALUATIONS OF PERFORMANCE

Scientific performance was measured in two broad ways: (a) by evaluations of scientists' work by knowledgeable peers and supervisors, and (b) by scientists' own reports of their five-year output of papers, patents, and reports. Details will be presented about how the evaluations were obtained, the qualities evaluated, the extent of agreement between judges, and how evaluations made by different judges were combined into a single score. (Appendix B considers the output measures.)

Method of Obtaining Evaluations

Approximately 20% of the respondents at each site were asked to act as judges. Roughly half of the judges at each lab were supervisors, the other half were not. All were respected members of their organization who were acquainted with the work of many people in it. Judges were selected by us from nominations proposed by the research director and other senior people in each laboratory.

Judges evaluated the performance of other members of their lab with respect to two qualities: first, according to their *contribution to general technical or scientific knowledge in the field* (within the past five years), and second, according to their *over-all usefulness in helping the organization carry out its responsibilities* (within the past five years).

Each judge was given a group of cards containing the names of the scientists in his lab. Judges were instructed to remove the card containing their own name and also the cards bearing the names of people with whose work they were not familiar. They were then asked to sort the remaining cards into as many rank-ordered piles as they could discriminate with respect to the men's contribution. After the first sorting had been recorded and the cards shuffled, each judge made a second sorting using the criterion of usefulness. Judges worked individually and they were told that only the project staff would ever see their rankings.

In an effort to give the judges maximum freedom and increase the validity of the resulting evaluations, they were asked not to consider people whom they felt unqualified to evaluate. In addition, decisions about the number of categories to use, and the number of respondents to place in each category, were made separately by each judge on the basis of what seemed appropriate to him. (The one restriction was that no more than one-third of the names should be placed in any one pile.) This method of obtaining evaluations caused some problems when the judgments were to be combined (as will be discussed later), but had some important advantages over the more conventional methods of *rating* each man on numerical scales from "outstanding" to "poor."

One trouble with ratings is that they tend to bunch at the high end of the scale since anything lower is considered a black mark. Furthermore, judges often have difficulty giving numerical ratings to individuals (how does one decide whether John Jones is "outstanding" or simply "excellent"?). Finally, judges tend to approach the rating task with different frames of reference, some giving top ratings to a higher proportion of people than others.

Under the procedures used in the present study, judges seemed to find it relatively easy to decide whether John Jones was a better scientific producer than Bill Smith, or poorer, or about equal. Without evidence of strain, judges could evaluate approximately 50 scientists on two dimensions in less than an hour.

Qualities Evaluated

Perhaps the best way to describe the qualities judges were asked to evaluate is to present the instructions given to the judges.

> [Technical or scientific contribution] refers to the man's *own work*—his own production at the desk or in the laboratory, whether he is working individually or as part of a team. [It] refers to contributions which help the *field* move forward—whether in basic research or technical applications. Anything by way of contributing new knowledge or techniques, creative ideas, or inventions, regardless of whether this organization benefits or not.
>
> [Usefulness] refers to his over-all value to the organization, whether by his own work or by stimulating others. [It] refers only to how useful the person's work is to this organization—regardless of whether he performs a research or service function, whether his assigned task is creative or routine, technical or administrative.
>
> Base your evaluations on *evidence known to you personally*. This could consist of such things as watching the person at work, participating with him in group discussions, reading reports about work to which he has contributed, etc. But do *not* base evaluations simple on opinions which some other person has expressed about him.
>
> Each scale refers to the individual's performance in the *present or immediate past* (within the last four years). Look at what the person has actually produced, or is now producing—work on which substantial progress has been made, even if not yet written up.

Ignore contributions beyond five years ago. If a man made substantial contributions earlier than that but has not since produced, rate him low.

Disregard *future promise.* If you are rating a very able person who will probably be outstanding in five years, but has not produced anything tangible to date, rate him low.

Disregard the fact that a piece of work may not have appeared in the literature owing to security restrictions, because not yet completed, etc. On the other hand, if a man has neglected to publish or patent something when he might have done so, this fact may lessen the value of the contribution.

Rate the actual contribution, regardless of the *individual's education or background.* A man may not have the training needed to make outstanding contributions; he will still rate low if he has not contributed. Compare him with all other persons on the list on the basis of output only.

Rate the actual contribution *regardless of job assignment.* If a supervisor spends all his time in supervision and none at the bench, rate his bench contribution low. If a man was assigned to a project which has not contributed to the field, rate him low.

Multiple Evaluations of Respondents

The performance of the average respondent was evaluated by about five judges. The median number of judges per respondent varied between laboratories from a low of 3.4 to a high of 9.5 (median across all labs = 4.9); 95% of the respondents were evaluated by at least two judges and 84% by three or more.

Where two judges evaluated the same men it was possible to examine the agreement in their evaluations. Spearman's rank-order correlation coefficients were computed between the evaluations provided by all possible pairs of judges in ten academic or industrial departments. The median correlation within departments ranged from .3 to .9; in half the departments the median was .7 or greater. (On the basis of a limited examination, there was no evidence that the agreement between judges was markedly different on the contribution and usefulness scales.) Thus, although different judges did not always give the same rankings to an individual, they did tend to show substantial agreement.

Combining Evaluations

Since the work of most respondents had been evaluated by several judges, it was necessary to combine these evaluations into a single score (on each quality) for each respondent. The great flexibility allowed the judges in their choice of respondents and use of categories posed certain difficulties in determining these scores.

To illustrate the difficulties and the method used for handling them, a simple set of artificial data may be used. Table A-1 shows how six judges might have evaluated ten scientists. One may note that all judges showed high, but not perfect, agreement in their rankings. Judges I, V, and VI happened to be acquainted with relatively mediocre people. Judges II and IV knew only high performers. Judge III knew some of each. The number of ranks used by the judges ranged from three (Judge I) to six (Judge IV). Although Judge III used only four ranks, he evaluated more people than any other judge. Although these data are artificial, their only important deviation from much of the actual data is the small number of scientists and judges.

Given the simplicity of Table A-1, one can identify the "true" rank order of the scientists. This is shown at the left of the table. Note that the order between Scientists B and C is undefined (Judge II placed C above B, but Judge IV placed B above C, and no other judge compared them). It is clear, however, that these two scientists fall below A and above D. In addition, there was minor disagreement among the judges concerning the relative positions of F versus G (two judges out of three placed F ahead) and I versus J (three out of four placed I ahead).

In our more complex real data, it would have been exceedingly difficult to identify the true rank order by inspection. Consequently, several other systems were tried. Before describing the system which gave the best results, some of the simpler ones which gave misleading results will be mentioned to illustrate the kinds of problems which had to be solved.

Some Unsatisfactory Combining Systems. In some kinds of data the average *rank order* assigned by several judges provides a satisfactory way of combining judgments. The fact that in the present data judges evaluated different sets of scientists, and used different numbers of categories, however, made this system inappropriate. Note in Table A-1 that this first method gives results very different from the true rank order.

Allowance could be made for the judges' using different numbers of categories by converting to a "percentile equivalent" for the rank (if one were willing to assume equal distances among ranks). A combined score could then be based upon an average of the "percentile equivalents." This second method was applied to the data of Table A-1. One may note that it also failed to approximate the true rank order. The reason is that this method, like the first, neglects the fact that judges ranked different sets of scientists.

A somewhat more sophisticated system is to set up a "win-loss matrix." For each possible pair of people, this matrix indicates the number of times the first person was ranked above the second, and vice versa. Such a matrix can be used to determine each person's "win percentage"—the number of times he was preferred to others relative to the total number of times he was compared. The "win percentages" derived by this third

TABLE A-1 *This shows artificial evaluations of ten scientists by six judges. Methods I to III for combining evaluations gave misleading results when judges were free to rank only scientists whom they knew, and were allowed flexibility in the number of categories to be used and the number of people assigned to each. The Ford Program (Method IV) gave results close to the true order.*

		Evaluations by judges						Method I		Method II		Method III		Method IV
		Judge												
True rank	Scientists	I	II	III	IV	V	VI	Mean rank	Order	Mean percentile	Order	Win %	Order	Order by Ford Program
1	A			1	1			1.0	1	89	1	100	1	1
2.5	B		2		2			2.0	3.5	72	3	78	3	2
2.5	C		1	1	3			1.7	2	78	2	86	2	3
4	D		3	2	4			3.0	6	50	6	50	6	4
5	E		4		5			4.5	10	28	9	22	8	5
6	F		5	2	6	2	1	3.2	7	47	7	45	7	6
7	G			3		1	2	2.0	3.5	63	4	61	4	7
8	H	1				3	3	2.3	5	61	5	60	5	8
9	I	2		3		5	4	3.5	8	30	8	20	9	9
10	J	3		4		4	5	4.0	9	16	10	6	10	10

method are also shown in Table A-1. This method gives an order somewhat different from either of the previous two, but still fails to approximate the true order.

The Ford Program. Another solution to the problem of combining evaluations in such data—and the one which we adopted—was proposed by Ford.[1] Ford's proposal (with certain modifications to insure that our data met his assumptions) was translated into a program for the IBM 704, 709, and 7090 computers. This program came to be known as the "Ford Program" and was used to achieve the needed rank orders (Method IV in Table A-1). It can handle up to 150 judges who may use up to 150 ranked categories in evaluating up to 150 subjects. Although no attempt will be made to present the details of the program, the rationale and basic formula of the technique will be described.

Ford proposed that one may attempt to assign to each object (in our data these were scientists) a number or "weight" (w) which could be interpreted as odds, in the sense that the probability of object i being preferred to object j in a future comparison would be $w_i/(w_i + w_j)$. With these probabilities, one could compute the a priori probability of obtaining exactly the win-loss matrix actually obtained. The problem, then, was to determine that set of numbers which maximized the likelihood of obtaining the given matrix.

The percentage of wins in the win-loss matrix was used as the initial estimate of w_i. From these, Ford showed that the desired set of weights could be obtained by an iterative technique.[2] His procedure was to solve the following equation for each object in the set until the resulting weight for each object had stabilized.[3]

[1] L. R. Ford, Jr., "Solution of a Ranking Problem from Binary Comparisons," *American Mathematics Monthly,* 1957, vol. 64, no. 8, Part II, pp. 28–33.

[2] Any arbitrary starting assumption for w might be used, but the procedure would take longer.

[3] Although the criterion of what constitutes "stability" is arbitrary, the primary criterion used by us was that no scientist's weight should change by more than 0.5% from one iteration to the next: that is, for all scientists

$$\frac{w_i^{n+1} - w_i^n}{w_i^n} \leq .005$$

It was discovered, however, that an exceedingly large number of iterations would be needed before the weights in some sets of data would reach this criterion. Therefore if the foregoing criterion had not been met after 100 iterations (which occurred in about 10% of the runs), the iterative process was arbitrarily stopped. Although the weights themselves might still be changing from one iteration to the next, their *rank order* generally stabilized after 20 to 30 iterations. Since we were interested only in the rank order, this phenomenon presented no problem for our purposes. (The data shown in Table A-1 provided one instance of this. After 150 iterations, only the weights of two scientists had stabilized, although the rank order of the weights did not change after the 25th iteration.)

$$w_i^{n+1} = \frac{\sum_j a_{ij}}{\sum_j \frac{a_{ij} + a_{ji}}{w_i^n + w_j^n}}$$

where a_{ij} = number of times object i was preferred to object j

a_{ji} = number of times object j was preferred to object i

w_i^n = the number (weight) assigned to object i on the nth iteration

For Ford's technique to yield a solution, the data must meet the following assumption: "In every possible partition of the objects into two non-empty subsets, some object in the second set has been preferred at least once to some object in the first set."[4] Thus if the objects were all baseball teams, this assumption would be violated if it were possible to divide the teams into a major and minor league, where the major teams had always defeated the minor teams.

In our data this basic assumption could be violated in four ways. (a) One scientist might be universally preferred by the judges, that is, he headed all lists on which he appeared. (In Table A-1, Scientist A is such an object.) (b) A scientist might be at the bottom of all lists on which he appeared. (c) Some scientists might be judged neither universally high nor universally low but, taken as a group, were simply not judged in relation to the other scientists. (d) Some scientists might fall in a subset such that the comparisons with another subset were all in one direction. (After the removal of Scientist A from Table A-1 (see next paragraph), the subsets B-C, B-C-D, B-C-D-E, and B-C-D-E-F-G constitute violations of this type.)

Identifying violations of the first and second types proved relatively easy. A procedure was incorporated in the Ford Program which would identify scientists who were "universal highs" or "universal lows" and remove them before the computation of the weights. (They were subsequently given ranks above or below the set of scientists for whom weights were computed, as appropriate.)

Violations of the third and fourth types, however, proved difficult to identify before computation. Our solution consisted of adding an extremely small constant (0.00001) to every cell of the win-loss matrix derived from the judges' evaluations. This constant insured that the matrix would not contain violations of the third and fourth types.[5]

[4] Ford, *ibid.*, p. 29.

[5] This solution was accepted only after experimentation with several other possibilities which included: adding 0.00001 only to the zero cells of the win-loss matrix, adding 1.0 to all cells, and adding 1.0 only to zero cells. In several trial sets the smaller constant produced more rapid stabilization of the weights, and whether the constant was added to all cells or only to zero cells proved to make little difference.

The Ford Program proved capable of yielding a combined rank order which closely approximated the "true" order in a wide variety of test cases. In Table A-1, for example, the only difference between the true order and that determined by the program was that the program produced a completely ordered set (that is, no ties) whereas in actuality the set contained a pair of tied scientists. The minor discrepancies were probably attributable to the addition of the small constant to the cells of the win-loss matrix.

Although the program proved satisfactory for our purposes, it disregarded certain data which an even more complex procedure might have considered. The program took no account of the fact that a judge might have placed two or more individuals in the same category. Furthermore, it did not separately maintain the identity of each judge, nor examine the extent to which his rankings were consistent with those of other judges. Although such refinements might be desirable from the standpoint of elegance, considering the large errors inherent in any judgment process, it seems unlikely that these refinements would have been particularly useful in the present situation.

Conversion to Percentiles

On the basis of the sizes of the weights generated by the Ford Program, scientists were rank ordered and their positions expressed in percentile form. Since evaluations had been made with respect to two qualities, it was necessary to perform the operation twice, once for each quality. All evaluations had been made relative to other scientists in the same laboratory. Accordingly, use of the Ford Program, and the subsequent percentiling of the weights, produced two sets of performance scores for each laboratory (one for contribution, one for usefulness). Of course, each set had a rectangular distribution and a mean of 50. For lack of better information, it was assumed that the meaning of a given percentile score was roughly the same in all laboratories. (Note that this assumption did not have to be made with respect to the output measures.)

Transformation of Percentiles

For analysis purposes it seemed desirable to group the scientists according to the classifications shown in Chapter 1 (the five "primary groups") rather than by their laboratories. A difficulty arose, however, when doctoral people were separated from nondoctorals. For example, almost all university scientists had their doctorates, therefore their mean score was 50. In government and industrial locations, however, both Ph.D's and non-Ph.D's had been included, and the doctorals tended to score substantially higher than the nondoctorals. In order to combine Ph.D's in

TABLE A-2 *This shows how the transformations equalized the means and standard deviations of the evaluation scores for each group.*[*] *For some groups the order of individuals changed, as intended, when the subgroups on which the transformations had been made were combined into the groups shown in the table. The magnitude of these changes is indicated by the correlations.*[†]

	Scientific contribution					Over-all usefulness				
	Pre-transformation		Post-transformation		Correlation between pre- and post-transformation scores	Pre-transformation		Post-transformation		Correlation between pre- and post-transformation scores
Group	mean	s.d.	mean	s.d.		mean	s.d.	mean	s.d.	
Ph.D's in development	70	23	50	29	.94	67	25	50	29	.95
Ph.D's in research	57	29	50	28	.94	57	29	50	28	.97
Engineers	49	29	50	28	1.00	49	28	50	28	1.00
Assistant scientists	33	22	50	28	.97	35	22	49	28	.99
Non-Ph.D scientists	43	30	46	30	1.00	45	30	47	29	1.00

[*] At the time this transformation was made, we had not decided to analyze engineers and non-Ph.D scientists separately, therefore the mean for the latter deviated slightly from 50. An appropriate constant was subsequently added to the scores of non-Ph.D scientists to move the mean for their group up to 50.

[†] The greater the deviation of the correlation from unity, the greater the changes in order of individuals between pre- and post-transformation. For example, "Ph.D's in research" include scientists from both government and university; those in government originally scored spuriously high (relative to those in university) because the former had been compared with non-Ph.D's, whereas the latter had not. Thus on the transformed scores—which compensate for this—some university Ph.D's would score above some government Ph.D's who had formerly been unbeaten.

government research labs with those in university research labs, (that is, to produce the group "Ph.D's in research labs") it seemed necessary to first equalize the distributions of their performance scores with respect to shape, mean, and standard deviation.

The required transformation was accomplished by dividing the scientists into eight subgroups and repercentiling the scores of each. This amounted to a slight shifting of scores—without changing their order—to produce a rectangular distribution ranging from 00 to 99 with a mean of 50. The eight subgroups were as follows:

Ph.D's in government development labs
Ph.D's in industrial development labs
Ph.D's in university research labs
Ph.D's in government research labs
Engineers and non-Ph.D scientists in government[6]
Engineers and non-Ph.D scientists in industry
Assistant scientists in government
Assistant scientists in industry.

The results of the transformation and subsequent combining of subgroups may be seen in Table A-2.

[6] At the time this transformation was made, we had not planned to analyze the engineers and non-Ph.D scientists separately. Their subsequent separation, however, did not require a retransformation of the evaluation scores since the distributions for both groups were closely similar.

Appendix B

OUTPUT OF SCIENTIFIC PRODUCTS

In addition to the evaluations described in Appendix A, scientific performance was also measured by the number of scientific products produced by each respondent over the preceding five years. Data on four types of products were collected: a) patents or patent applications, b) published papers, c) books, and d) unpublished technical manuscripts, reports, or formal talks.

This appendix will describe the source of this information, its accuracy, and a transformation which was applied. Appendix D shows the interrelationships among the various measures of performance.

Source of Data

Information about respondents' output of scientific products came from the question shown in the following box. Although "books" were included in the question, so few scientists wrote books that our analyses depended primarily on patents, papers, and reports.

Question 75. Over the past five years, about how many of the following have you had: FILL EACH SPACE

	Approximate number
Patents or patent applications	____
Technical papers accepted by professional journals	____
Technical books accepted for publication	____
Unpublished technical manuscripts, reports, or formal talks (either inside or outside this organization)	____

Respondents' accuracy in reporting their output was checked in two ways. First, answers given by one group of respondents were compared with records maintained by their laboratory. Second, we observed how closely the answers given by a group of respondents at one time agreed with answers given by the same people three months later.[1]

[1]Technically, the first was a check for "validity;" the latter, a check for "reliability."

Agreement with Laboratory Records

Output data were obtained from a group of 27 scientists responsible for advanced research in an industrial laboratory. Their answers were then compared with records maintained by their laboratory. In general, there was reasonably close agreement.

To assess the degree of agreement, Spearman rank-order correlations were calculated between laboratory records and respondents' claims. For patents, this correlation was .91, indicating a very high level of agreement. For papers, the correlation was .82, again indicating a high level of agreement. The coefficient for reports, however, was only .06.

Several sources of discrepancies were possible. Scientists could have forgotten the exact number of their patents, papers, or reports; they could have inflated their output; and, we observed, some appeared to round the number of their output to approximate figures such as 5, 10, or 20.

In addition, records from the laboratory were not perfect. Although the questionnaire definition of reports included *any* unpublished written or oral formal presentation, the laboratory counted only a limited set of written documents prepared for internal consumption. Nearly all respondents claimed more reports than had been recorded by the laboratory, suggesting the more restricted nature of the laboratory definition.

Of particular interest was the finding that respondents' own claimed output of reports correlated more strongly with judgments of their "technical contribution"[2] (rho = .58) than did the laboratory records (rho = .02). This finding suggested that the questionnaire measure was probably reasonably valid and the lack of agreement was primarily attributable to omissions (due to different definitions) from the laboratory records.

In summary, these scientists' claims of their output—as elicited by the question shown in the foregoing box—seemed to be reasonably accurate.[3]

Agreement at Different Times

A second test of the accuracy of reporting output was part of a larger examination of the "reliability" or "stability" of answers to the questionnaire. This study was conducted on 418 research and development engineers in an industrial laboratory. Approximately two months after completing the questionnaire, 52 of these people were given many of the same questions again.

Claimed output of patents and papers showed very high agreement

[2] These judgments were collected as described in Appendix A.

[3] A complete report of this investigation is available in F. M. Andrews, "A Check on the Validity of Reporting Scientific Output," Analysis Memo #19, September 1963, available as Publication #2134 from the Survey Research Center, University of Michigan.

between the two administrations of the questionnaire (Pearson product-moment correlations were 1.00 for patents, .91 for papers). For reports, there was moderate agreement ($r = .56$). Thus these scientists were very likely to give closely similar answers to these questions over a two-month interval.[4]

Logarithmic Transformation

All of the output measures were highly skewed: many scientists produced few papers (or patents, or reports), but few produced many. Since the statistical techniques we desired to use would have given misleading results if applied to highly skewed distributions, it seemed advisable to apply a transformation which would reduce the skew. What would be a satisfactory transformation?

It was discovered that the outputs of many subgroups of our respondents had a distribution closely approximating the lognormal curve. In other words, the distributions were such that they would produce the familiar bell-shaped normal curve if the logarithm of units of output were plotted against frequency. There was one exception to good fit between the theoretical lognormal distribution and the observed distributions: there often were more zero producers than would have been "predicted" by a lognormal curve. Since this was the only important exception, the output scores were normalized using a logarithmic transformation.[5]

Since the lognormal curve assumes a continuous variate, it seemed wise to think of people who had produced N units of output as being distributed between N and $N + 1$ units. Accordingly each person was arbitrarily credited with one-half unit of output more than the number he claimed—corresponding, in a rough fashion, to work in progress. The addition of a half unit to each person's output served an important practical function for the zero producers. It avoided the necessity of handling the awkward quantity *minus infinity* (the log of zero).

After the conversion to log scores, a constant of 1.0 was added to each score to avoid the inconvenience of negative scores.

[4] A discussion of other results from this reliability study is contained in Appendix F. The complete report is available in D. C. Pelz, "Reliability of Selected Questionnaire Items," Preliminary Report #9, August 1962, available as Publication #1991R from the Survey Research Center, University of Michigan.

[5] We are indebted to our former colleague S. S. West for the investigations which led to these findings. A description of them appears in D. C. Pelz, "Some Properties of the Measures of Scientific Output," Analysis Memo #1, December 1960, available in Publication #1741 from the Survey Research Center, University of Michigan.

The lognormal curve is characteristic of many types of data observed in nature and society, for example, distributions of income. Theoretical discussions appear in J. Aitchison and J. A. C. Brown, *The Lognormal Distribution,* Cambridge University Press, 1957.

Thus the completed transformation was as follows:

$$\text{transformed score} = 1 + \log_e (\text{raw score} + 0.5)$$

This produced output scores falling in a convenient range for computer processing, having distribution characteristics which did not grossly violate assumptions of normality implicit in many statistical techniques, and in no way disturbed the relative order of scientists with respect to their output of various products.[6]

Conversion to Percentiles

After analysis had been completed, we sometimes wanted to report results using charts which showed mean output for various subgroups. In these charts it was desirable to have a common scale which could be used for the several different output measures, as well as the performance evaluations. As the last step before plotting, therefore, the output measures were converted to percentile equivalents. (This conversion was not used when results were reported in the form of correlation coefficients.)

[6]The transformation process and some properties of the transformed scores are described in greater detail in F. M. Andrews, "Logarithmic Transformation of Output of Scientific Products," Analysis Memo #11, May 1961, in Publication #1826 from the Survey Research Center, University of Michigan.

Appendix C

ADJUSTMENT OF PERFORMANCE MEASURES

Since the purpose of this study was to explore the effects of contemporary social and motivational factors on scientific performance, it was desirable first to remove the effects of antecedent factors. Of course, it was impossible to know all antecedent conditions which might have influenced a scientist's performance, but several conditions of known importance were identified. These included: (a) The type of laboratory in which the scientist was working—people in government, for example, tended to publish more than those in universities or industry, due, in part, to needs for "letting the public know" by government, and for "security" by industry. (b) Length of working experience—scientists with substantial experience tended to produce more than relatively inexperienced ones (see Chapter 10). (c) The speed with which scientists completed their formal training—scientists who received their B.S. degrees early, or who earned Ph.D's soon after their B.S.'s, tended to outperform their slower colleagues. And (d) amount of formal education—the performance of Ph.D's tended to exceed that of scientists with master's degrees, and master's degree holders tended to outperform bachelor's degree people.

There is no attempt here to imply that these antecedent conditions should or should not be related to performance. It was simply observed that they were. The speed with which a scientist completed his formal education, for example, was probably itself a function of several other factors, such as ability, motivation, and receipt of financial assistance. Similarly, the kind of laboratory a scientist was in may have markedly affected his opportunities to publish. There were undoubtedly many factors which accounted for the relationships between performance and the antecedent conditions examined. It was not our purpose to identify the entire chain of causality. If we could at least remove the effects of several antecedent conditions of known importance, we could be sure that our findings were not due to these conditions.[1]

[1]Chapter 1 gives an example of the misleading results which could appear had antecedent conditions not been removed.

Possible Solutions

A typical procedure for ruling out unwanted factors is partial correlation. This would have been inappropriate in our data, since many relationships were known to be curvilinear. A second common practice is to analyse within subgroups; for example, study the relationships between motivation and performance within subgroups having roughly the same amounts of experience. This also would have been inappropriate since we wished to allow for several background factors simultaneously and the number of cases in the resulting subgroups would have been too small to permit analysis.

Therefore, we adopted a third strategy, which was to *adjust* the performance scores in such a way that the antecedent factors would have no relationship with the adjusted scores. This was accomplished by constructing a predicted score for each individual, based on his particular antecedent characteristics, and then subtracting his predicted score from his actual score. The resulting adjusted (or residual) score indicated how much better (or poorer) the individual actually performed than one would have expected on the basis of his experience, type of laboratory, etc. These adjusted scores were then used in later analyses with confidence that the antecedent factors could not produce any spurious component in the relationships observed.

ADJUSTMENT OF INDIVIDUAL SCORES

Method of Adjustment

The adjustment was accomplished in two phases. The first (Adjustment I) may be regarded as a preliminary equating of means for different types of laboratories. Adjustment II began with the scores which resulted from Adjustment I and removed effects of additional background factors.

Adjustment I. The goal of this adjustment was to provide scores with equivalent means in various subgroups so certain of these subgroups could subsequently be combined.

For the evaluations (contribution and usefulness), this had already been approximated by the transformation described in Appendix A. (See the discussion there for a full description of the rationale and method.) Therefore, Adjustment I produced only a minor refinement of these scores, necessitated by a slightly revised method of classifying scientists. For the output measures (papers or patents and reports), however, substantial shifts in scores were made.

Adjustment I was accomplished by subtracting an appropriate constant from the score of each person in a particular group so the mean score for this group would equal the mean score for all other groups.[2] The groups whose means were thus adjusted were as follows:

Ph.D's in government development labs
Ph.D's in industrial development labs
Ph.D's in university research labs
Ph.D's in government research labs
Engineers in industrial labs
Engineers in government labs
Assistant scientists in industrial development labs
Assistant scientists in government development labs
Assistant scientists in government research labs
Non-Ph.D scientists (all in government development labs)

Performance scores incorporating Adjustment I were used in Chapters 10 and 11. Chapters 2 through 9 and 12, however, used scores which reflected further adjustments as described next.

Adjustment II. This adjustment[3] started with the Adjustment I scores and removed effects of the following additional factors in the five groups shown:

Ph.D's in development labs Ph.D's in research labs	Engineers Non-Ph.D scientists Assistant scientists
a—time since receiving Ph.D	a—time since receiving B.S.
b—time since joining division	b—time since joining division
c—years between B.S. and Ph.D	c—age at which B.S. was received
	d—whether held M.S.

The relationships of these background factors to the performance measures were studied by use of a new computer program called Multiple Classification Analysis (MCA).[4] As an example of the type of results gen-

[2]The exact adjustment values used for each group appear in F. M. Andrews, "Revised Adjustment of Performance Measures to Hold Constant Some Background Factors," Analysis Memo #21, March 1966, available from the Survey Research Center, University of Michigan.

[3]Although we speak of an "adjustment," the reader should keep in mind that 20 parallel operations were carried out: one each for the four performance scores (papers or patents, reports, contributions, and usefulness), separately for each of five primary analysis groups.

[4]J. Sonquist, K. Goode, H. Hinomoto, and R. Hsieh, "Multiple Classification Analysis for the IBM 7090," Institute for Social Research Data Processing Section. This program is a revised and expanded version of an IBM 650 program developed by Vernon Lippitt for the General Electric Company. A monograph giving a full description of the MCA Program is being prepared by F. Andrews, J. Sonquist, and J. Morgan of the Institute for Social Research, University of Michigan.

erated by this program, Table C-1 presents complete data for one group —Ph.D's in development labs.[5]

To find the relationship between each background factor *considered alone* and the performance scores was not a problem. One had only to compare the mean performance of the various subgroups of the background factor being considered. These means could be presented either as absolute values or, equivalently, as deviations from the grand mean for all groups together. (In Table C-1 the columns labeled "Raw deviations" present means in this latter fashion. For example, the first raw deviation shown in the "Contribution" columns of the table (−08.5) indicates that among Ph.D's in development labs, those scientists 0 to 3 years beyond their Ph.D's were evaluated, on the average, 8.5 percentile points below the average for all Ph.D's in development labs.[6])

For adjusting, however, one cannot examine each background factor individually but must consider all simultaneously. Were they not all considered simultaneously, one would "over-adjust" the scores in situations where the background factors were positively correlated, as some of ours were. ("Underadjustment" would occur if background factors were negatively correlated.)

An example will illustrate the effect. We observed that scientists who had just recently completed their education tended to perform below the average for all scientists. Similarly, those who had just recently joined their division tended to perform below average. But, of course, many of the scientists who had just joined their divisions were the same people as those who had just completed their educations. Therefore, *part* of the low performance of newcomers to a division was attributable to lack of post-degree experience, whereas another part of the low performance was attributable to lack of experience in the division. Thus there were two separate but partially overlapping effects. Since they overlapped, the effect of the two together was less than the sum of the effects of each considered separately.

Thus it was necessary to determine how much to adjust each scientist's score for his particular combination of background characteristics, considering the fact that these background characteristics were expected to have overlapping effects. Since the background characteristics were known to have complex curvilinear effects on performance, traditional techniques of multiple regression were inapplicable. The MCA Program provided a powerful technique for determining the correct adjustments.

The MCA Program uses an iterative technique to find that set of

[5]The document described in Footnote 2 gives full details for all five groups.

[6]These are Adjustment I scores, of course.

TABLE C-1 *Since this study focused on the effects of current motivations and laboratory environments, it was desirable to remove effects of several "background" factors, such as length of experience. Shown below are the background factors examined for Ph.D's in development labs and the effects of each. The "raw deviations" show how the mean for each subgroup deviated from the over-all mean. The "fitted deviations" differ only in that they compensate for related effects of other listed background factors (see text for method of computation). The effects of these background factors were removed from the performance scores by determining the subgroups to which each person belonged and subtracting the fitted deviations for those groups from his performance scores. Adjustment II scores were the result.* (*Note: All scores shown below incorporate Adjustment I—see text.*)

		Data from Ph.D's in Development Labs							
		Papers		Reports		Contribution		Usefulness	
Mean (after Adjustment I)		3.0		3.0		50		50	
Cases with all data		178		176°		174°		174°	
Percent of variance attributable to listed background factors		14%		5%		8%		10%	
	N	Raw dev.	Fitted dev.	Raw dev.	Fitted dev.	Raw dev.	Fitted dev.	Raw dev.	Fitted dev.
Years since Ph.D									
0–3 years	15	−0.73	−0.34	−0.50	−0.49	−08.5	−01.5	−15.0	−05.4
4–5 years	22	0.08	0.36	0.05	0.02	06.2	−01.5	−06.5	−00.8
6–10 years	55	−0.20	0.02	−0.07	−0.11	−02.6	−00.3	00.8	05.2
11–15 years	28	0.20	0.03	−0.05	−0.06	00.5	−02.3	−01.4	−05.7
16–20 years	23	0.64	0.28	0.44	0.47	12.1	06.6	06.4	−01.7
21+ years	35	0.00	−0.31	0.06	0.13	03.3	−00.6	06.1	00.2
Years in division									
Under 2 years	24	−0.48	−0.46	−0.27	−0.11	−11.4	−12.2	−15.5	−16.5
2–4 years	38	−0.26	−0.34	0.04	0.11	−06.4	−06.0	−04.9	−05.1
5–9 years	56	−0.12	−0.11	0.02	0.09	00.8	01.4	00.1	−01.2
10+ years	60	0.47	0.50	0.07	−0.11	08.3	07.7	09.4	11.2
Years between B.S. and Ph.D									
0–3 years	25	0.15	0.02	0.20	0.14	−05.3	−08.0	−00.7	−02.9
4–7 years	114	0.03	0.05	0.00	−0.01	01.7	02.5	00.2	01.6
8+ years	39	−0.19	−0.16	−0.11	−0.07	−01.8	−02.6	−00.1	−02.9

° *N*'s in each category are very close to those in the first column.

"fitted deviations" which minimize the (squared) error associated with the following model:

$$Y_{ij\ldots\alpha} = \overline{Y} + a_i + b_j + \cdots + e_{ij\ldots\alpha}$$

where $Y_{ij\ldots\alpha}$ = the performance score of individual α who falls in category i of background factor A, category j of background factor B, etc.

$\overline{Y}$ = the grand mean of the performance scores for all people being considered

a_i = the fitted deviation from the grand mean associated with category i of background factor A

b_j = the fitted deviation from the grand mean associated with category j of background factor B

$e_{ij\ldots\alpha}$ = error term for this individual

The model is far from new, having been described by Yates in 1934.[7] A few of the fitted deviations generated by the MCA Program are also shown in Table C-1. It was these fitted deviations which were subtracted from the Adjustment I scores to produce the Adjustment II scores.

As an example, consider the case of a young doctoral scientist in a development lab whose score on contribution (after Adjustment I) was at the 35th percentile. At the time he took part in the study, it had been two years since he had received his Ph.D and one year since he had joined his division. It had taken him five years to earn his Ph.D after he had received his B.S. To obtain his Adjustment II score, one would start with his Adjustment I score (35) and subtract the fitted deviations associated with his particular category on each of the three background factors (−01.5, −12.2, and +02.5). This yields a score of 46.2. In a sense, this young scientist "moved up" when his score was compensated for his relative lack of experience. Put differently, although he was well below average when compared to all Ph.D's in development labs, he was only slightly below average when compared with other scientists with similar backgrounds.

Of course, when some of the background factors which related to—or "explained"—some of the differences observed in the scores were removed, there were fewer (or smaller) differences left to be explained by other factors. Thus it was of considerable interest to determine what portion of the differences (technically, what portion of the variance) in the Adjustment I scores had actually been explained and removed by the background factors used to produce Adjustment II. As may be seen in Table C-2, the Adjustment II background factors explained from 5 to 38% (median = 15%) of the variance in the Adjustment I scores.

[7]F. Yates, "The Analysis of Multiple Classifications with Unequal Numbers in the Different Classes," *Journal of the American Statistical Association*, 1934, Vol. 29, pp. 51–66.

TABLE C-2 *This shows the percent of variance in the Adjustment I scores attributable to Adjustment II background factors. This portion of the variance was removed in the process of producing Adjustment II.*

	Papers	Reports	Contribution	Usefulness
Ph.D's in development labs	14%	5%	8%	10%
Ph.D's in research labs	17%	10%	12%	15%
Engineers	16%	8%	7%	10%
Assistant scientists	27%	11%	20%	19%
Non-Ph.D scientists	38%	21%	35%	26%

As shown in Appendix D, one result of removing the effects of background factors from the performance scores was to reduce the extent to which they were intercorrelated. This was expected and desirable. Part of their interrelationship was spurious, in the sense that all measures were related in the same ways to the background factors. When these factors were removed, the resulting measures were "purer," that is, more distinct, indicators of the different forms of scientific performance.

Interpreting Adjusted Scores in Turns of Actual Output

After the various transformations and adjustments had been applied to the performance measures, we were in a position to examine the effects of contemporary social and psychological factors on scientists' performance. Furthermore, we could be sure that any effects found were not attributable to the antecedent factors whose effects had been removed. It was no longer possible, however, to specify a precise number of papers (or patents, or reports) which would correspond to a mean falling at any given percentile position. For example, how many papers, on the average, were written by the set of people with "semimonthly or less" contact with their colleagues—shown in Chapter 3, Chart 1-A to fall at the 50th percentile? This figure would vary according to the backgrounds of the people who composed the set.

For example, consider a set whose mean output of papers placed it at the 60th percentile after Adjustment II. If the set were composed of relatively inexperienced people, the actual output of papers would have been lower than if the group were composed of people with long experience. Thus, although any particular individual's score can be converted back to the actual number of papers, patents, or reports he produced over the five-year period,[8] in general it is not feasible to convert means for sets of people to raw output figures in any exact way.

[8]Using data shown in the document cited in Footnote 2, the reader can determine the raw score equivalents of Adjustment I and II values for individuals with any particular set of characteristics.

TABLE C-3 *This shows the approximate average output (per person) required for a set of people to score at designated percentile points after Adjustment II. Values after Adjustment I would be roughly the same.**

Group	Percentile on papers or patents†			Percentile on reports		
	30th	50th	70th	30th	50th	70th
Ph.D's in development	3–5	5–8	9–14	3–8	6–16	11–22
Ph.D's in research	4–6	6–10	9–16	3–6	6–14	9–21
Engineers	0–0	1–2	2–3	2–6	4–12	6–15
Assistant scientists	0–1	0–5	2–7	0–2	2–7	4–12
Non-Ph.D scientists	0–1	1–3	3–5	2–5	3–7	5–11

*The reader is cautioned that these data are approximate. The exact value will vary depending on the people who compose the set (see text). Under extreme conditions it is possible for average output to fall outside the ranges shown.

The table was prepared as follows: respondents were divided into numerous sets according to a variety of classification schemes. For each of these sets, the mean raw output and mean Ajustment II value were obtained. By plotting one against the other, the approximate range of mean raw outputs associated with a given mean on Adjustment II could be observed.

† Patents were used for engineers, papers for all other groups.

To provide an indication of the *approximate* average output (per person) required to place a group at the 30th, 50th, and 70th percentile points after Adjustment II, Table C-3 has been prepared. Although the table gives data for Adjustment II, the Adjustment I values would be roughly the same.

ADJUSTMENT OF GROUP SCORES

In Chapter 13, results of an analysis of *group* performance are shown. Of course, many of the problems previously outlined with respect to individual scores and the unwanted effects of various antecedent factors also applied to group scores. (The score for a group was simply the mean of the scores of the individuals who composed the group.[9]) For reasons already discussed, it seemed desirable to also "adjust" the group scores.

Method of Adjustment

The method of adjustment was similar to that employed for the individual scores. After an initial exploration of the effects of various background factors, three that seemed most important were included in a Multiple Classification Analysis. These were: the average age of members, the proportion of members with Ph.D's, and the type of setting (industrial,

[9] Percentile evaluations—prior to the transformation described in Appendix A—were used to compute these means.

government research, or government development). Separate analyses were performed for groups from departments characterized by high amounts of individual autonomy and for groups from more coordinated departments.[10] As before, the resulting "fitted deviations," shown in Table C-4, were subtracted from the original mean scores to produce a set of residual or adjusted group scores. The sum of the adjustments for groups from autonomous departments ranged from +4.8 to −4.9 for contribution, and from +5.5 to −5.7 for usefulness. The groups from coordinated departments tended to score lower than those from autonomous departments and there was more variation among their scores. To equalize both the means and standard deviations of the adjusted performance measures, scores for groups from coordinated departments were multiplied by an appropriate constant to reduce their variation (.66 for contribution, .77 for usefulness), and were added to an appropriate constant to raise their means (19.8 for contribution, 13.8 for usefulness).

The resulting scores were then ready for use in the analyses described in Chapter 13.

[10] The method of classifying departments is described in Appendix H.

TABLE C-4 *This shows the fitted deviations resulting from the MCA Program which were subtracted from group scores to produce the adjusted measures of group performance. Groups from autonomous and coordinated departments were analyzed separately.*

I = Groups from autonomous departments
II = Groups from coordinated departments

	Contribution		Usefulness	
	I	II	I	II
Average age of members				
27–33 years	−2.4	−3.7	−2.7	−3.1
34–39 years	−0.7	−0.2	−0.6	0.8
40+ years	2.3	7.3	2.5	3.7
Percent of members with Ph.D				
0–7%	1.1	−4.1	2.8	−2.9
8–33%	−1.4	2.9	−2.7	1.3
34+%	−0.3	6.1	−2.1	5.1
Type of setting				
Industrial	1.5	1.1	−0.1	−1.0
Government development	−2.7	−3.2	0.4	3.0
Government research	−0.5	°	−0.2	°

° No groups fell in this category.

Appendix **D**

INTERRELATIONSHIPS AMONG PERFORMANCE MEASURES

TABLE D-1 *As shown in these matrices, all performance measures were positively correlated. Highest correlations were between the two evaluations (contribution and usefulness). Adjustment I had little effect on the correlations. After background factors had been removed by Adjustment II, the correlations declined but were still positive. (Note: Data shown are Pearson correlation coefficients.)*

Upper figures = Unadjusted output scores (log transformation—see Appendix B) or evaluations (transformed—see Appendix A)
Lower figures = Scores after Adjustment II (see Appendix C)

I. Ph.D's in development labs

	Papers	Reports	Contribution
Reports	.14 .12		
Contribution	.40 .35	.16 .14	
Usefulness	.26 .19	.21 .20	.72 .70

II. Ph.D's in research labs

	Papers	Reports	Contribution
Reports	.42 .35		
Contribution	.43 .39	.33 .30	
Usefulness	.23 .15	.23 .19	.57 .55

III. Engineers

	Patents	Reports	Contribution
Reports	.19 .16		
Contribution	.35 .32	.27 .22	
Usefulness	.39 .34	.23 .17	.69 .67

IV. Assistant scientists

	Papers	Reports	Contribution
Reports	.22 .16		
Contribution	.38 .33	.26 .18	
Usefulness	.25 .12	.20 .12	.65 .59

V. Non-Ph.D scientists

	Papers	Reports	Contribution
Reports	.18 .01		
Contribution	.40 .22	.30 .11	
Usefulness	.29 .02	.26 .15	.79 .75

Appendix **E**

JOB GRADES AND CAREER LEVELS

On several occasions we wished to classify respondents according to their relative position in their laboratory's organizational hierarchy. Was the man a recent college graduate with little experience, the director of research, or somewhere in between?

Although the government labs shared a common classification scheme (the Civil Service System), the industrial and university settings each had its own system. We desired a set of categories which would have roughly the same meaning in the different settings. This appendix describes two codes we used.

Job Grades

By examining titles, job descriptions, salaries, and supervisory responsibilities, it was possible to derive a set of nine categories—"job grades"—which were reasonably comparable in the various settings. The categories and some typical characteristics are shown in Table E-1.

Career Levels

For purposes of analysis it was desirable to have a smaller number of "career levels" than the nine job grades shown in Table E-1.

It also seemed desirable to consider length of working experience. Because of market conditions, a few individuals might have been hired at excessive salaries relative to their experience. Other individuals with much experience might be making only modest salaries if in an organization with tight salary ceilings.

The "career level" code consisted of four categories derived from the job grade code described previously and the respondent's experience. All persons with the title of section head or higher (in the university, full professor or higher) was assigned to the "supervisor" category provided they had completed their education (Ph.D or B.S.) at least six years ago. Then the remaining nonsupervisory personnel were classified "senior," "junior," or "apprentice," according to the scheme shown in Table E-2. (Note that slightly different definitions were used for Ph.Ds and non-Ph.Ds. Among the latter less experience was required to occupy high positions

TABLE E-1 *System of job grades*

Job grade	Typical job description in industry or government	Title in university	Gov't. civil service level*	Typical industrial salary†	Percent who supervise‡
9	Director of research	—	Supergrade	?	100%
8	Head of major department, branch or division	Department chairman	15	$18,000	100%
7	Head of large section; senior consultant	Full professor	14	$15,000	85%
6	Head of small section; investigator of broad stature	Associate professor	13	$12,000	75%
5	Mature investigator; engineer with substantial responsibility	Assistant professor	12	$10,000	33%
4	Engineer or scientist (B.S.) with ten or more years experience OR new Ph.D	Instructor	11	$9,000	10%
3	Engineer or scientist (B.S.) with about five years experience	—	9	$8,000	5%
2	Beginning engineer or scientist, recent B.S.	—	7	$6,300	0%
1	Experienced technician or assistant	—	5	$5,200	0%

*Government data were collected in the spring of 1960.

†Industrial data were collected in the spring of 1959.

‡Shown are the percent who have one or more subordinates (varies widely in different locations).

than was the case for Ph.Ds.) Table E-2 also shows the overlap between the senior and supervisor categories with respect to grade and experience.

After respondents had been classified by career level, some typical characteristics of scientists or engineers at each level were ascertained. These data are shown in Table E-3. At the junior and senior levels, one may note the close similarity between Ph.Ds and non-Ph.Ds with respect to median age and time in organization. However, the typical non-Ph.D who attained the level of supervisor did so several years sooner than the typical Ph.D.

TABLE E-2 *Four career levels.*

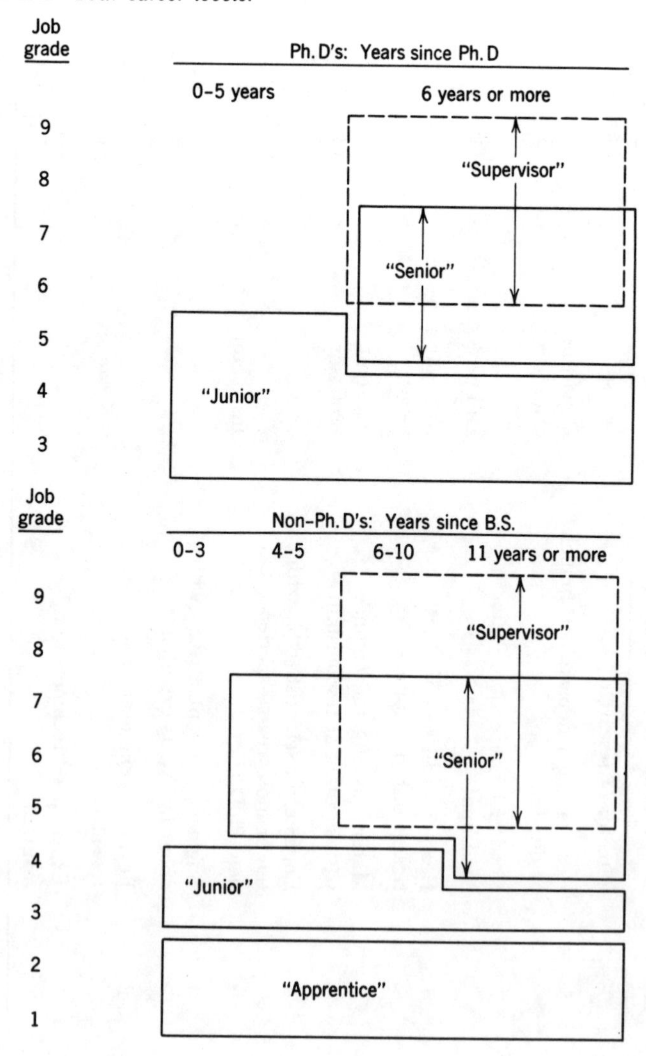

TABLE E-3 *Some typical characteristics of individuals at each career level.*

	Median age		Median years since		Median time in org'n.	
Level	Ph.D's	Non-Ph.D's	Ph.D	BS	Ph.D's	Non'Ph.D's
Supervisor	47	39	19	13	13	8
Senior	39	38	10	13	8	7
Junior	33	33	4	8	3	4
Apprentice	-	29	-	5	-	2

Appendix **F**

THE QUESTIONNAIRE

Since all questionnaire items used in generating final results have been quoted elsewhere, this appendix will not reproduce the entire questionnaire. Included here are short descriptions of the questionnaire, of the procedures used in its administration, and of a study designed to examine its reliability.

Two Questionnaire Forms

Two forms of the questionnaire were used. The long form consisted of approximately 230 items of information and required an average of 2.5 hours to complete. This was administered to a 42% cross section ($N = 552$) of the respondents. Since we were reluctant to ask all respondents to spend this much time, the other 58% ($N = 759$) completed the short form. The short form included about half of the long-form items, principally to compare motivations of the long-form sample with their colleagues. Except where questions asked for the names of people, work units, research specialties, and the like, all questions were of the closed-end (that is, fixed-alternative) type.

Questionnaire Administration

After it had been agreed that a certain organization would participate in the study, preliminary meetings were held with section and division heads. These meetings were used to inform them of the study, and answer the questions which arose. Participants were assured that the questionnaire was based on careful preliminary interviewing, and that their answers would be treated as confidential. These meetings were generally held several weeks before the questionnaire was administered.

During the following period, a memo was sent to each person selected to participate, informing him about the study and giving the time and place for questionnaire administration. Although all prospective respondents knew the study had the support of their management, no one was forced to take part. Response rates, however, were high (71 to 99%; median across 11 sites = 94%). Although a few people refused, more frequent reasons for not taking part were temporary assignment away from the site, vacation, illness, etc.

The questionnaire was administered to groups of 20 to 50 respondents at a time. Each group was given a short introduction to the study and an opportunity to ask questions. The introductory front page of the questionnaire is reproduced on page 292.

Reliability of Questionnaire Items

Any study must be concerned with the quality of measurements it uses. One way of assessing quality is to examine the extent to which one gets consistent measurements at different times.[1]

If the questionnaire was measuring stable psychological and organizational conditions, answers given at one time should have been similar to those given by the same person at another time. For example, a person who answered that he experienced considerable autonomy at one time should have answered in roughly the same way at a later time. This repeatability or consistency of results is technically termed the "reliability" of the measuring instrument.

An opportunity for examining the reliability of the questionnaire arose when an industrial R & D department participated. The management agreed to let us ask a sample of the respondents to answer some of the same questions a second time two months later. Of the sample selected, 96% ($N = 52$) cooperated.

Each person was given a blank copy of the original questionnaire. Approximately every third item was selected for answering again (94 items in all). Respondents were instructed: "Do not try to remember how you answered the questions before. Simply answer them as you would if you were seeing them for the first time."

In general, answers were satisfactorily stable over the two-month interval.

One way of examining the stability was to compute the average response for this group of 52 respondents on each item at each administration.[2] The two sets of averages correlated .97, indicating that the relative standing of the group on these items at the first administration was an almost perfect predictor of the relative standing at the second administration. Since our analyses involved comparisons among groups rather than individuals, an examination of the stability of group means seemed an appropriate way to assess stability.

Another way of examining stability was to compare the way each indi-

[1]Appendices A and B describe other checks on the quality of measurements used in this study.

[2]Of the 94 items answered a second time, 89 had response categories consisting of a 5- or 7-point scale where it was appropriate to compute a mean response for the group. The other five items were not included in this analysis.

UNIVERSITY OF MICHIGAN
INSTITUTE FOR SOCIAL RESEARCH
ANN ARBOR MICHIGAN
RENSIS LIKERT, DIRECTOR

SURVEY RESEARCH CENTER
ANGUS CAMPBELL, DIRECTOR
RESEARCH CENTER FOR GROUP DYNAMICS
DORWIN CARTWRIGHT, DIRECTOR

<u>Short Form</u>

MOTIVATIONS AND WORKING RELATIONS
OF SCIENTISTS AND ENGINEERS

To respondents:

The attached questionnaire is part of a continuing series of studies on scientific and engineering personnel, which we have carried on at the Institute since 1952. The present questions are taken from a longer form that was developed on the basis of intensive personal interviews with technically trained people.

We are attempting to obtain reliable quantitative data on many factors -- both within the individual and in his working environment -- which may serve to promote or inhibit technical performance.

We hope that these studies may help to find out the kind of working situations in which scientists and engineers can do their best work.

<u>Confidentiality</u>

As with all of the studies at the Institute, answers of individual respondents are kept in strict confidence. For purposes of reporting, individuals are grouped together in meaningful categories. No names of units or organizations will be used in published reports.

<u>Comments</u>

It is difficult to design questionnaire items which can capture the details of all possible situations. We therefore invite your comments and qualifications as you go along, either in the margin or on the back cover.

We are very grateful for your assistance.

Sincerely,

Donald C. Pelz

Donald C. Pelz
Project Director

Form 202

vidual answered the items at the two administrations. For each of 89 items,[3] we correlated answers given at the first administration with those given at the second. The median correlation (reliability coefficient) was .62; 83% of the correlations were .5 or higher.

We were not surprised to find reliability higher when we looked at

[3]Five items were omitted because their response categories were inappropriate for this analysis.

group means than when we looked at individual scores. The group means changed over time only when individual shifts were noncompensating, that is, if one member of the group increased and another decreased an equal amount, their shifts would cancel each other. When we looked at individual scores, however, all shifts acted to lower the reliability coefficient. Nevertheless, a reliability of about .6 for individual scores was deemed an acceptable degree of stability.[4]

[4]Complete details on this reliability study appear in D. C. Pelz, "Reliability of Selected Questionnaire Items," Preliminary Report #9, August 1962, available as Publication 1991R from the Survey Research Center, University of Michigan.

Appendix G

CONSTRUCTION OF INDICES

At numerous points in the analysis, sets of questionnaire items were combined to provide measures of stimulation from designated sources, preferred styles of working, desirability of certain goals, etc. This appendix provides information about the construction of these composite measures.

Technically, composite measures of the type used here are known as indices. In some circumstances an index has several advantages over a measurement based on just a single questionnaire item. An index provides a way of measuring a concept which may be too broad to be tapped fully by a single question. A related advantage is that an index emphasizes those aspects which two (or more) items have in common. In addition, an index may provide a better measure than a single item because measurement errors associated with one item may be reduced or cancelled out by other items.

An example will help to clarify the following discussion. Consider the index called "stimulation from own ideas." This index was used in Chapters 2, 6, 9, 12, and 13. It consists of an additive combination of three questionnaire items: 13E + 13J + 62L. These items were, respectively, stimulus received from "my own previous work or plans, stimulus received from "my own curiosity," and importance attached to having "freedom to carry out my own ideas." Here were three different items all of which seemed to measure the importance of inner sources of motivation. It was discovered that scientists who strongly endorsed one item also tended to endorse the other items (that is, the items were positively correlated—correlations are shown in Table G-1). Thus it seemed reasonable to combine these items into a single index.

Two criteria had to be met before items were combined into an index: (a) the content of the items had to be similar, and (b) there had to be a correlation between them which was reasonably consistent across the several major groups of scientists. Sometimes not only two, but three or more items were found to cluster together in these ways and were combined into a single index.

Items were combined so that each item was about equally weighted in the composite score. This was achieved by recoding items, where necessary, so that each had about the same range or variation. The actual com-

TABLE G-1 *This shows intercorrelations (Pearson r's) among items which were combined to form the more important indices described in Chapters 6, 7, 10, 12, and 13. Intercorrelations are shown separately for each of the five groups of scientists.*

Name of index and items included*	Pair correlated	Correlations				
		Ph.D's, devel.	Ph.D's, res.	Engineers	Ass't. scients.	Non-Ph.D's, res.
Stimulation from	13E & 13J	.4	.4	.2	.5	.3
own ideas	13E & 62L	.2	.2	.2	.2	.3
13E + J + 62L	13J & 62L	.3	.3	.2	.3	.4
Desire for self-	62A & 62B	.3	.4	.3	.4	.3
actualization	62A & 62J	.3	.3	.3	.3	.4
62A + B + J + L	62A & 62L	.3	.3	.2	.3	.1
	62B & 62J	.4	.3	.4	.4	.4
Professional orien-	62B & 62L	.3	.2	.3	.4	.2
tation	62J & 62L	.4	.4	.4	.5	.2
62A + B + I + J + L	62A & 62I	.3	.2	.1	.3	.2
+ M	62A & 62M	.4	.4	.2	.3	.2
	62B & 62I	.4	.2	.2	.3	.1
Science orientation	62B & 62M	.3	.3	.3	.3	.3
62A + L + M	62I & 62J	.4	.3	.3	.3	.1
	62I & 62L	.2	.2	.2	.3	.3
	62I & 62M	.4	.3	.4	.5	.4
	62J & 62M	.5	.5	.4	.5	.4
	62L & 62M	.5	.4	.5	.5	.3
Provision for self-	63A & 63B	.6	.5	.5	.7	.6
actualization	63A & 63J	.6	.7	.5	.4	.6
63A + B + J + L	63A & 63L	.4	.6	.4	.3	.6
	63B & 63J	.6	.5	.5	.4	.5
	63B & 63L	.5	.4	.5	.2	.6
	63J & 63L	.4	.6	.5	.4	.7
Desire for advance-	62C & 62D	.3	.2	.4	.4	.4
ment in status	62C & 62H	.2	.2	.3	.2	.4
62C + D + H	62D & 62H	.6	.4	.5	.6	.6
	19B & 19C	.6	.6	.6	.6	.7
Status orientation	19B & 62C	.2	.3	.2	.2	.3
19B + C	19B & 62D	.2	.2	.3	.3	.3
+ 62C + D + H	19B & 62H	.1	.2	.3	.2	.3
	19C & 62C	.2	.3	.3	.3	.4
	19C & 62D	.5	.3	.5	.5	.5
	19C & 62H	.3	.2	.4	.4	.3
Provision for advance-	63C & 63D	.5	.1	.5	.2	.2
ment in status	63C & 63H	.3	.3	.2	.1	.4
63C + D + H	63D & 63H	.6	.4	.5	.5	.3

*Chapter 6 gives the exact wording for each of the items indicated here.

bining was done separately for each respondent, of course, and was achieved by adding together his (recoded) scores on each item.

Items were combined into the indices described in this book only after many thousands of correlations had been examined. In some instances, it was not clear whether a given item should be included in a certain index. In these instances, we sometimes scored several different versions of the same index and waited to see which worked best. (At one point in Table G-1 the reader will see three versions of roughly the same index. The one we made heaviest use of was called "desire for self-actualization;" related versions were "professional orientation," and "science orientation.")

The table which follows shows the correlations between each pair of items composing the more important indices used in this book. Results are shown separately for each of the five major groups of scientists. The items themselves all appear in Chapter 6.

Appendix H

CLASSIFICATION OF DEPARTMENTS

Scientists from a wide range of laboratories participated in this study. For purposes of analysis, it seemed desirable to classify the type of environment in which they worked so that those experiencing different conditions could be examined separately.

In addition to the distinction between university, industry, and government, two other classifications proved useful: research versus development, and coordinated versus autonomous. This appendix describes how the latter two classifications were determined.

Examination of Whole Departments

In attempting to classify environments, it seemed important to concentrate on perceptions which were shared among members of an organizational unit. The unit which seemed appropriate consisted of two or more work groups (for example, sections) reporting to one chief.[1] The name of this unit could be department, laboratory, branch, or sometimes, division. We arbitrarily called all these units "departments," and identified 53 of them.

In most of these departments, between ten and 45 professional members answered the questionnaire (median = 22). Industry and government departments generally contained two levels of supervision. Often the head of the department reported directly to the director of research.

To arrive at department scores, answers from department members were averaged together. Thus, although a department might receive one classification (for example, highly coordinated), it was possible that a few individuals in that department might experience considerable autonomy. This was as intended, for we wished to measure characteristics of the immediate working environment rather than the individual's own job. Accordingly, we concentrated on what most people perceived as management's goals, values, and methods of operation.

[1]Occasionally two or three small groups similar in function but not under the same supervisor were combined as one unit.

Research versus Development

There has been much use of the terms "basic research" and "applied research." Although no consensus has emerged on the meaning of these terms, they remain an intuitive basis for distinguishing some laboratories from others. Clearly there is a difference between the laboratory devoted to writing scholarly articles and the one dedicated to improving commercial products.

About 30 items in the questionnaire seemed as if they might be relevant to the basic-applied dimension. For each of these a department score was computed by averaging members' responses. Then, with each department treated as a single "case," intercorrelations among the scores were obtained. Two parallel analyses were done: one based on 33 industrial and university departments; the other based on 20 government departments.

Six items clustered together strongly in both the university-industry and government analyses. These items are shown in the following box. Their intercorrelations appear in Table H-1. The reader will note that these six all reflect the distinction between management's emphasis on useful products or processes and discovering general knowledge.[2] Thus we have called it the research versus development dimension.[3]

Question 15. In your estimation, how important do executives in this technical division regard each kind of activity as an objective for the division? [For each item respondent checked a five-point rating scale of importance.]

A. Discovering general knowledge relevant to a broad class of problems
C. Improvement of existing products or processes
D. Invention of new products or processes

Question 56. To what extent do you feel that each experience (if it occurred) would help you to *get ahead in your technical organization?* [Respondent checked a five-point scale on helping to get ahead.]

A. Contributing to a product with high commercial success
B. Contributing to a product of distinctly superior quality
D. Publishing a paper which adds significantly to the technical literature

[2]Included in the 30 items examined were other distinctions such as likelihood of short-run usefulness, source of activity (own curiosity versus assignments from supervisor), and length of time for which support was assured. Items tapping these dimensions proved not to cluster together as well as the six which were selected. Full details appear in F. M. Andrews, and D. C. Pelz, "Dimensions of Organizational Atmosphere," Analysis Memo #7, March 1961, available in Publication #1825 from the Survey Research Center, University of Michigan.

[3]Descriptions of the scientist's own work, that is, amount of time spent on research or development, also tended to correlate with the six selected items. However, since the index was intended to measure perceptions of *management's* goals, descriptions of particular jobs were not included in it.

TABLE H-1 *This shows intercorrelations (Pearson's r's) among department means for six items used to characterize departments on the research versus development dimension. (Upper figure based on 33 industrial and university departments; lower figure on 20 government departments.)*

	15C	15D	56A	56B	56D
15A. Discovering general knowledge —stressed by executives	−.7 −.3	−.7 −.4	−.8 −.3	−.7 −.4	.8 .6
15C. Improvement of products —stressed by executives		.8 .8	.8 .2	.8 .5	−.7 −.3
15D. Invention by new products —stressed by executives			.9 .6	.8 .7	−.7 −.3
56A. Commercially successful product —helps one get ahead				.9 .7	−.8 −.3
56B. Superior quality product —helps one get ahead					−.8 −.3
56D. Publishing paper —helps one get ahead					

The sum of the (averaged) scores on these items—with the signs of 15A and 56D reversed—constituted the department score on the research-development dimension.

Table H-2 shows how departments in various locations scored on this index. One may note that university departments tended to be heavily research oriented, those in industry heavily development oriented, and those in government midway between university and industry. (Since these data do not constitute a representative sample of scientists, Table H-2 is not intended to describe anything other than the particular departments included in the study.)

TABLE H-2 *This shows how the 53 departments represented in this study scored on the research versus development dimension. (Figures indicate number of departments.)*

	University	Government	Industry
Strong research emphasis	6	0	0
Mild research emphasis	1	7	0
Mild development emphasis	0	10	8
Strong development emphasis	0	3	18

Autonomy versus Coordination

To what extent was coordination of effort expected in each department? In some departments, investigators worked independently; in others, tight groups worked within a highly structured organizational pyramid.

Fifteen items had been included in the questionnaire which promised to tap various aspects of autonomy versus coordination. Before obtaining department averages, however, we faced the possibility that perceptions about autonomy and coordination might differ at different levels within the organization. A chief, for example, might feel free to determine his own goals, but might impose direction on his subordinates. Therefore, two scores were obtained for each item: the department average for supervisors, and the department average for nonsupervisors. (The latter, of course, might supervise some technicians.)

These scores were then intercorrelated across 48 industry and government departments.[4] (University departments were omitted since many of the questions on supervision were not relevant for university scientists.)

Two clusters were observed in the data—one for supervisors, a different one for nonsupervisors. Items in the nonsupervisory cluster concerned the amount of autonomy or self-direction which the scientist felt he had over his own work goals. These items are shown in the following box; intercorrelations appear in Table H-3.

Nonsupervisors' Answers to the Following Items Clustered:

Question 26C. To what extent does [your own group or work] have autonomy to determine its own technical program, in contrast to having work suggested or assigned by higher authorities, clients, etc.? [Respondent checked a five-point scale indicating degree of autonomy.]

Question 29. Consider the choice of *goals or objectives* of the various technical activities for which you are responsible (either your own work, or work which you supervise or coordinate). Who has weight in deciding on these goals and objectives? [Respondent estimated the relative percent of weight exerted by eight sources, including:]

A. Myself
E. Higher-level technical supervisors in this organization

Question 31. To what extent do you feel you can influence [the person or group with most weight in choice of your work goals] in his recommendations or decisions concerning your technical goals? [Respondent checked a six-point scale of influence.]

[4] Three small government departments were subsequently merged together.

TABLE H-3 *This shows intercorrelations among department means for four items used in a preliminary index of nonsupervisors' estimates of intragroup autonomy. (Based on data from nonsupervisors in 48 industrial and government departments.**)

	29A	29E	31
26C. Autonomy of own group (or self) to set program	.5	−.6	.6
29A. Weight of self in deciding goals		−.5	.7
29E. Weight of higher technical supervisors in deciding goals			−.6
31. Influence on person having most weight in setting goals			

*Three small government departments were subsequently merged together.

The cluster for supervisors included items concerned with the amount of coordination between the respondent and heads of other groups. These items are shown in the lower part of the following box; all intercorrelations were .5 (data not shown).[5]

Supervisors' Answers to the Following Items Clustered:

Question 25. To what extent do members of these groups coordinate their efforts for some common objective? [Respondent checked a five-point scale indicating degree of coordination.]

A. Most significant group*
B. Group headed by my chief

Question 39. To what extent do you and each colleague coordinate your activities toward some common objective? [Respondent checked a five-point scale indicating degree of coordination with each of five colleagues. Answers were subsequently averaged across colleagues to produce a variable indicating mean coordination with colleagues.]

*If the respondent's "most significant group" was headed by his chief or chief's chief, we substituted his answer to Question 25C: coordination in group headed by the respondent himself.

Two preliminary indices were formed by adding the departmental scores for each cluster (the sign of 29E was reversed). We thought it might be worthwhile to distinguish different combinations of these characteristics. However, the two indices correlated −.62, indicating that departments with high intergroup coordination tended to have low intragroup autonomy. It therefore seemed best to combine the two indices into a single index of coordination. The following formula was used, which resulted in weight-

[5]The document described in Footnote 2 gives full details of our explorations in this area.

ing the two preliminary indices approximately equally in the final index:

$$C = 2S - N$$

where: C = department score on coordination versus autonomy dimension.
S = preliminary index of supervisors' estimates of intergroup coordination.
N = preliminary index of nonsupervisors' estimates of intragroup autonomy.

Table H-4 shows that our industrial departments tended to be mildly or strongly coordinated, university departments autonomous, and government departments midway between industry and university.

Relationships between Research-Development and Autonomy-Coordination Dimensions

A final analysis examined the interrelationships between the two indices described in this appendix. Within industrial departments, and also within government departments, the two indices seemed to be measuring rather different things. Across the industrial departments, the correlation between the two indices was −.10; the comparable figure for government was +.25.[6]

TABLE H-4 *This shows how the departments represented in this study scored on the autonomy versus coordination dimension. (Figures indicate number of departments.)*

	University	Government	Industry
Autonomous	7	3	2
Mild coordination	0	11	12
Strong coordination	0	6	12

[6]In this analysis, each department was weighted by its size.

Appendix I

SOURCES OF ADDITIONAL INFORMATION

This appendix lists references which people in search of additional information may find helpful. The appendix has been organized into two sections. Section A lists some general works, including bibliographies, relevant to research or R & D and/or creativity. Section B lists documents cited in this book.

A. SOME GENERAL WORKS RELEVANT TO RESEARCH ON R & D AND/OR CREATIVITY

Listed here are some general works relevant to research on R & D and/or creativity. Inquiries should be directed to the respective authors or publishers.

Barber, B. and Hirsch, W. (eds.), *The Sociology of Science,* The Free Press, Glencoe, Illinois, 1962. Some of the 38 selected readings deal with relations between science and society, communication among scientists, the social image of the scientist, and the social process of scientific discovery.

Bush, G. P., *Bibliography on Research Administration, Annotated,* The University Press of Washington, D.C., 1954, 146 pp.

Conferences: Since 1950 there have probably been at least three or four conferences a year on research administration and creativity, with proceedings published. One series, the Annual Conference on Administration of Research, has been held since 1947 at several universities, principally Pennsylvania State University. Other series have been held at Columbia University, the American University in Washington D.C., and the University of Buffalo. Other sponsors of conferences include the Industrial Research Institute, New York City; the Institute of Contemporary Art, Boston; and Foundation for Research on Human Behavior, Ann Arbor, Michigan. See also the reference to Taylor and Barron in this section.

Creativity Research Exchange, Educational Testing Service, Princeton, New Jersey. Periodic summaries of current studies.

Deutsch and Shea, Inc., "Creativity—A Comprehensive Bibliography on Creativity in Science, Engineering, Business, and the Arts," *Industrial Relations Newsletter,* March 1958, 16 pp.

Evans, Jill, *Factors Related to the Performance of Scientists and Engineers: Synopses of Research Findings,* Center for Research on Utilization of Scientific Knowledge, Institute for Social Research, University of Michigan, Ann Arbor, Michigan, (in press), 228 pp. Contains synopses of research findings and bibliographies of studies on R & D.

Foundation for Research on Human Behavior, *Human Factors in Research Administration,* University of Michigan, Ann Arbor, Michigan, 1956. Proceedings of a conference with this title sponsored by the Foundation.

———, *Creativity and Conformity,* Ann Arbor, Michigan, 1958. Proceedings of a conference with this title sponsored by the Foundation.

Guilford, J. P., Christensen, P. R., and Wilson, R. C., *A Bibliography of Thinking, Including Creative Thinking, Reasoning, Evaluation, and Planning,* the Department of Psychology, University of Southern California, Los Angeles, July 1953.

McPherson, J. H. and Rapucci, L. C., *The Creativity Review,* The Dow Chemical Company, P.O. Box 632, Midland, Michigan. Quarterly summaries of current studies since 1955.

National Science Foundation, *Current Projects on Economic and Social Implications of Science and Technology,* U. S. Government Printing Office, Washington D.C. Yearly summary.

Parnes, S. J. (ed.), *Compendium of Research on Creative Imagination,* The Creative Education Foundation, Buffalo, New York, 1958, 8 pp.

———, *Compendium #2 of Research on Creative Imagination,* The Creative Education Foundation, Buffalo, New York, 1960, 11 pp.

———, *Bibliography on the Nature and Nurture of Creative Behavior,* The Creative Education Foundation, State University of New York at Buffalo, 1964, 39 pp. Lists major books, journal articles, and doctoral theses relevant to creative behavior appearing between 1960 and mid-1964. See also previous two references.

——— and Harding, H. (eds.), *A Source Book for Creative Thinking,* Charles Scribner and Sons, New York, 1962. Contains 29 articles, annotated compendium of 60 other studies, list of 27 institutions with investigators currently doing research, and a central list of books on creativity.

Rubenstein, A. H., *A Directory of Research on Research,* College on Research and Development of the Institute of Management Sciences, Northwestern University, Evanston, Illinois, May 1964. The Primary emphasis in the directory is on studies directly related to the management of R & D. Studies in related areas are included, such as scientific information and economics of innovation.

Stein, M. I. and Heinze, Shirley J., *Creativity and the Individual,* The Free Press, Glencoe, Illinois, 1960. Summary of more than 300 articles and books from psychology, psychiatry, and various other disciplines, bearing on creativity.

Taylor, C. W. and Barron, F. (eds.), *Scientific Creativity: Its Recognition and Development,* John Wiley and Sons, New York, 1963. Selected papers from a series of conferences on the Identification of Creative Scientific Talent sponsored by the University of Utah in 1955, 1957, and 1959.

Weislogel, Mary H. and Altman, J. W., *Abstracts of Literature Concerning Scientific Manpower,* American Institute for Research, Pittsburgh, 1952, 85 pp.

B. DOCUMENTS CITED IN THIS BOOK

Listed below are the documents cited in this book, together with the chapter in which the citation occurs. The initials SRC indicate that the document may be obtained from the Publications Department, Survey Research Center, University of Michigan, Ann Arbor, Michigan. Inquiries about all other publications should be addressed to the respective authors or publishers.

Adams, J. S., "Toward an Understanding of Inequality," *Journal of Abnormal and Social Psychology,"* 1963, vol. 67, pp. 422–36. (Chapter 7)

Aitchison, J. and Brown, J. A. C., *The Lognormal Distribution,* Cambridge University Press, Cambridge, 1957. (Appendix B)

Allison, D., "Engineer Renewal," *International Science and Technology,* June 1964, pp. 48–54, 109–110. (Chapter 10)

Andrews, F. M., "An Exploration of Scientists' Motives," Analysis Memo #8, 1961, SRC Publication #1825, $.75. (Chapters 6, 11)

———, "Logarithmic Transformation of Output of Scientific Products," Analysis Memo #11, 1961, SRC Publication #1826, $.75. (Appendix B)

———, *Creativity and the Scientist,* doctoral dissertation, University of Michigan, 1962. (Chapter 9)

———, "A Check on the Validity of Reporting Scientific Output," Analysis Memo #19, 1963, SRC Publication #2134, $.35. (Appendix B)

———, "Scientific Performance as Related to Time Spent on Technical Work, Teaching, or Administration," *Administrative Science Quarterly,* 1964, vol. 9, pp. 182–193; SRC Publication #2132, $.50. (Chapter 4)

———, "Factors Affecting the Manifestation of Creative Ability by Scientists," *Journal of Personality,* 1965, vol. 33, no. 1, pp. 140–152. (Chapter 9)

———, "Revised Adjustment of Performance Measures to Hold Constant Some Background Factors," Analysis Memo #21, 1966, SRC Publication #2400, $.50. (Appendix C)

———, Sonquist, J. A. and Morgan, J. N., *The Multiple Classification Analysis Program,* 1966, SRC Monograph (Appendix C), (in press)

——— and Pelz, D.C., "Dimensions of Organizational Atmosphere," Analysis Memo #7, 1961, SRC Publication #1825, $.75. (Chapter 1, Appendix H)

Argyris, C., *Personality and Organization,* Harper, New York, 1957. (Chapter 7)

Barron, F., *Creativity and Psychological Health,* Van Nostrand, Princeton, New Jersey, 1963. (Chapter 9)

———, "The Needs for Order and Disorder as Motives in Creative Activity," in Taylor, C. W. and Barron, F. (eds.) *Scientific Creativity: Its Recognition and Development,* John Wiley and Sons, New York, 1963. (Chapter 9)

Bayley, Nancy, "On the Growth of Intelligence," *American Psychologist,* 1955, vol. 10, pp. 805–818. (Chapter 10)

Brehm, J. W. and Cohen, A. R., *Explorations in Dissonance Theory,* John Wiley and Sons, New York, 1962. (Chapter 7)

Burrey, S., "The Question of Creativity," *Industrial Design,* January 1957, p. 32. (Chapter 8)

Davis, R. C., *Commitment to Professional Values as Related to the Role Performance of Research Scientists,* doctoral dissertation, University of Michigan, 1956. (Chapter 6)

Evan, W. M., "Conflict and Performance in R & D Organizations: Some Preliminary Findings," *Industrial Management Review,* 1965, vol 7, no. 1, pp. 37–46. (Chapters 8, 13)

Farris, G. F., "Congruency of Scientists' Motives with their Organizations' Provisions for Satisfying Them: Its Relationships to Motivation, Affective Job Experiences, Style of Work, and Performance," Department of Psychology, University of Michigan, 1962. (Chapter 7)

Fiedler, F. E., *Leader Attitudes and Group Effectiveness,* University of Illinois Press, Urbana, 1958. (Chapter 13)

Ford, L. R., Jr., "Solution of a Ranking Problem from Binary Comparisons," *American Mathematics Monthly,* 1957, vol. 64, no. 8, part II, pp. 28–33. (Appendix A)

Gerard, R. W., *Mirror to Physiology: A Self Survey of Physiological Science,* American Physiological Society, Washington, D.C., 1958. (Chapter 4)

Ghiselli, E. E., "The Validity of Management Traits in Relation to Occupational Level," *Personnel Psychology,* 1963, vol. 16, pp. 109–113. (Chapter 12)

Glazer, B. G., "The Local-Cosmopolitan Scientist," *American Journal of Sociology,* 1963, vol. 69, pp. 249–259. (Chapter 6)

Gordon, G. and Marquis, Sue, "Effect of Differing Administrative Authority on Scientific Innovation," Working Paper #4, 1963, from project on Organizational Setting and Scientific Accomplishment, Graduate School of Business, University of Chicago. (Chapter 2)

Guilford, J. P., "Three Faces of Intellect," *American Psychologist,* 1959, vol. 14, pp. 469–479. (Chapters 10, 11)

Houston, J. P. and Mednick, S. A., "Creativity and the Need for Novelty," *Journal of Abnormal and Social Psychology,* 1963, vol. 66, pp. 137–141. (Chapter 9)

Kornhauser, W., *Scientists in Industry: Conflict and Accommodation,* University of California Press, Berkeley and Los Angeles, 1962. (Chapter 6)

Kuhn, T. S., "The Essential Tension: Tradition and Innovation in Scientific Research," in Taylor, C. W. and Barron, F. (eds.), *Scientific Creativity: Its Recognition and Development,* John Wiley and Sons, New York, 1963. (Chapter 11)

Lehman, H. C., *Age and Achievement,* Princeton University Press, Princeton, New Jersey, 1953. (Chapter 10)

———, "The Chemist's Most Creative Years," *Science,* 1958, vol. 127, pp. 1213–1222. (Chapter 10)

———, "The Age Decrement in Scientific Creativity," *American Psychologist,* 1960, vol. 15, pp. 128–134 (Chapter 10)

Likert, R., *New Patterns of Management,* McGraw-Hill Book Co., New York, 1961. (Chapters 2, 12)

Longenecker, E. D., "Perceptual Recognition as a Function of Anxiety, Motivation, and the Testing Situation," *Journal of Abnormal and Social Psychology,* 1962, vol. 64, pp. 215–221. (Chapter 13)

Mednick, S. A., "The Associative Basis of the Creative Process," *Psychological Review,* 1962, vol. 69, pp. 220–232 (Chapter 9)

——— and Mednick, M. T., *Manual: Remote Associates Test, Form I,* Houghton Mifflin, Boston, 1966. (Chapter 9)

Meltzer, L. and Salter, J., "Organization Structure and Performance and Job Satisfaction of Scientists," *American Sociological Review,* 1962, vol. 27, pp. 351–362. (Chapter 4)

Morton, J. A., "From Research to Technology," *International Science and Technology,* May 1964, pp. 82–92. (Chapter 2)

Oberg, W., "Age and Achievement in the Technical Man," *Personnel Psychology,* 1960, vol. 13, pp. 245–259. (Chapter 10)

Patchen, M., *The Choice of Wage Comparisons,* Prentice-Hall, Englewood Cliffs, New Jersey, 1961. (Chapter 7)

———, "A Conceptual Framework and Some Empirical Data Regarding Comparisons of Social Rewards," *Sociometry,* 1961, vol. 24, pp. 136–156. (Chapter 7)

Pelz, D. C., "Some Social Factors Related to Performance in a Research Organization," *Administrative Science Quarterly,* 1956, vol. 1, pp. 310–325. (Preface)

———, "Motivation of the Engineering and Research Specialist," *Improving Managerial Performance, AMA General Management Series, No. 186,* 1957, pp. 25–46; SRC Publication #1213, $.50. (Preface, Chapters 8, 10)

———, "Uncertainty and Anxiety in Scientific Performance," 1960, SRC Publication #1588, $1. (Chapters 11, 13)

———, "Some Properties of the Measures of Scientific Output," Analysis Memo #1, 1960, SRC Publication #1741, $.75. (Appendix B)

———, "Intensity of Work Motivation as Related to Output," Analysis Memo #2, 1960, SRC Publication #1741, $.75. (Chapter 5)

———, "Congruence between Personal and Organizational Values, as Related to Output," Analysis Memo #4, 1960, SRC Publication #1741, $.75. (Chapter 7)

———, "Satisfaction with the Work Situation as Related to Output," Analysis Memo #5, 1960, SRC Publication #1741, $.75. (Chapter 7)

———, "Self-Determination and Self-Motivation in Relation to Performance," (prepublication draft), 1962, Survey Research Center, University of Michigan. (Chapter 12)

———, "Self-estimates of Motivation Strength as Indicators of Scientific Performance," Analysis Memo #17, 1962, SRC Publication #1922, $.50. (Chapter 5)

———, "Reliability of Selected Questionnaire Items," Preliminary Report #9, 1962, SRC Publication #1991R, $1. (Chapter 1, Appendices B, F)

———, "Time and Influence Factors in Laboratory Management as Related to Performance," Analysis Memo #18, 1962, SRC Publication #1993, $.75. (Chapter 2)

———, "Relationships between Measures of Scientific Performance and Other Variables," in Taylor, C. W. and Barron, F. (eds.), *Scientific Creativity: Its Recognition and Development,* John Wiley and Sons, New York, 1963, pp. 302–310. (Preface)

———, "Freedom in Research," *International Science and Technology,* February 1964, pp. 54–66. (Chapter 2)

———, "The 'Creative Years' and the Research Environment," *IEEE Transactions in Engineering Management,* 1964, vol. EM-11, no. 1, pp. 23–29. (Chapter 10)

——— and Andrews, F. M., "How Motives Relate to Three Kinds of Output in Various Types of Laboratories," Analysis Memo #10, 1961, SRC Publication #1826, $.75. (Chapter 1)

——— and ———, "Organizational Atmosphere as Related to Types of Motives and Levels of Output," Analysis Memo #9, 1961, SRC Publication #1826, $.75. (Chapter 1)

——— and ———, "Organizational Atmosphere, Motivation, and Research Contribution," *American Behavioral Scientist,* 1962, vol. 6, no. 4, pp. 43–47; SRC Publication #1944, $.35. (Chapter 1)

——— and ———, "Diversity in Research," *International Science and Technology,* July 1964, pp. 21–36; SRC Publication #2213, $.35. (Chapter 4)

——— and ———, "Autonomy, Coordination, and Stimulation in Relation to Scientific Achievement," *Behavioral Science,* 1966, vol. 11, pp. 89–97, $.35. (Chapter 12)

Roe, Anne, *The Making of a Scientist,* Dodd Mead, New York, 1953. (Chapter 9)

Selye, H., *From Dream to Discovery: On Being a Scientist,* McGraw-Hill Book Co., New York, 1964. (Chapter 11)

Shepard, H. A., "Creativity in R & D Teams," *Research and Engineering,* October 1956, pp. 10–13. (Chapter 13)

Stein, M. I., "Creativity and the Scientist," in Barber, B. and Hirsch, W. (eds.), *The Sociology of Science,* The Free Press of Glencoe, New York, 1962, pp. 329–343. (Chapter 7)

——— and Rodgers, R., "Creativity and/or Success?" paper delivered at 1957 Convention of the American Psychological Association. (Chapter 7)

Tannenbaum, A. S., "Control in Organizations: Individual Adjustment and Organizational Performance," *Administrative Science Quarterly,* 1962, vol. 7, pp. 236–257. (Chapters 2, 12)

Taylor, C. W., "A Tentative Description of the Creative Individual," in Parnes, S. J. and Harding, H. F., *A Source Book for Creative Thinking,* Charles Scribners and Sons, New York, 1962. (Chapter 9)

Torrance, E. P., "Some Consequences of Power Differences on Decision Making in Permanent and Temporary Three Man Groups," in Hare, A. P., Borgatta, E. F., and Bales, R. F. (eds.), *Small Groups,* Knopf, New York, 1955, pp. 482–492. (Chapter 13)

Weaver, W., "Dither," *Science,* 7 August 1959, vol. 130, p. 301. (Chapter 8)

Wells, W. P., *Group Age and Scientific Performance,* doctoral dissertation, University of Michigan, 1962. (Chapter 13)

Yates, F., "The Analysis of Multiple Classifications with Unequal Numbers in the Different Classes," *Journal of the American Statistical Association,* 1934, vol. 29, pp. 51–66. (Appendix C)

INDEX

Where "performance" appears in this index, it refers to performance of individuals unless otherwise noted. The word "and" designates a relationship between two measures—e.g., "autonomy and performance." The word "by" indicates that a relationship between two measures is examined separately for subgroups on some third measure—e.g., "autonomy and performance, by coordination." The third measure is also indexed, either: "coordination—effect on relationship between performance and autonomy," or (more tersely) "coordination—mediating performance and autonomy." This index does not cover references in Appendix I, nor does it include mediating effects of primary analysis groups, which are scattered throughout.

PART TWO

PROBLEM SOLVERS VS. DECISION MAKERS

Donald C. Pelz

MOST thoughtful managers of R&D laboratories have solid convictions about the goals of the organization they run, but they are less certain about how to motivate their technical men toward these goals: If you insist that your staff tackle only the problems you see as essential to the organization, you may squelch their enthusiasm—maybe even lose them to competitors; but if you allow them undue leeway to pursue their own leads, they may go off on irrelevant tangents.

I'd like to put forward some ideas on how something as seemingly simple as the climate of communication among researchers can motivate them to undertake the right problems and solve them effectively.

My thinking is partly data-based, partly buttressed by the ideas of others on the nature of creative people, and partly the result of speculation based on my own experience with and observation of such men. I'm hoping that you'll react to these ideas, for the ultimate test of their validity is how they sit with men who manage R&D day in and day out. To this end, there's a feedback mechanism discussed at the conclusion of my remarks. I hope you will take advantage of it.

Inasmuch as the argument is a bit complex, I'd like to preface it with a few words about the structure of what is to follow.

First, I shall review two broad concepts labeled "security" and "challenge," which have proved useful to me in understanding a number

Reprinted by permission of the publisher from *Managing Advancing Technology*, Vol. 2, © 1972 by American Management Association, Inc. This article originally appeared as "Problem Solvers Versus Decision Makers" in *Innovation*, January 1970, © 1970 by Technology Communication, Inc., and has subsequently been included in *Organizing the Organization for Better R & D, An AMA Management Briefing*, ©1975 by AMACOM, a Division of American Management Associations.

of seemingly contradictory findings from a study that my colleague, Frank Andrews, and I did a few years ago on eleven R&D organizations. Second, I would like to compare these concepts with some that other writers have used—particularly the notion of communication "bonds" and "barriers." In the third section I shall explore further the question of how both challenge and security, as well as communication bonds and barriers, are linked in the problem-solving process.

Moreover, I shall put forward some principles by which working managers might shape the communication climate in their laboratories so as to affect their basic components and thus get their researchers to become better problem solvers.

LET me begin with the matter of autonomy among researchers. More than three decades ago the distinguished head of research at Eastman Kodak, C. E. K. Mees, wrote: "The best person to decide what research work shall be done is the man who is doing the research, and the next best person is the head of the department, who knows all about the subject and the work; after that you leave the field of the best people and start on increasingly worse groups, the first of these being the research director, who is probably wrong more than half the time; then a committee, which is wrong most of the time; and, finally, a committee of vice-presidents, which is wrong all the time."

This view is widely shared among scientists. Autonomy or self-direction is felt to be an essential condition for scientific achievement—perhaps the most essential one. A close second, of course, is resources. When my colleague and I began our research on scientific organizations, with the National Institutes of Health being the first, we asked laboratory directors how they managed research. They were likely to answer: "All you can do is find a good man, give him the facilities he needs, and leave him alone."

Hence when we undertook the comparative study of eleven industrial, government, and university laboratories, we asked several questions about freedom and autonomy. In one question, for example, we asked the individual researcher to estimate how strong a voice various people had in deciding his technical assignments: What percentage of the total weight was exerted by himself, his immediate chief, his colleagues, research directors, nontechnical executives, and outside sponsors?

The more weight a man claimed for himself, and the less for other people, the more he was likely to regard himself as autonomous or free. We then plotted this measure of autonomy against several criteria of performance—such as the judgment of technical colleagues, and the number of recent publications, patents, or unpublished reports.

Upon examining our data, we found (with an important exception) the expected trend: As autonomy increased to a high level, performance also increased. But only among scientists and engineers without a doctoral degree. Among men with a doctoral degree, we found a surprise: Performance increased with autonomy up to the point where half the weight was exerted by the man himself; after that, when he felt largely or wholly autonomous, his performance dropped to mediocre.

These results were puzzling. Why should non-Ph.D.s perform better when they had considerable autonomy, but not Ph.D.s? We proceeded to dig further. We ascertained how many decision-making sources exerted at least a slight weight on the man's assignments. Was this choice concentrated in one or two sources? Three? More? What we observed was unexpected: As more decision-making sources were involved, performance of the Ph.D.s also rose.

If Mees was right in his skepticism about research directors and vice-presidents, how could we account for the finding that scientists who allowed these gentlemen some voice in selecting their problems were more effective by scientific standards?

Our research results contained other puzzles. The conditions that accompanied achievement often appeared contradictory. In searching for a framework to accommodate these inconsistencies, I was led to the idea of "creative tensions" in the research and development climate—the idea that a sense of security and a sense of challenge can combine to spur a scientist to creative problem solving. In the paragraphs that follow, I shall outline just enough of this concept to advance my hypothesis concerning communication among researchers.

In looking over our research data, it became apparent that technical achievement was high under several conditions in the laboratory which served to *protect* the individual researcher from the demands of the environment, or which promoted continuity or stability. To designate any such protecting condition, I used the term "security."

Personal autonomy is such a source of protection, and there are many others. Possession of a doctoral degree, for example, makes it easier to say no to a department head. Evidence of this in our study showed up in the fact that Ph.D.s were less often reorganized into new groups. Length of time in the same group, or length of time on the same project, can also provide security. Among the younger scientists and engineers in our sample, performance was positively correlated with length of time on one's main project. Longevity on a group or a project enables one to build up specialized knowledge, and the specialist is better able to resist a disrupting assignment outside his specialty.

Perhaps the most important source of security, in the sense in which I'm using it, is self-confidence or self-esteem. The better performers in our sample preferred to rely on their own ideas. Of course, it's debatable whether achievement generates confidence or whether confidence stimulates achievement (I shall come back to this matter of causality later on). But in either case, it is undeniable that a self-assured person is better able to ignore disrupting demands.

Now in examining our data further for conditions under which technical achievement proved high, we found several which appeared to be the opposite of security—conditions which served to *expose* the individual to demands of the environment, or to disrupt his ongoing patterns. As a general label for conditions like these I used the term "challenge."

One example is the involvement of several other people in selecting assignments. Here, the technical man is exposed to other people's ideas or criticisms. Challenge can arise from facing an unfamiliar problem, from encountering approaches different from his own, from having flaws in his solution pointed out.

Diversity of activity can provide challenge. Our data showed that the most effective scientists and engineers, both in research and in development labs, were not those who concentrated on research only or on development projects only, but those who did some of each. Nor did these same high performers concentrate wholly on technical activity as such. Rather, they spent up to one-quarter of their time on administration or teaching.

Challenge can also arise from dissimilarity among or disagreement with colleagues. In older groups, whose members had been together several years, the more effective groups were those in which the men differed in their technical approach to problems and engaged in intellectual dispute.

In pondering this evidence of achievement flourishing under conditions that seemed antithetical, I began to wonder whether beneath the apparent disorder there might lie a more basic order. Did achievement flourish not in spite of the contradictions but because of them? Was problem solving stimulated by a "creative tension" between some conditions that gave security and others that provided challenge? Let us see how this might be so.

Consider the case of independence versus interaction. A dominant trait of first-rate scientists is their self-reliance, their insistence on their own ideas, faith in their own judgment. To measure the strength of this trait among our sample, Andrews and I constructed an index of "motivation from one's own ideas," using items from our questionnaire in which the scientist reported a stimulus from his own previous work and his own curiosity.

It turned out that our high performers were strongly motivated by their own ideas, by stimulation from within. But at the same time they did not avoid stimulation from without. They interacted more vigorously with colleagues than did less effective scientists. The same trend has been noted by other investigators. Tom Allen at M.I.T. and Schilling and Bernard at George Washington University observed, in industrial and government labs, a positive relationship between the performance of engineers and scientists and the extent of their communication with other members of the organization. Effective performers, in short, seem to be men open to both internal and external stimuli.

Are the two types of stimuli incompatible? Not logically, of course, but psychologically each tends to weaken the other. We know from everyday experience something that psychological experiments have verified—that in the face of social consensus it is difficult to maintain one's independence. Yet creative scientists are able to do this and to flourish.

Herbert Shephard, a management consultant wise in the ways of R&D people, has borrowed a term from personality theory to shed further light on this phenomenon. The creative man, he says, must be able to act alone, to compete or rebel, when that is what the task requires. But he must also possess what O. J. Harvey and his colleagues at the University of Colorado call "autonomous interdependence." That is, the creative man must also be able to depend on others and to join with them in intimate teamwork, when that is what the task requires.

INTERESTINGLY, one can find this same dissonant blend in the worlds of letters and of common sense. Take Emerson's essay on self-reliance, for example, in which he used "the world" to mean one's social milieu: "It is easy in the world to live after the world's opinion; it is easy in solitude to live after our own; but the great man is he who in the midst of the crowd keeps with perfect sweetness the independence of solitude." For Emerson's "great man" substitute "creative scientist," and you have the tension between independence and interaction.

Or take the commonly held adage that necessity is the mother of invention. There is certainly some pertinence here, if "necessity" is taken to mean not merely adversity but rather the perception of a problem, coupled with the belief that a solution can be found. In fact, awareness of a problem is among the most essential forms of challenge. It sounds commonplace to say that problem solving requires that a problem be perceived, but the point is easily missed by many who proclaim the virtues of idle curiosity and serendipity.

Necessity, then, is a form of challenge which can spur invention. But invention in my conceptual framework has more than one parent. Necessity is better called the father of invention. Challenge in my view is a masculine attribute. The mother of invention is, rather, security. When the masculine and feminine components are joined, the creative tension between them can give birth to technical achievement.

In concluding this part of my discussion, I should like to make more explicit two features of this creative tension concept. First, I do not consider the optimal climate to lie halfway between extremes—at some compromise between security and challenge. Rather, the creative scientist needs a lot of both. He should be exposed to disrupting demands from his organizational environment, and at the same time have the means to filter these demands.

Very broadly, my major hypothesis is that for creative problem solving to occur there must coexist conditions both of strong security and of strong challenge. But coexistence need not mean strict simultaneity; the two conditions can occur in succession—periods of intense exposure followed by periods of withdrawal, with the cycle periodically repeated.

A second hypothesis is a plausible corollary; namely, that the intensity of the two components must be in balance. The stronger the security, the stronger must be the challenge if creativity is to flourish. Otherwise the individual or the group will stagnate. On the other side, if security is weak then challenge must also be mild; too much challenge in this case will arouse anxiety and rigidity.

Some hints of this corollary appeared in our data. Among men of lower status in the organization, or among those who felt they lacked influence—that is, men low in security—maximum performance occurred when their assignments were affected by relatively few other people.

At the other extreme—that of high security—we pressed the question of why scientists with maximum autonomy were only average performers. One clue emerged from our measurements on the tightness or looseness of coordination within the department where such men resided. Now a loose organization does not make demands on its members. We found that when scientists were both autonomous and in loose departments, they withdrew from contact with colleagues, and they specialized in narrow areas. They were even less involved in their work! In short, they minimized their challenge.

But—and here's the point—it was precisely under these conditions of high security (as defined by looseness of coordination) that challenge was shown to be most essential. We found performance to be most strongly correlated with stimulation from the man's environment. It would seem, therefore, that a nondemanding organization permits an autonomous member to withdraw into an ivory tower of maximum security and minimum challenge, where he can grow comfortably stale.

A little while ago I said I would take up the question of causal sequence, and right here is where it should be examined: If we do observe that creative performance is strong in the presence of both security and challenge, how is this association to be interpreted?

One view is that the individual creates his own conditions. An outstanding scientist can insist on autonomy and stability, and he thereby generates his own security. His achievements can also attract attention from colleagues and top management, who then become eager to seek his help; thus he generates challenge.

But what about the reverse sequence? Can the right combination of externally generated security and challenge stimulate a technical man to perform above his natural level of competence?

Many technical readers will subscribe to the first sequence since it happens all the time. It is consistent, too, with Mees's view quoted earlier, or with the philosophy that all a lab director can do is to recruit good men, give them facilities, and leave them alone. Implicit in this viewpoint is the conviction that you can smother a good man with rejection, or starve him with poor equipment, but you can do nothing to boost his achievement beyond certain natural limits of his competence or training.

But I believe that the second sequence can also work. Or, to change the slant slightly, I favor a circular or feedback interpretation of creative performance in which both sequences operate: achievement engenders conditions which, in turn, stimulate achievement. But, most important, I would argue that the cycle need not start only with the individual and his given abilities. A research manager can, I believe, promote conditions which will help a man to achieve, and can thus cause the circular process to operate more intensely.

As a prelude to constructing a model of how such a circular process might work, I want to draw upon the thinking of Jack A. Morton, vice-president of Bell Telephone Laboratories. As you will see, his notion of communication "bonds" and "barriers" ties in rather nicely with the ideas about creative tensions between security and challenge that I have been discussing.

The focus of my own studies, as discussed thus far, has been on the individual and his interactions with other individuals and groups. Morton looks rather at the departments within a research and development structure and their relation to each other and with the rest of the company. His approach is that of the systems engineer, and his analogy is an electronic device or system in which "the thing being processed is information . . . that goes from one person to another. . . . Just as in an electronic circuit, you use insulators, conductors, semiconductors to build barriers and bonds to the flow of electrical information." He argues that both barriers and bonds are needed to keep the total R&D system productive.

Now an information or communication barrier is intended to buffer the individual from outside stimuli; hence the parallel with security. A communication bond, of course, ensures exposure to outside stimuli; hence the parallel with challenge.

Morton defines two kinds of bonds and barriers. One kind he calls "organizational," meaning linkages or separations created by lines of authority and responsibility in the organizational structure. The second kind of bond or barrier is "spatial," arising from physical closeness or separation.

Information must be transmitted between basic and applied research, between applied research and design development, between design and manufacturing. The people in each must be able to understand the others, and be able to work together if the total organization is to operate. Yet if the design or engineering groups can dictate to the research groups, this will stifle the latter's freedom. How do you accomplish the first but avoid the second?

The answer adopted by Bell Labs is not total separation. If you separate the groups physically as well as organizationally, there will be too great a barrier to the forward flow of new knowledge and designs and to the feedback flow of evaluation. Says Morton, "Now we know we should never have a space barrier and an organizational barrier on top of one another. We use organizational and spatial links in complementary relations—wherever we have a space barrier we also have an organizational bond, and vice versa."

A nice example of the same philosophy was given not too long ago by Jack Goldman, then head of the Ford Scientific Laboratory and now in charge of R&D at Xerox. Ford Motor Co. had acquired the Philco Corporation, and the Scientific Laboratory wanted to establish a basic research group in electronics. The Scientific Lab was in Detroit, Philco in California. Goldman created a group of basic researchers, which he made organizationally responsible to himself. But he located the group physically at Philco's California plant, so that contact between research and engineering could stimulate discovery. Between them, that is, he placed an organizational barrier but a spatial bond.

THERE is an obvious parallel between this strategy of Morton's (and Goldman's) and my central hypothesis—that conditions of challenge and security should be complementary and balanced for creative achievement to occur. With this parallel in mind, I want to move on now to the circular mechanism I spoke of earlier when talking about the matter of causality. Through what sequence of events does a combination of security and challenge lead to creative achievement, and how does achievement in turn strengthen security and challenge? Further, what kinds of communications climates will enhance or inhibit these reciprocal processes?

If the several causal linkages can be clarified, I believe we can find ways to modify the climate of communication—the mix of barriers and bonds. To this end, I'm going to outline a model of the problem-solving process, incorporating as I go the notions of security and challenge. Then I shall focus on what seem to be the critical linkages between events in the hypothesized network, points at which managers might effectively intervene. The manner of the intervention will be suggested by additional findings from our study of R&D organizations. Finally, I shall comment on aspects of the climate in which this intervention would take place, again drawing upon research data where it is helpful.

As the adjacent diagram suggests, there are two main components in this model of the problem-solving process: the qualities of the individual problem solver (left half of the diagram), and the technical environment in which he works (right half). Let's start with the first of these components.

There are really four key qualities of the individual himself, but the three shown at the far left group logically together. The first of these is simply *competence,* arising from the man's intellectual ability and from his training and experience. A second is *self-confidence,* which sometimes appears as dominance or even arrogance, and is widely found in assessments of creative individuals. A third characteristic I have labeled *curiosity.* This may appear in several forms, such as zest for new experience, or enjoyment of puzzles.

A little apart from these three qualities I have placed the fourth—*involvement*—the capacity to become absorbed in the problem-solving activity. In our study, the more effective scientists in all settings were deeply involved in their work. I've located this quality midway between personal qualities on the left and the technical environment on the right because it depends partly on a personal capacity for enthusiasm, and partly on the nature of the work itself.

Some of these personal qualities reinforce others, as the arrows suggest. Competence usually increases self-confidence, which in turn sustains curiosity—especially when one is probing into unfamiliar territory. The combination of self-confidence (a security factor) and curiosity (a challenge factor) will heighten involvement.

The connection between competence and curiosity is less clear. On the one hand, you have to know something about a field to realize where its puzzles lie. But if you know a great deal you will take much

Model of the Problem-Solving Process

The model of how problems get solved is a complex interaction between qualities of the researcher himself (at left) and characteristics of the environment in which he works (at right). Many of these qualities and characteristics can be interpreted, as explained in the text, in terms of "security" for the research or "challenge" to him. Creative tension between these two factors tends to spur the researcher to achieve solutions to problems.

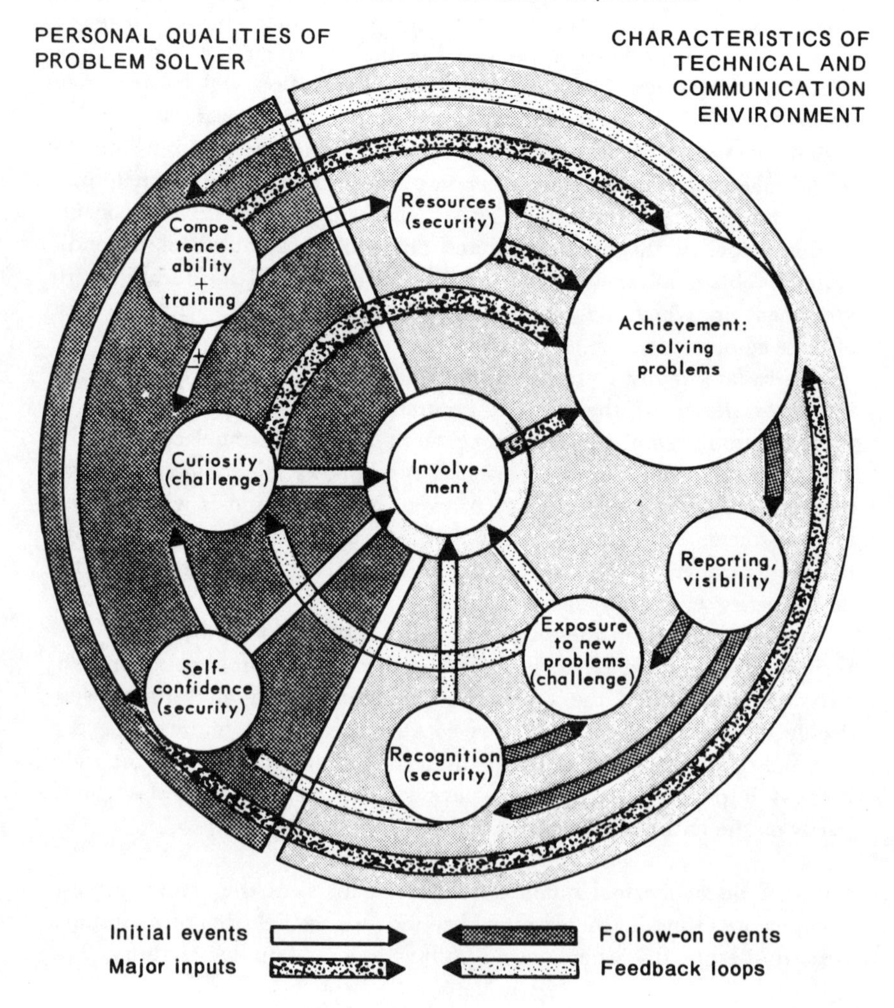

for granted, and curiosity is inhibited. This is why the arrow between them is marked (±). To emphasize the point, I cite the comment of a physics professor renowned for his gift for turning out Ph.D. aspirants, many of whom went on to high achievement: "My method? I spend as much time talking about what the field doesn't know as about what it does."

Each of these four personal characteristics is essential to problem solving. The absence of any one can block the effect of the others. The roles of competence and curiosity are obvious; they are necessary, but by themselves not sufficient. Involvement serves to keep the mind absorbed and the energy flowing, for problem-solving is tough work.

The role of self-confidence is less appreciated than that of the other three qualities. Problem solving is a hazardous enterprise, beset with frustration, failure, and hostility. Creativity can be threatening to an organization, which often tends to react by suppressing it. So it can take enormous confidence in one's own ideas, sustained over years sometimes, before one is able to prove them out and win acceptance.

The next step in the model network is technical *achievement*—attacking and solving significant problems. As the arrows indicate, all the personal qualities contribute here, and you must also have adequate *resources* in the form of equipment and assistants. (There are other inputs, of course, which I have not tried to picture.)

Achievement is likely to trigger follow-on events, provided that the achievement becomes known through some form of *reporting* or visibility—a most important step, whether it be done by publication, seminar, personal dialog, or whatever.

Hopefully, there follows *recognition*—and I mean this in the basic sense of awareness and appreciation rather than monetary reward. One preferably gains recognition not only from one's immediate colleagues or boss, but also from executives at higher levels, and perhaps from the lay public. Whatever the source, recognition acts as a security factor, of course.

As the individual's work becomes known, other people begin to exchange ideas with him, perhaps to seek his help on their problems. These contacts lead to another step in the network by increasing his *exposure to new problems,* to other areas of inquiry. These are all a natural source of challenge.

With the basic elements of the model in place, the feedback loops can now be completed: Achievement means more experience, and this directly increases competence. Recognition reinforces self-confidence; exposure to new problems nourishes curiosity; and both of them strengthen involvement. Achievement also gains the problem solver access to additional resources.

Sustained by heightened self-assurance, intrigued by bigger problems, armed with added knowledge and resources, the individual is able to accept tasks of increased difficulty and risk. And thus the conditions are set for further achievement in the next cycle.

Now I want to examine some of the points in this model where the manager can intervene to good effect, for I believe the R&D manager can do far more than simply provide assignments and resources and then wait for the man's personal qualities to produce results.

One of the manager's essential functions, it seems to me, is to prod the younger technical man into early achievement, and then to push its visibility and recognition. Too often this function is obstructed by typical company practice: When a new man is hired, he is assigned to first one department, then another, and another—all this to give him a well-rounded picture, to diversify his interests.

But the younger man is more in need of a solid foundation of success. We found in our study that among scientists and engineers under 30 years of age, performance was strongly correlated with having worked on his main project for one or two years. Of course, the area in which the young man focuses should be compatible with his interests. But he should be urged to focus soon on a task that is challenging but within his reach, and to stay with it. His manager should give him every assistance in producing a creditable outcome, and should see that the outcome is publicized.

Behind all this is the fact that confidence in your own ideas is a fundamental security factor. A dominant personality helps, to be sure, but even more helpful is achievement you can claim as your own. That's why the research manager should see to it that, at least once or twice a year, each man generates a product or part of a product—whether a publication, a technical report, a patent application, a design—which he helped create, and which bears his name.

I have said that achievement is likely to increase exposure to new problems, which in turn will stimulate curiosity. These feedback loops offer excellent opportunities for managerial intervention. Let me take a couple of examples.

In listening to laboratory administrators recount successful developments, one hears many anecdotes in which the manager brought together a research man and some applications people to talk about a problem on the applications front. At first the research man could see no relevance to his own endeavors. As they talked, he became aware of possible connections, and saw exciting possibilities of translating the practical problem into a form which his theory could attack. Often, in the translation process, questions would arise for which the theoretical background provided no answer, and the researcher was stimulated to devise one. This is a common enough tale, when you think back on your own experience. But how often are such confrontations deliberately set up so as to speed up the feedback loops? Not often enough I suspect.

My second suggestion for intervention arises from an important finding from our research study. We wanted to find out how the performance of a scientist or an engineer related to the amount of time he spent, by his own estimate, on strictly technical activity as against administration or teaching. As you might expect, the more time spent on research or development work, the higher the performance—but only up to a point.

It turned out that the men who spent about three-quarters of their time on technical matters and one-quarter on teaching were more effective technically than those who spent full time at the bench. This result makes sense, since students provide challenge by forcing the teacher to test and expand his ideas. But we were surprised to find that technical men who spend about one-quarter of their time in administration were also more effective technically than full-time researchers. Why?

It could be simply because the best men are given administrative assignments, but I think we must also consider the possibility that some forms of administrative activity can serve as a challenge to technical creativity.

For example, serving on a department-wide committee (whatever its ultimate effectiveness) can bring a man into contact with others he might not ordinarily encounter. It can help to build communication bonds where spatial and organizational bridges are lacking. Now such committee assignments normally go to the senior staff. But what if a deliberate effort were made to involve junior men as well? What if each technical man, new as well as old, were involved in some sort of cross-organizational administrative activity a few times a year? I argue that technical performance would improve, rather as our study data indicated.

As you think about my rough model of the problem-solving process, and check it out against your own experience, I am sure you will see other specific points of effective managerial intervention. Rather than pursue such specific actions further, I should like to examine briefly several aspects of the general climate for communication over which managers have control, and to suggest some attitudes which may lead to more creative achievement.

The first aspect arises from the question of just who should decide what tasks the researcher should undertake. Our study data suggest that multiple involvement is needed in such decision making. When the chief alone had the main voice in determining a man's assignments, performance was lower than under any other condition. By contrast, in development-oriented labs, weight shared jointly by the investigator and his chief was a favorable condition, while in research-oriented labs joint weight shared between the man and his colleagues proved favorable.

Now what this says to me, in terms of the communication climate in the lab, is that the smart manager will assure that multiple channels exist for spreading news of the individual's work and for recognizing his achievements. Not one man but several must know what he is doing—including people outside his own section, and those at higher levels. Then, whenever the investigator allows other people to have a voice in deciding his assignments, he is also letting them appreciate his achievements. This provides security as well as challenge.

Connected with this aspect of multiple channels for communication is the matter of the personal interaction that takes place through them. Effective technical men, we know, communicate often with many other people in a variety of roles. What creative functions are served by frequent and diverse communication?

I have already suggested that communication not only provides challenge in the form of unsolved problems, but also security in the form of recognition for accomplishment. Glancing once more at the model diagram, it's clear that interaction can also stimulate curiosity, and can build self-confidence.

Another major function is assuring relevance. Does the technical man attack problems that are central or peripheral to the organization's concerns, or to the discipline's state of knowledge? Communication with superiors helps to assure organizational relevance; communication with colleagues solidifies scientific relevance.

Creative thinking is said to occur when previously known but unassociated elements are brought together in combinations that are both novel and useful. If so, interaction among persons with different approaches can provide a diversity of inputs and thus help creative problem solving.

And lastly, we saw from the model that achievement depends in part on the personal quality of involvement. Is the technical man gripped by what he is doing? Or is it just a job, one of several interests? Enthusiasm is contagious. If supervisors or colleagues are interested in what you are doing, and express this interest, your own involvement is heightened. Thus the strengthening of motivation is an important function of interaction.

EARLIER I suggested that challenge can be provided by dissimilarity between the researcher and his colleagues, and by disagreement on technical strategies. How much conflict, and of what type, is desirable? How much is harmony needed for creative problem solving?

The ultimate answer may depend on where we focus on the continuum from the generation of a new idea to its final incorporation in a changed technology. When the necessity is to originate or to invent, perhaps more disagreement is needed and more disharmony can be tolerated. When the task is rather to execute designs already agreed upon, the tolerance is lower, and disagreement can become disruptive.

It is important to distinguish between two forms of disagreement—technical and personal. Our study data suggest that intellectual disagreement and conflict can facilitate problem solving. On the other hand, personal conflict or hostility probably inhibits it. Hostility will usually block the channels of communication—by preventing people from talking together who should be talking, or preventing them from saying what they should be saying.

When we examined groups of researchers who had performed well together over a considerable period, we found that they attached great value to intellectual disagreement (a challenge factor) in conjunction with personal agreement or attraction (a security factor). Such groups remained effective if, on the one hand, they maintained social cohesion—if the members valued one another and voluntarily sought contact with each other. On the other hand, group effectiveness continued if the members maintained different technical strategies, and (surprisingly) were somewhat hesitant to share their technical ideas freely with colleagues. They seemed to be intellectually wary of each other—respectful but argumentative. Here again is a creative tension between security and challenge.

Critical evaluation of a proposed solution to a knotty technical problem is essential at some point, but how can it occur without blocking communication? My own view is that if an atmosphere of trust and confidence can be generated, a high level of intellectual conflict can be tolerated without damage to the communication channels.

Precisely how this can be accomplished is a subject in itself. Nevertheless, we can certainly say that having the right leader plays a big part—a leader who believes in what John Stuart Mill called the "morality of public discussion."

There is great potential, too, in the introduction of sensitivity training into more R&D organizations. If it's done carefully, one can gain valuable insights from an open discussion of how one's actions affect other people, and how others in turn affect him. Ultimately, as interactive skill is increased, and along with it one's security in communicating with others, the prevailing trust among members of the organization will rise. And as this occurs, greater intellectual conflict can be permitted.

But whatever the wellsprings of a more trustful climate, it seems to me that the end objective should always be to nurture Emerson's great man "who in the midst of the crowd keeps with perfect sweetness the independence of solitude." Among such men, I am sure, are to be found the creative problem solvers of this world.

CREATIVE PROCESS

Frank M. Andrews

An Orientation to the Creative Process

ONE CAN DIFFERENTIATE between a person's *creative ability* and the *innovativeness* of his output.[1]

The first is a quality of the person—different from, but analogous to, intelligence. Mednick (1962) has proposed that the creative process involves bringing together ideas which are not usually associated with one another —that is, making "remote associations." "Creative ability" is this capacity for making remote associations which are *useful.*

Creativity ability is assumed to be reasonably stable for any individual over a 5–10 year period, though it may show short-run fluctuations—perhaps related to factors such as fatigue, motivation, depression, or colleague stimulation. While one can imagine that differences among individuals in their creative ability might be related to early learning experiences and/or physiological factors, these are not topics for this investigation. From the standpoint of the creator, or of an administrator of an organization responsible for producing innovations, a person's creative ability represents a resource which may or may not be well used. Essentially, creative ability is an *input* to the creative process.

Outputs of the creative process, on the other hand, are products—scientific papers, artistic drawings, musical compositions, reports, devices, processes, substances. Such products vary in their "innovativeness" and "productiveness." Highly innovative outputs open *new* possibilities for further research, appreciation, development, or utilization; *productive* outputs, on the other hand, permit significant advances along *established* lines (Ben David, 1960). On the other hand, some outputs of the creative process may be low in innovativeness, productiveness, or both. Creative ability, we

This article originally appeared as "Social and Psychological Factors which Influence the Creative Process" in *Perspectives in Creativity* edited by Irving A. Taylor and J. W. Getzels (Chicago: Aldine Publishing Company, 1975).

expect, should be relevant for understanding the innovativeness of a person's output, though not for understanding its productiveness.

The formation of new, useful combinations of ideas—creative acts—does not happen in a vacuum. The person must be aware of a specific problem, task, or technological "gap," and he must be motivated to work on it. Furthermore, he must have at his command the discrete bits of knowledge and skills which, in combination, can contribute to its solution.

But a new idea, if it occurs, is only the first step in the creative process. The resulting association must be evaluated, communicated, and developed before it can contribute to an innovative product.

Let us sketch some of the stages a creative idea might go through in becoming an innovative product. One can imagine that the person who achieves a remote association must first evaluate that "new idea." Does it seem likely that this idea can contribute to solving some particular problem about which the person has been thinking? Implicit in the evaluation is the notion that there exists a set of criteria against which the person can assess his idea. This evaluation probably occurs both consciously and subconsciously. It may require anywhere from a fraction of a second to months or even years. While some ideas may be evaluated solely by thinking about them, others may be the objects of extended efforts at "private" investigation.

Once a technical idea has been judged appropriate or useful, or at least "worth trying," it will usually need to be communicated to others. Several factors suggest the need for such communication. Often additional resources will be needed for the further evaluation and developmenet of new ideas—additional information, computers, new tools, working space, time, and the like. Furthermore, often one "new idea" is not sufficient, and can result in an innovative product only if combined with several other new ideas.

Crucial issues, then, are whether the person is willing to communicate a new idea and whether he is able to. Relevant factors include his willingness to take the risks inherent in possible failure; his need for, and sense of, security in his job, family, and community; his perception of his own role and the appropriateness of his suggesting something different; the ease with which he feels he can approach other people—friends, colleagues, supervisors; his skills at making himself understood; and the willingness of others to listen to this man make new proposals.

Once an idea has been successfully "sent" by its originator, and "received" by other people, further action will often need to occur for the idea to be fully implemented. Whether this will occur will depend in part on the nature of the other activities already underway in the external environment into which the idea is sent, and the ease of accumulating or shifting the resources needed for a new effort.

Thus the creative process can be conceived as an input of new, potentially useful ideas, a series of developmental stages or hurdles which those ideas must pass, and an output of innovative products. The present research focuses on factors which affect the likelihood of a new idea crossing the hurdle-filled gap and being developed into an innovative output.

Design of the Study

GENERAL PROCEDURES

To test these ideas about social and psychological factors which may affect the creative process, data were obtained from 115 scientists, each of whom had been the director of a research project. It was reasoned that the conduct of scientific research provides a setting of some social significance where the creative process occurs, and where outputs can be evaluated for innovativeness and productiveness using criteria about which there is reasonable consensus.

As is described in more detail below, the scientists completed one questionnaire several years after the initiation of their projects to provide information about social and psychological conditions present during the course of the research. After termination of their projects, they were asked to identify the principal report of the research, a copy of which was subsequently obtained, abstracted, and rated for several scientific qualities by expert judges. Five years after the initial contact, the scientists were contacted again for administration of psychological tests of creative ability and intelligence and a follow-up questionnaire.

RESPONDENTS AND THEIR RESEARCH PROJECTS

The respondents were all medical sociologists who had directed a project dealing with social psychological aspects of disease. The 115 scientists on whom full data were available constitute about half (47 percent) of all directors of such projects which were listed in the *1953–60 Inventory of Social and Economic Research in Health.* A careful check showed no evidence of marked differences between these 115 respondents and the entire set of all listed directors with respect to 11 variables describing demographic characteristics, research role, and qualities of scientific performance.[2] Thus the project directors analyzed here can be assumed to be reasonably representative of all directors of such projects active during the period.

The median age of the respondents was 38 at the time they began the specified projects; 86 percent were male; and two-thirds had a doctoral degree (nearly all the rest had master's degrees). The most common professional fields were sociology, psychology, and medicine. The typical project director had had his degree for about 7 years at the time the project was undertaken and had directed 2 or 3 previous projects. Fifty-eight percent claimed their primary activity was as a researcher; others mentioned teaching or administration in about equal proportions. Just over three-quarters, however, said their *preferred* activity was research. Typically, they spent half to two-thirds of their working time directly on research, and roughly half of the research time was devoted to work on the specified project.

Nearly all respondents (94 percent) felt they had "considerable" or more influence over the people who made decisions about their work goals; 87 percent felt their colleagues and superiors had "considerable" or "complete" confidence in their abilities; and 70 percent claimed their sense of involvement with the specified project had been "strong" or "very strong." Four out of five said they had "much" or "very much" responsibility for initiating new activities.

As a group, the respondents scored very high on a measure of verbal intelligence. The mean score for respondents was: 41.2, standard deviation: 8.0; by comparison, the mean for all employed Americans is 21.5, standard deviation: 9.4 (U.S. Dept. of Labor, 1967). While no national norms exist for the Remote Associates Test (a measure of creative ability), these respondents scored slightly higher than a large group of undergraduates at the University of Michigan, in which the mean score for respondents was: 17.8, standard deviation: 5.4; mean for 2,786 undergraduates: 16.4 (Mednick and Mednick, 1966).

As for the research projects themselves, about two-thirds were "problem oriented" (the others were mainly "theoretically oriented"); the median length was 2–3 years; and they typically had a staff of one or two professional people in addition to the project director. Findings from most projects were published as an article in a professional journal or as a health agency report. The most frequently mentioned laboratory sites were health agencies, academic social science departments, and medical schools.

MEASURES EMPLOYED

The measures to be discussed in this chapter fall into three broad types: (1) two mental abilities—creative ability and verbal intelligence; (2) a large number of factors characterizing the social and psychological environment in which the research took place; and (3) two qualities of the project's output—innovativeness and productiveness.

Creative Ability. The measure of creative ability was the Mednicks' Remote Associates Test (RAT). This is a timed, 40-minute, paper-and-pencil test which closely operationalizes Mednick's associational theory of creativity (Mednick, 1962). The person taking the test is presented with a series of items each containing three words. He is asked to think of a fourth word which can be associatively linked with each of the three given words. The following is an example of the type of item used:

ITEM: rat blue cottage ANSWER: cheese

Studies on architecture students, psychology graduate students, suggestion award winners in a large business firm, and children have shown that people scoring high on the RAT tended to receive higher ratings for the creativity of their products (architectural designs, research projects, suggestions, and drawings, respectively) than low scorers (Mednick and Mednick, 1966). Furthermore, on a sample of scientists Gordon and Charanian found that those who scored high on the RAT tended to write more research proposals, to win more research grants, and to win bigger grants than low scorers (Gordon and Charanian, 1964). A study by Andrews (1967), however, found scientists' RAT scores to be unrelated to their output of papers, patents, or reports, and unrelated to the judged quality of their technical contribution or organizational usefulness. This lack of overall relationship seemed attributable to two opposite effects which were self-canceling; scientists in flexible settings showed positive relationships between RAT scores and performance; those in inflexible settings, negative relationships.

Taking a somewhat different approach, which may also be relevant for scientific innovation, Mendelsohn and Griswold (1966) found that high scorers on the RAT were more likely than others to make effective use of information which at first seemed irrelevant, but which in fact could be helpful in solving the problems presented.

The RAT has been shown to have very satisfactory inter-item reliability. For several different groups, the reliability has been consistently above .9 (Mednick and Mednick, 1966).

Although the Remote Associates Test has high reliability and appeared promising in the validity studies cited above, there remain some doubts as to whether it "really" measures creative ability.[3] We make no claim that the RAT is a perfect measure of this ability, but in view of its close ties to an explicit theory of association and incorporation of a usefulness criterion it seemed an appropriate measure to use in this situation.

Since the Remote Associates Test requires a close familiarity with colloquial American English, it does not yield valid scores for people who grew up learning a different language. As part of the data collection procedures, respondents were asked to indicate their childhood language. Those who did not check "American English" were excluded.[4]

Intelligence. The psychological test used to measure intelligence was the V scale of the General Aptitude Test Battery, Form B-1001, Part J, (GATB) (United States Department of Labor, 1967).[5] This test consists of 60 items, each containing four words. The test taker is asked to identify that pair of words which have similar *or* opposite meanings. One sample item is:

ITEM: big large dry slow ANSWER: big and large

The test has a five-minute time limit.

The test is a widely used measure of verbal intelligence, is appropriate for adults, and is sufficiently difficult that very few of these scientists—a very bright group compared with the general population—hit its maximum score.[6]

The Laboratory Environment. Data about the social and psychological setting in which the specified project was conducted are based on respondents' answers to two questionnaires. They included items assessing decision-making power, independence from colleagues and supervisors, the quality of communication channels available, sense of professional security and environmental support, level of motivation, opportunities for initiating new activities, research roles, the size and duration of the project, and the adequacy of facilities and resources.

Since some (but not all) projects had been completed several years prior to the time respondents were asked to answer the first questionnaire, and since five additional years elapsed before the administration of the second questionnaire, opportunities for errors in recall were substantial. To check the extent of such errors, ten ordinally scored items which had been part of the first questionnaire were repeated in the second. The reliabilities of these ten items ranged from $\gamma = .81$ to $\gamma = .37$, with a median gamma of .55. Although reliabilities in this range are far from perfect, they do serve to show that memory errors did not grossly distort the data in the five-year interval between the administration of the two questionnaires.[7] It seems reasonable to infer that memory errors similarly did not markedly obscure the data during the interval between the time the project was underway and the time it was described by the respondent on one of the questionnaires.[8]

Quality of Output. Each respondent was asked to indicate what was the principal report or major publication of a designated project he had directed. Copies of these reports were obtained and abstracted, and the abstracts were then rated on the following two qualities within the context of disease control and treatment.[9]

Innovativeness: The degree to which the research represents additions to knowledge through new lines of research or the development of new theoretical statements of findings which were not explicit in previous theory.

Productivity: The extent to which the research represents an addition to knowledge along established lines of research or as extensions of previous theory.

The judges who performed the ratings were persons chosen as leaders in medical sociology by members of the Section on Medical Sociology of the American Sociological Association. The number of judges independently rating each project ranged from one to seven with a median of 4.5 per project.The scores of the different judges evaluating a project were averaged to determine a final score for that project on each of the two qualities.

Although a measure of agreement among these judges has not been computed, their ratings were compared with ratings of the same reports obtained by similar methods from two other groups of judges, medical doctors, and administrators of medical sociology research. Agreement coefficients (Pearson r's) for innovation and productivity ranged from .42 to .53. Since the medical sociologists probably agreed more among themselves than they did with other groups of judges, it seems probable that the interjudge agreement among the medical sociologists was at least .5. Application of the Spearman-Brown Formula would thus suggest these quality ratings have a reliability of at least .8.

Derivation of Adjusted Innovation Scores. A preliminary examination of the innovativeness and productivity ratings showed a possible "halo" effect: projects that received a high rating on one quality rating tended to also receive a high rating on the other ($r = .76$).

Since our interest focused on the extent a project was innovative, we sought a way of removing the effects of extraneous qualities from the innovation scores. This was achieved by computing a residual innovation score (hereafter called the "adjusted" innovation score) which consisted of the deviation of the project's actual innovation score from that score which would have been predicted solely on the basis of the productiveness of the project.[10] Thus, the adjusted innovation scores indicate the innovativeness of the project *after* "allowing for" or "holding constant" the productiveness of that project.

Parallel analyses were subsequently carried out for both the raw and adjusted innovation scores. In general the same conclusions emerged from each, though the adjusted scores tended to show sharper effects.

Findings I: Relationships among Creative Ability, Intelligence, and Qualities of Creative Output

CREATIVE ABILITY AND QUALITY OF SCIENTIFIC OUTPUT

As is shown in the top row of Table 5.1, creative ability, as measured by the Remote Associates Test, was virtually unrelated to either the innovativeness or the productiveness of the scientists' output.

This is in direct conformity with previous findings which showed RAT scores to be unrelated to several qualities of scientific output (Andrews, 1967). Nevertheless, the present finding came as a surprise because in the writer's previous study innovation had not been one of the qualities of scientific performance specifically examined. Here it was, and although numerous factors were expected to intervene between a scientist's creative ability and his producing innovative output, a stronger relationship was

TABLE 5.1. *Relationships among Creative Ability, Intelligence, and Two Qualities of Scientific Output.*

	Intelligence	*Innovativeness (adjusted)*	*Innovativeness (raw)*	*Productiveness*
Creative ability	.41	.05	.07	.05
Intelligence		−.11	−.09	−.01
Innovativeness (adjusted)			.65	.00
Innovativeness (raw)				.76
Productiveness				

NOTE: Each correlation is based on about 100 cases.
Coefficients are Pearson *r*'s.

expected than the .05 or .07 shown in Table 5.1. (The low relationship between creative ability and productivity was predicted: productivity had been defined as progress along *established* lines, and was not expected to require creative ability.)

The lack of even a modest relationship between creative ability and innovation might be explained in at least two ways. One is that creative ability or innovation (or both) may have been inadequately measured. However, other relationships in Table 5.1 indicate that these variables clearly consisted of more than just "random noise." The correlation between creative ability and verbal intelligence (.41) was similar to results obtained from other groups which have taken both the Remote Associates Test and a verbal intelligence test (Mednick and Mednick, 1966). Similarly, the fact that the raw innovation scores correlated substantially with the productivity scores further suggests that the innovation scores were not merely "noise." The validity studies cited previously for the RAT, and the fact that different judges showed moderate agreement on innovation, suggest that these scores should at least be related to the concepts they were intended to measure.

The other explanation—which we tend to favor—is that social and psychological factors may so affect the translation of creative ability into innovative performance that there is no general effect which one can describe or identify. This is what seemed to account for the lack of an overall relationship observed previously by Andrews, and, as will be shown below, can also explain the phenomenon here.

Before turning to the examination of the social and psychological factors, however, several other comments may be made about results shown in Table 5.1.

INTELLIGENCE AND QUALITY OF SCIENTIFIC OUTPUT

The second row in Table 5.1 shows relationships between verbal intelligence and the several qualities of scientific output. One can see that verbal intelligence was virtually unrelated to these qualities.

Verbal intelligence was included in this study because it might have confounded relationships between the Remote Associates Test (with which it was known to be correlated) and the quality measures. Finding intelligence unrelated to the quality measures removed our initial concern, but was contrary to the folklore which says that the research output from the brightest scientists will be seen as the best.[11] After an extended and unsuccessful search for possible contingency effects, our data suggest this bit of folklore is simply wrong. Among successful researchers (all of the respondents were project directors), verbal intelligence was not a useful predictor of these qualities of output.

RELATIONSHIPS AMONG OUTPUT CHARACTERISTICS

Also included in Table 5.1 are the interrelationships among the several qualities of the project reports. The substantial relationship shown at the bottom of the table, the .76 correlation between raw innovation and productivity, was what suggested the advisability of computing an adjusted version of the innovation score.

Correlations of these adjusted innovation scores with the other quality measures appear in the third row of the table. One can note that the adjusted scores correlated substantially (.65) with the original raw scores and were completely independent ($r = .00$)—as intended—from the productivity scores.

DISCUSSION

This first set of findings contains two key results: (1) the lack of relationship between creative ability and innovativeness, and (2) the lack of relationship between verbal intelligence and either innovativeness or productiveness. Given that at least moderate relationships might have been expected, that the measures themselves seemed to show convergent and discriminant validity, and that there was adequate variability among the respondents for some relationships to emerge if they were present, these nonrelationships called for further exploration.

For both creative ability and intelligence it was suspected that the apparent lack of relationship with certain output qualities might be the result of opposite, self-canceling effects.[12] An extensive analysis of possible social and psychological factors was undertaken and it was concluded that these did in fact have an important influence on the process by which creative ability was translated into innovative outputs. These results are detailed in the following section.

On the other hand, a similar analysis conducted with respect to intelligence showed that such factors could not explain the observed lack of relationship between verbal ability and output qualities. Convinced that differences between respondents in intelligence would not confound other analyses, we made no further use of the intelligence measure.

Findings II: Social and Psychological Effects on the Relationship Between Creative Ability and Innovation

OVERVIEW OF RESULTS

Prior to examining the specific analyses which lead to the conclusions of this section, it will be helpful to sketch these conclusions very briefly.

Although the magnitude of any social-psychological effect on the creative process was not very great, it was observed with reasonable consistency that creative ability tended to pay off more (in the sense of being more positively related to innovativeness of output) in the following kinds of settings: (1) when the scientist perceived himself as responsible for initiating new activities; (2) when the scientist had substantial power and influence in decision-making; (3) when the scientist felt rather secure and comfortable in his professional role; (4) when his administrative superior "stayed out of the way"; (5) when the project was relatively small with respect to number of professionals involved, budget, and duration; (6) when the scientist engaged in other activities (teaching, administration, and/or other research) in addition to his work on the specified project; and (7) when the scientist's motivation level was relatively high. These findings were in line with our expectations. We were surprised, however, to find that various communication factors and several indications of environmental constraints showed no consistent effect on the payoff from creative ability.

When several of these factors were considered simultaneously, very substantial and statistically significant effects emerged, though the number of scientists in the specified environments was sometimes extremely small.

SPECIFIC SOCIAL AND PSYCHOLOGICAL EFFECTS

Responsibility for Initiating New Activities. One of the psychological factors which was expected to influence the payoff from creative ability was the extent the person saw himself as having opportunities and responsibilities for innovating.

Three items in the questionnaire were relevant here. One asked directly "In the setting in which you did this project, to what extent were the following present . . . responsibility for initiating new activities?" The respondent checked a five-point scale ranging from "none" to "very much." The second question posed the matter oppositely and asked to what extent the resepondent saw himself as "effective as a 'right-hand man' carrying the ball for a more experienced advisor." The third item concerned the extent the project director had been involved in determining the focus of his project.

Table 5.2 shows that all of these items affected the relationship between creative ability and innovativeness in the expected direction. Scientists who saw themselves as having relatively high responsibility for initiating new activities, those who placed themselves relatively low with respect to acting as a "man Friday," and those who had been partially or wholly responsible for determining the research problem were the ones for whom mild positive correlations between RAT scores and innovativeness emerged. For other scientists, the correlations were consistently negative.[13]

TABLE 5.2. *Correlations between Creative Ability and Innovativeness (adjusted) According to Scientist's Role in Initiating New Activities.*

	*RAT scores and Innovativeness**	
	r	*difference*
Responsibility for initiating new activities:		
"much" or "very much" ($N = 84$)	.10	.32
"none" to "some" ($N = 22$)	−.22	
Effectiveness as a "right-hand man" for more experienced advisor:		
relatively low ($N = 70$)	.11	.35
relatively high ($N = 34$)	−.24	
Determination of research problem:		
determined by *R* himself ($N = 57$)	.12	.23
partially determined by *R* ($N = 19$)	.22	
determined before *R* became involved ($N = 19$)	−.11	

* Adjusted for productivity level

Thus the opportunity or responsibility for innovation was one factor which influenced the creative process. Without this, creative ability did not "pay off" in innovative output.

Influence Over Decision Making. Another effect which emerged with substantial consistency was related to the creator's role in decision making. Since a creative idea often requires development before it can result in an innovation, it was expected that people who were in positions to promote such development would be more likely than others to see their creative ideas carried through to innovative outputs.

As shown in Table 5.3, scientists who said they exercised relatively large amounts of influence in decisions about project-relevant matters tended to show stronger positive correlations between creative ability and innovativeness than less influential scientists. Out of eight items dealing with influence in decision making, seven generated differences in the expected direction.

As with the data shown in Table 5.2, this one effect was not very strong, but occurred with fair consistency.[14] It looked as if an ability to influence

TABLE 5.3. *Correlations between Creative Ability and Innovativeness (adjusted) According to the Scientist's Influence over Designated Decisions.*

	*RAT scores and Innovativeness**	
	r	*difference*
R's influence in hiring personnel:		
relatively high ($N = 56$)	.12	.18
relatively low ($N = 42$)	−.06	
Who had power to hire a research assistant?		
R alone ($N = 68$)	.12	.25
others above *R* ($N = 22$)	−.13	
R's influence over allocation of research funds:		
relatively high ($N = 48$)	.11	.07
relatively low ($N = 54$)	.04	
Who had power to purchase a $500 machine?		
R alone ($N = 45$)	.14	.17
others above *R* ($N = 53$)	−.03	
Who had power to requisition under $50 of supplies?		
R's subordinates ($N = 29$)	.26	
R alone ($N = 52$)	.12 }	.61
others above *R* ($N = 20$)	−.49 }	
Who had power to publish a controversial finding?		
R alone ($N = 66$)	.11	.21
others above *R* ($N = 26$)	−.10	
Ability of *R* to influence decisions about his technical goals:		
"great" or more ($N = 66$)	.06	.03
"no influence" to "considerable" ($N = 43$)	.03	
Exception		
R's influence in formulation of research design:		
relatively high ($N = 54$)	−.09	−.27
relatively low ($N = 55$)	.18	

* Adjusted for productivity level

decisions was a social factor which could reduce the "gap" between creative inputs and innovative outputs. Interestingly, this finding replicates a result obtained in a previous study (Andrews, 1967).

Of course, faced with the data in Table 5.3 one wonders whether it might simply be a seniority effect. Table 5.4 brings together data relevant to this issue and shows that older, more experienced or more educated scientists did *not* consistently get higher payoff from their creative ability. (Note the lack of consistent direction and low values shown in the "difference" column.) Thus it appeared that what was important was ability to influence research decisions, not just age or experience or formal education.[15]

TABLE 5.4. *Correlations between Creative Ability and Innovativeness (adjusted) According to Level of Seniority and Experience.*

	RAT scores and Innovativeness*	
	r	*difference*
Years of prior research experience:		
0–3.0 years ($N = 36$)	.15	
3.1–5.0 years ($N = 21$)	−.02	.10
More than 5 years ($N = 40$)	.05	
Numbers of projects previously directed:		
0–2 projects ($N = 47$)	.03	−.07
3 or more projects ($N = 49$)	.10	
Highest degree:		
Ph.D., M.D., or other doctoral degree ($N = 64$)	.09	.05
nondoctoral degree ($N = 31$)	.04	
Years since receiving highest degree:		
0–5.0 years ($N = 42$)	.11	
5.1–10.0 years ($N = 22$)	−.35	−.09
More than 10 years ($N = 32$)	.20	
Age at time study began:		
35 years or less ($N = 37$)	−.01	
36–40 years ($N = 21$)	.34	.07
41 years or more ($N = 38$)	−.08	

* Adjusted for productivity level

Security. The creative process may be a risky operation for the would-be creator. Hence his sense of security would be expected to influence his willingness to discuss and act upon new ideas.

Table 5.5 brings together 14 items relevant to the scientist's sense of

TABLE 5.5. *Correlations between Creative Ability and Innovativeness (adjusted) According to the Scientist's Sense of Professional Security.*

	RAT scores and Innovativeness*	
	r	*difference*
Confidence of colleagues and superiors in *R*'s abilities:		
"complete confidence" ($N = 40$)	.18	.23
"considerable confidence" or less ($N = 67$)	−.05	
Stability of employment:		
"very much" ($N = 53$)	.22	
"much" ($N = 30$)	−.03	.57
"none" to "some" ($N = 22$)	−.35	
Basis on which *R* worked on project:		
"permanent member of organization" ($N = 64$)	.16	.42
not a permanent member of organization ($N = 28$)	−.26	

TABLE 5.5 *(continued)*

	*RAT scores and Innovativeness**	
	r	*difference*
Ease of moving to a comparable position elsewhere:		
"rather easy," "no trouble at all" ($N = 69$)	.16	.29
"impossible" to "so-so" ($N = 38$)	−.13	
Extent roles were defined for professionals on project:		
relatively well defined ($N = 47$)	.27	.41
relatively undefined ($N = 45$)	−.14	
Extent administrative superior got along with *R* personally:		
relatively well ($N = 39$)	.22	.31
relatively poorly ($N = 35$)	−.09	
Extent of disagreement with administrative superior on methods used:		
none or not discussed ($N = 62$)	.21	.88
some ($N = 13$)	−.67	
Extent of disagreement with administrative superior on study's purpose:		
none or not discussed ($N = 64$)	.17	.61
some ($N = 11$)	−.44	
Extent of disagreement with administrative superior on definition of problem:		
none or not discussed ($N = 63$)	.23	.55
some ($N = 12$)	−.32	
Extent of disagreement with administrative superior on interpretation of findings:		
none or not discussed ($N = 52$)	.27	.72
some ($N = 15$)	−.45	
Extent first-named colleague understood *R*'s role on the project:		
relatively well ($N = 39$)	.16	.12
relatively poorly ($N = 41$)	.04	
Exceptions:		
Quality of professional relationship with first-named colleague:		
relatively good ($N = 53$)	−.02	−.22
relatively poor ($N = 28$)	.20	
Extent of disagreements about work activities with first-named colleague:		
little disagreement ($N = 46$)	−.06	−.26
relatively much disagreement ($N = 36$)	.20	
Risk-taking—what would happen if a year's activity led nowhere?		
"nothing, told to continue" ($N = 61$)	−.08	−.25
"severe criticism" to "mild" ($N = 44$)	.17	

* Adjusted for productivity level

professional security and shows their effect on the relationship between creative ability and innovativeness of output. Considered here are the scientist's relationship with his colleagues and administrative superior, the stability of his employment, the ease with which he could find a comparable position elsewhere, and the consequences of taking a risk that failed. Out of these 14 items, 11 showed effects in the expected direction: higher payoff from creative ability tended to occur among scientists in the more secure settings.[16]

It has already been shown that several seniority factors did not consistently or markedly affect the relationship between creative ability and innovativeness (Table 5.4). Thus it follows that the "security effect," present in Table 5.5, was not attributable simply to seniority.

From the magnitudes of the effects, it appeared that disagreement with the administrative superior on such fundamental aspects as a study's goals, methods, or interpretation of results was particularly debilitating. Conversely, there was inconsistent evidence about how the relationship with the first-named colleague (usually, the man's most important colleague) affected payoff from creative ability. Two of the three items which specifically mentioned this person suggested that a less than maximally comfortable relationship enhanced payoff from creative ability. (The third item was mildly opposite and favored the security notion.)[17]

Table 5.5 suggests that, with the possible exception of relationships with the first-named colleague, a sense of security in a scientist's professional life seemed to promote effective utilization of creative ability.

Role of the Administrative Superior. Our orientation to the creative process emphasizes the existence of various "hurdles" which a creative idea must pass in the course of its development into an innovative output. One such hurdle may be an administrative superior—a person who, in some instances, has much to say about the goals and methods of the creator's work, and the resources available to him.

As shown in Table 5.6, project directors whose administrative superiors "stayed out of the way"—at least with respect to the actual conduct of the research—were the scientists who tended to obtain higher payoff from their creative abilities. The first three items shown in Table 5.6 were different ways of tapping general involvement by the administrative superior. Thereafter, the items refer to his influence over funds, hiring, research design, and decisions about goals and objectives. Consistently, the higher payoff occurred when the superior was less involved or exercised less influence.[18] It seemed as if creatively able people needed to run their own show if their efforts were to result in innovative outputs.

Several cautions need to be added lest the results of Table 5.6 be interpreted to mean that the superior has *no* role to play. Recall, first, that all the respondents were directors of their own projects. Presumably they were reasonably competent scientists with at least some administrative

TABLE 5.6. *Correlations between Creative Ability and Innovativeness (adjusted) by Role of the Scientist's Administrative Superior.*

	*RAT scores and Innovativeness**	
	r	*difference*
Extent administrative superior involved himself in the research:		
relatively little ($N = 29$)	.32	.47
relatively much ($N = 45$)	−.15	
Extent administrative superior interfered in the research:		
relatively little ($N = 54$)	.15	.32
relatively much ($N = 20$)	−.17	
Extent administrative superior limited *R*'s research activities:		
relatively little ($N = 37$)	.23	.35
relatively much ($N = 37$)	−.12	
Influence of administrative superior over all allocation of funds:		
relatively little ($N = 33$)	.32	.55
relatively much ($N = 35$)	−.23	
Influence of administrative superior over hiring of personnel:		
relatively little ($N = 31$)	.21	.31
relatively much ($N = 34$)	−.10	
Influence of administrative superior over research design:		
relatively little ($N = 31$)	.25	.34
relatively much ($N = 42$)	−.09	
Influence of administrative superior on deciding goals and objectives:		
relatively little ($N = 63$)	.13	.19
relatively much ($N = 46$)	−.06	
Scientist has no administrative superior ($N = 23$)	.15	

* Adjusted for productivity level

experience. The appropriate role for the administrative superior of a person at this level may involve encouragement, facilitation, friendly criticism, and administration of the laboratory, rather than close involvement with the details of others' research. By hindsight, we regret not having included questionnaire items to measure these aspects of an administrative superior's role.[19]

Size and Time Factors. In line with the reasoning that high creative ability will be more likely to be translated into innovative outputs if the creative process occurs in a flexible setting, one would expect that short, small projects would provide better settings for innovation than massive projects involving many professionals and/or lasting many years.

TABLE 5.7. *Correlations between Creative Ability and Innovativeness (adjusted) by Size and Duration of Project.*

	*RAT scores and Innovativeness**	
	r	*difference*
Number of professionals on project staff:		
R worked alone ($N = 14$)	.13	
R worked with one other ($N = 17$)	.36	
R worked with 2–3 others ($N = 35$)	.07	
R worked with 4+ others ($N = 28$)	−.07	
R's preference for individual or team research:		
R prefers to work alone ($N = 33$)	.28	.29
R prefers to work as part of a team ($N = 53$)	−.01	
Size of project budget:		
Median size or below	.15	.19
Above median size	−.04	
Duration of project:		
Under 1 year ($N = 18$)	.44	
1–2 years ($N = 32$)	.20	
2.1–3 years ($N = 20$)	−.09	
3.1 years or more ($N = 17$)	−.16	
Did the costs of the study exceed initial estimates?		
no ($N = 51$)	.16	.53
yes ($N = 23$)	−.37	
Did project exceed original time limit?		
no ($N = 31$)	.24	.24
yes ($N = 41$)	.00	

* Adjusted for productivity level

The results in Table 5.7 are in line with these expectations and replicate earlier findings (Andrews, 1967) with respect to the advantage of flexible situations for effective utilization of creative ability. Note that projects in which the scientist worked either alone or with one other professional showed the most positive correlations. As the number of professionals became larger than this, the correlations receded toward zero and then turned mildly negative.[20] Similarly, respondents who preferred to work alone showed more positive correlations than those who preferred to be part of a team. Unfortunately, no item had been included to distinguish between preferences for small teams versus large teams.

Table 5.7 also shows that projects with relatively small budgets and short durations were the ones where the director's creative ability showed the highest correlations with innovativeness of output. Finally, it was important that the project stay within its initial time and money estimates.[21]

Allocation of Effort. Turning next to the range of activities involved in the scientist's job, Table 5.8 shows that project directors who had considerable diversity in their work tended to have the more positive correlations between creative ability and innovativeness. Based on these data, it would appear that for optimizing payoff from creative ability a project director should spend up to three-quarters of his time on research (but not all of it devoted to a single project), some time on teaching, and/or some time on administrative duties not directly related to his own research.

TABLE 5.8. *Correlations between Creative Ability and Innovativeness (adjusted) by the Scientist's Allocation of Effort.*

	*RAT scores and Innovativeness**	
	r	*difference*
Time allocation to all research activities:		
1–50% ($N = 45$)	.19	
51–75% ($N = 29$)	.13	.37
76–100% ($N = 34$)	−.18	
Time allocation to this project:		
1–25% ($N = 57$)	.19	.33
26–100% ($N = 40$)	−.14	
Time allocation to teaching:		
some ($N = 64$)	.20	.28
none ($N = 32$)	−.08	
Time allocation to unrelated administrative duties:		
some ($N = 46$)	.27	.36
none ($N = 51$)	−.09	

* Adjusted for productivity level

Although our theoretical orientation had not specifically predicted the diversity effect which is consistently present in Table 5.8, one can well imagine that some diversity in work roles may be another aspect of the overall flexibility phenomenon discussed previously. Through such diversity, a person may receive useful stimulation, increased knowledge, and may also see opportunities for ways of translating creative ideas into innovative outputs. Furthermore, diversity may facilitate a work schedule which includes legitimated "incubation periods" for creative ideas. The notion that having time away from the task is a requisite part of the creative process has been mentioned by various eminently creative people.[22]

Motivation. It was expected that people who were more motivated by their projects would be more likely to develop their creative ideas to the point where they could result in innovative outputs.

Two attempts were made to tap the scientist's level of motivation. One item asked about his sense of involvement with the project, the other asked how important he felt the project was when the study began. As shown in the upper portion of Table 5.9, both items produced mild effects in the expected direction.

TABLE 5.9. *Correlations between Creative Ability and Innovativeness (adjusted) According to the Scientist's Level of Motivation.*

	*RAT scores and Innovativeness**	
	r	*difference*
Sense of involvement with project:		
relatively high ($N = 45$)	.13	.15
relatively low ($N = 64$)	−.02	
Sense of study's importance when study began:		
relatively high ($N = 36$)	.10	.09
relatively low ($N = 56$)	.01	
Sense of involvement with project:		
Doctoral scientists:		
relatively high ($N = 27$)	.27	.39
relatively low ($N = 37$)	−.12	
Nondoctoral scientists:		
relatively high ($N = 13$)	−.02	−.06
relatively low ($N = 20$)	.04	

* Adjusted for productivity level

An earlier analysis by the writer on how motivation level affected the payoff from creative ability had suggested that high motivation enhanced payoff only among doctoral scientists who worked in "research" labs (as distinguished from "development" labs) and that among nondoctoral scientists the effect was opposite (Pelz and Andrews, 1966). Would the same results occur in the present body of data? As is shown in the lower portion of Table 5.9, the answer is "yes." Clearly, the effect for these doctoral scientists (all of whom worked in research labs) was stronger than for nondoctorals, and among the nondoctorals the effect was virtually nonexistent and slightly negative.

While it is not clear why motivation level should have a greater influence over the creative behavior of doctoral scientists than nondoctorals, the explanation may lie in the greater independence that doctoral scientists customarily enjoy. Perhaps the work of nondoctorals is more affected by external stimuli in their laboratory, whereas that of doctorals is influenced to a greater degree by their own involvement and interest in a project.

Constraints. It seemed reasonable to expect that the scientists who perceived fewest organizational constraints in their laboratory would be most likely to succeed in translating creative ideas into innovative outputs.

Seven items from the questionnaire were relevant here; however, no consistent set of results emerged from their analysis. (No table shown.) The most direct item, "How adequate were the facilities and resources?" produced a moderate effect in line with expectations. However, three other items which asked in a general way about the role of the scientist's institution and discipline produced virtually no effects, and the three remaining items, which dealt with adequacy of time, money, and office space, produced moderate effects contrary to expectation. Thus the present data provide no consistent indication about how the creative process is affected by environmental constraints.

Communication. Ease of communication was another organizational factor which had been expected to facilitate the translation of creative ability into innovative outputs. Of six items relevant to this area, however, four showed little or no effects, and the other two showed inconsistent effects.[23] A previous study also examined several communication factors, but, as here, only weak and inconsistent interaction effects emerged (Andrews, 1967). Thus the two studies agree in suggesting that there is no massive, general effect of communication on the creative process—at least within the range of communicative phenomena measured in these laboratories.

SIMULTANEOUS PRESENCE OF SEVERAL ENVIRONMENTAL FACTORS

Up to this point social and psychological factors have been examined one at a time. Our orientation, however, suggests that one should consider the total impact of the environment on the creative process.

With only 115 cases there were insufficient data to examine in great detail the simultaneous presence or absence of numerous factors. As a step in this direction, however, and to see whether the several effects identified above would be cumulative in their influence—as our orientation suggested they should be—the analysis shown in Table 5.10 was carried out.

One "indicator item" was chosen from each of the following four areas: opportunity/responsibility for innovation, influence over decision making, noninterference by supervisor, and security. The specific items were chosen on the basis that they produced at least a moderate interaction effect when investigated separately, and that there existed an adequate number of cases in each of the split categories.

As can be seen in the table, the correlations between creative ability identified factors were "prerequisites" to achieving payoff from creative ability. If even one of these factors was absent, high creative ability apparently failed to get translated into innovative output.

With several factors absent, creatively able project directors actually produced *less* innovative outputs than other directors. Why might this be so? If a scientist has potentially creative ideas which repeatedly fail to get translated into innovative outputs because of the nature of his laboratory setting, one can imagine this man losing self-confidence, becoming dejected with his work, experiencing an intense sense of frustration, and eventually performing in maverick ways or perhaps hardly performing at all. While the idea that creative ability could be negatively related to innovation seemed surprising, this clearly occurred under certain circumstances. The oft-cited need to consider the "fit" between a scientist's abilities and the setting in which he works receives must support from this analysis.

DISCUSSION

This section has provided an explanation for the surprising finding of the preceding section that creative ability was apparently unrelated to innovativeness of output. The answer, it seems, lies in the social and psychological characteristics of the environment in which the creative process occurs. Under "favorable" conditions the higher one's creative ability the more innovative one's output. Under unfavorable conditions, however, the reverse was the case, and creative ability was negatively related to innovation. This is strong evidence for the proposition that it is important to consider the social and psychological setting in which the creative process occurs if one wants to understand the dynamics of that process.

Results presented here provide indications about the Specific social and psychological factors which have an influence on the creative process. These have been summarized in the "Overview of Results" which opens this section and need not be repeated here. It is worth noting, however, that while the impact of any one factor considered alone was often not very large, the impacts seemed to be cumulative, so when several factors were considered together they produced a very substantial influence on the process by which creative ability became manifested as innovative outputs. Only when each of the several factors being analyzed were simultaneously "favorable" did a strong positive relationship emerge between creative ability and innovation.

Findings III: Relationships of Social and Psychological Factors to Creative Ability and Innovation

Although the main thrust of this chapter has been to examine the effects of social and psychological factors on the relationship between creative ability and innovation, the data also permit examination of two other

TABLE 5.10. *Correlations between Creative Ability and Innovativeness (adjusted) by Number of Factors Present in Research Setting.**

	*RAT scores and Innovativeness***
4 factors present (0 absent) ($N = 26$)	+.55
3 factors present (1 absent) ($N = 32$)	−.07
2 factors present (2 absent) ($N = 21$)	−.19
1 factor present (3 absent) ($N = 8$)	−.40
0 factors present (4 absent) ($N = 4$)	−.97

* The four factors considered in this analysis and their indicator items were:

Initiation: Responsibility for initiating new activities—high.
Influence: Power to hire research assistant—respondent alone.
Role of superior: Interference from administrative superior—none.
Security: Stability of employment—high.

** Adjusted for productivity level

NOTE ON STATISTICAL SIGNIFICANCE: Given the numbers of cases in the subgroups, the −.97 and +.55 correlations are statistically significant beyond the .05 level. Consistent progression among five correlation coefficients in the predicted direction (high to low) would occur with probability = .06 under random conditions.

and innovativeness of output ranged from +.55 (where all four factors were present) to −.97 (in settings where none of the four factors was present). Furthermore, despite the small number of cases in some categories, the progression of the correlations was perfectly consistent with the decrease in the number of factors present.

Of course we have "stacked the cards in our favor" by selecting items which we knew had individual effects. Table 5.10 demonstrates, however, that the several individual effects were *cumulative,* and that the cumulated effect was of substantial size.

It is worth noting that the results shown in Table 5.10 were closely parallel to results obtained previously in an analysis of a different body of data. Among scientists who were involved in initiating new activities, and who were simultaneously high in influence, self-confidence, status, and motivation, there emerged a +.37 correlation between creative ability and "technical contribution." Among other groups of scientists correlations were close to zero or negative (Andrews, 1965).

Three of the four factors included in Table 5.10—responsibility for initiation, influence, and security—closely match factors in this previous analysis, and a marked similarity in overall trends emerged. On at least two occasions, therefore, a cumulative impact of several social and psychological factors on the creative process has been demonstrated.

It is interesting that a positive correlation occurred in Table 5.10 only when all four factors were present. This suggests the possibility that the interesting questions: 1) How did social and psychological conditions (as perceived) differ between scientists high and low in creative ability? and 2) How did these factors relate to innovativeness of output?

Either of these question could appropriately be the topic of a major analysis, and our discussion will only highlight some of the stronger relationships. Even these, however, provide interesting insights into the creative process.

CREATIVE ABILITY AND SOCIAL AND PSYCHOLOGICAL FACTORS

In general, scores on the Remote Associates Test were not strongly related to the kinds of social and psychological conditions tapped by the questionnaires. Among the stronger trends, however, were indications that scientists high in creative ability tended to describe themselves as "loners": they preferred to work alone, saw themselves as relatively isolated from their discipline, had less contact with colleagues and other professionals, and received less aid from their institutions. They also tended to feel relativey insecure and uncomfortable in their organizational settings. Specifically, they were less likely to be permanent members of their organizations, they believed the consequences of taking a risk that failed would be more dire, and they got along less well with their supervisors. These scientists also tended to spend more time on research and were more likely to have their primary role in research rather than in education or administration. To summarize: high creative ability tended to be accompanied by aloneness, insecurity, and greater focus on research.

These relationships agreed surprisingly well with the folklore which describes the creatively able person as a rather lonely, isolated, and often frustrated worker. Of particular interest was the relationship between creative ability and insecurity, for security was one of the environmental factors which affected the payoff from creative ability. This may pose one of the true dilemmas of managing an organization which includes creative individuals—the flow of risky new ideas, which may be unsettling to organizational stability, may elicit responses from the organization which erode the creator's sense of security. But that sense of security needs to be high if there is to be an effective utilization of those ideas. The contradiction between the kinds of environments which seem to promote payoff from creative ability and the kinds of environments in which creatively able people find themselves might be called the "security dilemma."

INNOVATION AND SOCIAL AND PSYCHOLOGICAL FACTORS

Among the more interesting relationships between innovation and the social and psychological factors was the tendency for people who had relatively *poor* relationships with their administrative superiors to produce somewhat more innovative outputs. These were scientists who said they got along poorly with their administrative superior, had difficulty communicating with their superior, had disagreements with their superior on the study's methodology and purpose, believed their superior held their work in relatively low esteem, and felt their superior had a poor understanding of their work.

DISCUSSION: THE "SECURITY DILEMMA"

It cannot be determined, from the present data, whether scientists had poor relationships with their superiors and felt insecure *because* of their creative abilities and innovative output, or whether the direction of causality was just the opposite. (Our theoretical orientation suggests the former.) One can, however, imagine a kind of homeostatic model which would parsimoniously encompass the several relationships which have been identified.

One could speculate that as a creatively able person puts forth numerous new, untried, perhaps risky ideas, he tends to "make problems" for his administrative superior; as a consequence, his superior may reduce (or fail to increase) the security aspects of the person's job; whereupon negative feedback occurs, and the person becomes less willing to propose risky new solutions in the future, and his rate of innovation falls until it is "in balance" with the level of security he encounters in his professional environment. The model might look something like this:

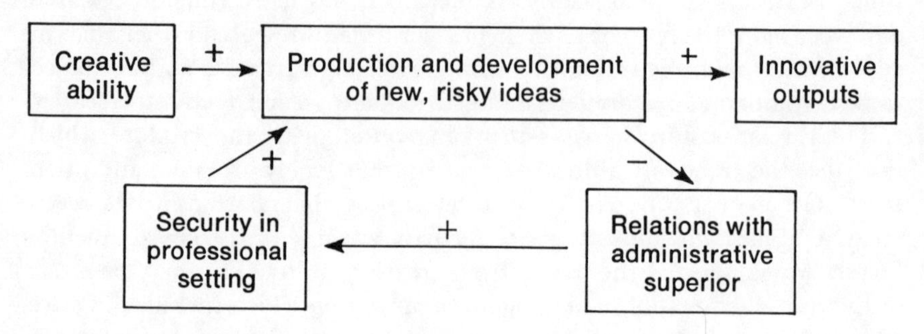

If the administrative superior could reduce the impact of the cycle, by providing relatively high independence for the person and/or by ensuring his security, the payoff from high creative ability might be enhanced.

Summary and Implications

One can conceive of the creative process as consisting of an input of new, potentially useful ideas, a series of developmental hurdles over which those ideas must pass, and an output of innovative products. The present research focuses primarily upon social and psychological factors which affect the likelihood of a new idea successfully becoming developed into an innovation.

The setting is the scientific laboratory. Creative ability (measured by

Mednick's Remote Associates Test) is taken as the input variable. The innovativeness of scientific reports (assessed by panels of expert judges) is the output variable. And the scientist's report about conditions in the laboratory where he conducted a specified research study provides information about the developmental "hurdles" and the process by which they were overcome.

The data about creative ability and organizational setting were obtained from slightly over 100 directors of research projects which investigated social psychological factors and disease. An abstract of the principal report of each project was independently assessed by 4.5 judges (on the average) for its innovativeness and productiveness. (The innovativeness judgments were subsequently adjusted to remove variance attributable to productiveness.) The adjusted innovativeness scores form the primary dependent variable.

Creative ability was found to be virtually unrelated to the measured qualities of scientific outputs when all project directors were considered together. (A measure of the project director's verbal intelligence was also unrelated to these.) However, mild positive relationships emerged for directors who: (1) had responsibilities or opportunities for innovation; (2) exercised considerable influence over decisions affecting the research; (3) felt a sense of professional security; (4) were allowed considerable independence by their administrative superior; (5) were strongly motivated toward the project; (6) worked on small, short projects; and (7) had diverse activities included as part of their work role. (Two other sets of factors—the presence of institutional constraints, and the ease of communication—were also examined, but produced inconsistent results.)

When the first four factors listed here were considered simultaneously, the correlation between creative ability and innovativeness of output varied from +.55 (all four factors present) to −.97 (all factors absent), showing that the different social and psychological factors could, cumulatively, exert a very substantial influence on the creative process. The nature of this multiple relationship suggested that these various factors might be acting as "prerequisites" for obtaining innovative payoff from creative ability.

Finally, a brief examination was made of the direct relationships between the various aspects of the research setting and creative ability and innovativeness. It was found that scientists with higher creative abilities tended to be relatively isolated from their colleagues and institutions, relatively low in their sense of professional security, and more exclusively research oriented. Those who produced more innovative outputs seemed to have more trouble than others in relating to their administrative superior. These results suggested that organizations seeking innovation may face a fundamental "security dilemma." Creative activities may be incompatible with organizational stability; as a consequence the creator's professional security erodes; but without that security he is unlikely to be able to effectively utilize his creative ability.

This research has both theoretical and practical implications. On the theoretical side, it shows the need to include social and psychological factors in theories about the dynamics of the causal process. Apparently it is not safe to assume that people with high creative abilities will generally produce highly innovative products. Some of the specific conditions which may enhance or block such a translation of input into output have just been described. Further investigation of the impact of these factors on the creative process in settings other than research labs would be in order, as would be attempts to identify additional factors which may also influence the creative process.

If it is determined by subsequent research that social and psychological factors have widespread influences on the creative process similar to those indicated here, this will have important implications for the validation of proposed measures of creative ability. Given a trait which may have either "positive" or "negative" effects, depending on the situation, and where the average effect may be close to zero, it would be inappropriate to assess validity by simply relating a measure of this trait (e.g., creative ability) to some criterion measure (e.g., innovativeness).[24] In assessing the validity of such a trait, appropriate social and psychological factors would have to be taken into account.

One of the central difficulties with the field of "creativity research" has been its failure to develop a set of unequivocal measures of creative ability which show high convergent validity among themselves. It is possible that this difficulty stems from the general neglect of social and psychological factors which may critically influence the creative process. Paying increased attention to such factors might result in both improved understanding of the creative process, and also a clearer conceptualization and measurement of the fundamental concept in the field—creative ability.

The present findings also have practical implications. They show that individuals concerned with promoting innovation within organizational settings (administrators of research and development, scientists, and certain members of many organizations devoted to activities other than research and development) need to ensure that the organizational climate facilitates the translation of new, potentially useful, ideas into innovative outcomes. The first part of this section summarizes some of the specific social and psychological factors which merit attention.

Notes

1. This research makes use of data collected previously by Dr. Gerald Gordon, my collaborator in a more general project of which just one part is reported here. His stimulation and support are gratefully acknowledged, as is the assistance received from Dr. Don-

ald Pelz, Frances Eliot, Ann Smith, and Lia Kapelis, who overcame many difficult problems in collecting new data for this study. The National Institute of Health, through grant GM-13507-01, provided financial support. Statements made herein are the responsibility of the author and not of the Public Health Service. This chapter is based on an unpublished paper titled "Social and organizational factors affecting innovation in research," presented at the 1970 Annual Convention of the American Psychological Association.

2. The following variables were examined: age, education, professional experience, principal professional activities, time allocation to project, role on project, size of project staff, project duration, involvement of administrative superior on project, and two qualities of performance: innovativeness and productiveness.

3. Goodman, et al. (1969), for example, found RAT scores did not correlate substantially with several other attempts to measure creative ability.

4. There were 13 people in this group. A check showed that the average score they obtained on the RAT was a full standard deviation lower than the mean for those who learned American English in childhood. (The mean for this latter group was 17.8; the standard deviation, 5.4.)

5. Used with permission from the Michigan Employment Security Commission.

6. To assess scoring reliability a 25 percent sample of the completed RAT's and GATB's were rescored by a second test scorer. The percentage of items where the two scorers agreed exceeded 99 percent, a highly satisfactory level.

7. Somewhat comparable retest reliabilities of questionnaire items answered by scientists have been reported by Pelz and Andrews (1966, p. 292). For 89 items over a two-month interval, they report a median correlation coefficient of .62.

8. As a further precaution against unreliability in data about the laboratory environment, separate parallel analyses have been conducted for each questionnaire item relevant to a given concept. Conclusions have been drawn only when nearly all such analyses point to the same finding.

9. Gordon (1963) compares ratings based on abstracts of project reports with ratings based on the full report of some of the projects included in the present study. From data he presents one can determine that the magnitude of agreement between the two ratings was $\gamma = .50$ for innovativeness and $\gamma = .67$ for productivity.

10. This prediction was achieved by the simple regression equation: $X = 11.88 + 0.77Y$, where X was predicted innovation and Y was observed productiveness (scaled to have a mean of 50 and a standard deviation of 10).

11. For other instances in which intelligence was found to be unrelated to occupational performance, see Super and Crites (1962) and Kraut (1969).

12. In statistical terminology these would be known as an "interaction." Other terms sometimes used for these phenomena are "moderator effects," "contingency effects," or "conditioning effects."

13. The fact that these three environmental factors were only modestly related to one another (gammas ranged from .11 to .31) made the finding of a consistent effect particularly interesting.

14. As before, the demonstration of a consistent effect becomes more meaningful if the items that produced that effect represent somewhat different ways of tapping an underlying concept. The gammas across the 28 possible pairs of these 8 items had a median value of .43 and ranged from .07 to .83.

15. Lest the reader be tempted to conclude, on the basis of data shown for age in Table 5.4, that the late 30s were an especially propitious time for translating creative ability into innovative output ($r = .34$), note that the data for years since receiving highest degree—most often a doctoral degree among these scientists—were directly contrary. People with 5–10 years research experience, most of whom would be in their 30s, showed a *negative* correlation ($r = -.35$). The lack of consistency suggests we have not identified any real effect.

16. Omitting the three items referring to the scientist's first-named colleague, the median gamma among the 55 pairs of remaining items was .27 (range: $-.20$ to .91). Thus the generally consistent interaction effect was not attributable to a uniformly high relationship among these items.

17. Pelz and Andrews (1966) have described a possible "dither" effect: that some mental

"shaking" may be required to keep innovating scientists keen and fresh. The finding that creative ability related more positively to innovativeness when a scientist tended to disagree with his main colleague and felt their professional relationship was relatively poor could be attributed to such a dither effect.

18. Once again the reader is reminded that the effect apparent in Table 5.6 was not simply a matter of seniority or prior experience (see Table 5.8). Nor was the consistent effect attributable to very high relationships among the items—median gamma was .34 (range: .16 to .87).

19. In a different study involving some of the same concepts, Andrews and Farris (1967) found that innovativeness in scientific teams tended to be higher when the team supervisor was seen as effective in such "task" functions as exercising technical skill, critical evaluation, and influence on goals. Two important distinctions are to be noted between the present study and the Andrews and Farris study: first, the present study focuses on the relationship between creative ability and innovativeness, not the absolute level of innovativeness; second, the superior-subordinate relationship investigated in the present study exists between people relatively high in a laboratory hierarchy, whereas the Andrews and Farris study examined the superior-subordinate relationship among people relatively low in the hierarchy.

20. This is one of the few places where the unadjusted innovation scores were more affected by one of the control factors than the adjusted scores. The basic trend, however, was similar to that described here.

21. A subsequent section on "constraints" discusses the effects of insufficient time or money.

22. A discussion of diversity and its *direct* relationship to scientific productivity appears in Pelz and Andrews (1966, chapter 4).

23. The six items tapped the following areas: sense of isolation from parent discipline, frequency of contact with colleagues, ease of getting across new ideas, problems in communicating with administrative superior, extent that work got discussed with people on other related projects, and whether the scientist maintained contact with other studies.

24. As an example, see Baird's (1972) review of the Remote Associates Test. As evidence questioning the RAT's validity, he cites previous research of the present writer showing that the RAT did not generally relate to several qualities of scientific performance. However, Baird does not go on to add that the relationships were affected by environmental conditions and that positive relationships emerged under certain conditions.

References

Andrews, F. M. Factors affecting the manifestation of creative ability by scientists. *Journal of Personality*, 1965, *33*, 140–152.

Andrews, F. M. Creative ability, the laboratory environment, and scientific performance. *IEEE Transactions on Engineering Management*, 1967, *14*, 76–83. (Note: same material also appears in Pelz and Andrews, 1966, chapter 9.)

Andrews, F. M., and Farris, G. F. Supervisory practices and innovation in scientific teams. *Personnel Psychology*, 1967, *20*, 497–515.

Baird, L. L. Review of Remote Associates Test. In O. K. Buros (Ed.), *The seventh mental measurements yearbook*. Vol. 1. Highland Park, N.J.: Gryphon Press, 1972.

Ben David, J. Scientific organization and academic organization in nineteenth century medicine. *American Sociological Review*, 1960, *25*, 828–843.

Goodman, P., Furcon, J., and Rose, J. Examination of some measures of creative ability by the multitrait-multimethod matrix. *Journal of Applied Psychology*, 1969, *53*, 240–243.

Gordon, G. The problem of assessing scientific accomplishment: a potential solution. *IEEE Transactions on Engineering Management*, 1963, *EM-10*, 192–196.

Gordon, G., and Charanian, T. Measuring the creativity of research scientists and engineers. Working paper, Project on Research Administration, University of Chicago, 1964.

Kraut, A. I. Intellectual success and promotional success among high-level managers. *Personnel Psychology,* 1969, *22,* 281–290.

Mednick, S. A. The associative basis of the creative process. *Psychological Review,* 1962, *69,* 220–232.

Mednick, S. A., and Mednick, M. T. *Manual: Remote Associates Test.* Form I. Boston: Houghton-Mifflin, 1966.

Mendelsohn, G. A., and Griswold, B. B. Assessed creative potential, vocabulary level, and sex as predictors of the use of incidental cues in verbal problem solving. *Journal of Personality and Social Psychology,* 1966, *4,* 423–432.

Pelz, D. C., and Andrews, F. M. *Scientists in organizations: Productive climates for research and development.* New York: Wiley, 1966.

Super, D. C., and Crites, J. O. *Appraising vocational fitness.* New York: Harper & Row, 1966.

U.S. Department of Labor. *Manual for General Aptitude Test Battery.* Washington, D.C., October 1967.

TIME PRESSURE

Frank M. Andrews and George F. Farris

Time pressure experienced by scientists and engineers predicted positively to several aspects of performance including usefulness, innovation, and productivity. Higher time pressure was associated with above average performance during the following five years, even when supervisory status, education, and seniority were controlled. Performance, however, did not predict well to subsequent reports of time pressure, suggesting a possible causal relationship from pressure to performance. High performing scientists also desired more pressure. Innovation and productivity (but not usefulness) were low if the pressure experienced was markedly above that desired. The five-year panel data derived from approximately 100 scientists in a NASA laboratory. Some theoretical and practical implications of the results are discussed.

Time pressure is often cited as a problem experienced by members of formal organizations. Moreover, it is an administratively interesting factor, since it is one over which management may have substantial influence.

The folklore about managing scientific laboratories includes two competing approaches to the management of time pressure: (1) provide professional staff with an unhurried "academic" environment, and (2) establish tight schedules and deadlines to avoid the Parkinsonian nightmare of work expanding to meet the time available.

What is the relationship between time pressure and scientific performance? Does time pressure tend more to predict performance or to be predicted by past performance? What characteristics of a scientist's working

[1] This research was supported by Grant NGR23-005-395 from the National Aeronautics and Space Administration. The authors are grateful for helpful comments from Donald C. Pelz, John R. P. French, Jr., and Raymond Faith, and for the technical assistance of Marita Di Lorenzi.

This article originally appeared as "Time Pressure and Performance of Scientists and Engineers: A Five-Year Panel Study" in *Organizational Behavior and Human Performance*, Vol. 8, No. 2, October 1972 and is reprinted here with the permission of the publisher.

environment are associated with his sense of time pressure? It is to these questions that the present article is addressed.

Despite the acknowledged importance of time pressure in organizations, surprisingly little research has been directly devoted to it. In group interviews about job pressures, Hall and Lawler (1971) found that a sense of time pressure was mentioned in more than three-quarters of the 22 scientific laboratories they studied. Of all the different kinds of pressures mentioned by the scientists and engineers in these interviews, "by far the most widely felt pressure was time (p. 67)." Although some of the other pressures did relate to laboratory performance as rated by the laboratory manager, no significant relationship was found between time pressure and performance.

The Hall and Lawler results were foreshadowed to some extent by two sets of findings reported by Pelz and Andrews (1966). One set indicated that scientists and engineers were especially likely to be low performers if they worked under conditions of loose coordination and high autonomy—a situation under which many kinds of job pressures might be expected to be minimal. Pelz and Andrews suggested that the low performance might be a result of low stimulation and/or motivation. The notion agrees well with that advanced by Hall and Lawler, who also suggested that motivational factors might account for the observed relationships between pressure and performance. Another set of findings by Pelz and Andrews indicated that scientific performance tended to be greater for those scientists and engineers who worked a nine- or ten-hour day, on the average; those who averaged only a standard eight-hour day or an eleven-hour day tended to perform at lower levels. To the extent that working hours are dictated by time pressure, these findings suggest a curvilinear relationship between time pressure and performance.

Related to time pressure is the concept of "overload." Kahn, *et al.* (1964) consider overload as "one of the dominant forms of role conflict . . . , which can be thought of as a conflict among legitimate tasks, or a problem in the setting of priorities (p. 380)."

> Overload could be regarded as a kind of inter-sender conflict in which various role senders may hold quite legitimate expectations that a person perform a wide variety of tasks, all of which are mutually compatible in the abstract. But it may be virtually impossible for the focal person to complete all of them within given time limits (p. 20).

Recent studies of colleague roles in a scientific laboratory indicate that role overload may be one characteristic of a scientist's working environment which is related to a sense of time pressure. Farris (1971) and Swain (1971) found that scientists who were named by more of their colleagues as helpful in their technical problem solving (a situation likely to engender inter-sender role conflict) experienced a greater feeling of time pressure.

Miller (1960) considers responses to information input overload, another factor which may be related to feelings of time pressure. Some of these are clearly dysfunctional in the organizational context—failing to process some of the information, processing some of the information incorrectly, or escaping from the task. Others may be functional or dysfunctional, depending on other factors—queuing, filtering, approximation, or employing multiple channels. As Katz and Kahn (1966) point out,

> people are likely to process the familiar elements in a message, which they readily understand and which do not constitute major problems for them. Under time pressures the parts of the communication difficult to decode are neglected for the more easily assimilated parts, even though the former may be more critical for the organization (p. 232).

Taken together, this research and theory on overload suggest that time pressure and overload are related. Role overload may be a source of time pressure, and responses to information overload, experienced by the scientist or engineer as time pressure, may well be dysfunctional for his performance.

Given this slim body of research and theory related to time pressure and performance, it is not surprising that the folklore about managing the time pressure of scientists is so contradictory. In the present study we shall attempt to resolve some of these contradictions. Specifically we shall consider the following questions:

1. How much time pressure is experienced by scientists and engineers in a government laboratory? How much time pressure do they consider optimum?
2. How does experienced time pressure relate to the scientist's performance as measured by his usefulness to his organization? Is time pressure related more strongly to the scientist's past usefulness or his subsequent usefulness?
3. How does experienced time pressure relate to the scientist's performance as measured by the innovation and productiveness of his work? Is time pressure related more strongly to the scientist's past innovation and productiveness or his subsequent innovation and productiveness?
4. How does experienced time pressure relate to these five characteristics of the scientist—freedom provided by his supervisor, preferences for working alone, involvement in technical work, time spent on administrative duties, and number of close colleagues?
5. How does performance relate to three other aspects of time pressure—optimal time pressure, the difference between experienced and

optimal pressure, and the "span" of different pressures experienced during a typical month's work?

METHOD

The present study was conducted in a NASA research division where scientists and engineers were exploring the effects of extreme physical conditions on various materials. Their work involved a mixture of research, development, and technical services.

The first wave of data was collected in 1965 (Time 1) from 117 scientists and engineers. The second wave occurred five years later (Time 2), and was based on 118 professional personnel, 78 of whom had also participated in 1965. At both Time 1 and 2, each participant completed a lengthy questionnaire and his performance was evaluated by judges selected from among other professionals in the lab.

Performance Measures

The performance criteria included the following:

Innovation—the extent the man's work had "increased knowledge in his field through lines of research or development which were useful and new,"

Productiveness—the extent the man's work had "increased knowledge along established lines of research or development or as extensions or refinements of previous lines," and

Usefulness—the extent the man's work had been "useful or valuable in helping his R & D organization carry out its responsibilities."

These qualities were independently assessed by an average of 4.4 judges at Time 1, and 7.6 at Time 2, each of whom claimed to be familiar with the man's work. Each judge ranked the scientists with whom he was familiar on the basis of their work over the preceeding five years. Approximately two-thirds of the judges were supervisors (the man's own chief might be among them), and one-third were senior-level nonsupervisors. Since the judges showed reasonably good agreement, their evaluations were combined into a single percentile score (on each quality) for each respondent.[2]

As is usually found for scientists and engineers, these performance measures varied according to the respondent's length of experience, seniority, and formal training (Pelz & Andrews, 1966). Since these effects might mask the relationships of interest, all performance measures were adjusted by adding or subtracting appropriate constants to remove such back-

[2] Based on the average inter-judge agreement and the average number of judges, the reliability of the performance ratings was estimated to be .95 at Time 1 and .88 at Time 2, using the Spearman–Brown formula (Guilford, 1954).

ground effects.[3] Thus, the final performance measures expressed how well or poorly each person performed *relative* to others with similar experience and training.

The interrelationships among these various criteria of performance were about as expected. All were substantially related to one another (correlations ranged .7 to .8 at both Times 1 and 2), with innovation being least similar to the others.

Measures of Time Pressure

The time pressure experienced by the respondent was assessed by the following question:

> Technical jobs sometimes involve working under time pressures exerted by other people—results are needed urgently, there are deadlines to be met, etc. In a typical month about what proportion of your time is spent working under the following amounts of pressure? (Five categories of pressure were listed, from "Relaxed—no pressure at all" to "Extreme pressure—I'm behind on important deadlines." The respondent entered the percentage of his time spent under each amount of pressure.)

Optimum time pressure was measured by a subsequent question which asked the respondent to indicate what he thought would be the optimum proportion of his time spent under each level of pressure in order for him to make his best contributions. From this basic information four scales were constructed:

(1) Typical level of time pressure experienced;[4]
(2) Amount of time pressure the respondent felt would be optimal;[4]
(3) Difference between the actual and optimum;[5] and
(4) "Span" or "range" of time pressures actually experienced.[5] (A person who said nearly all his work occurred under a single level of pressure had a low span; those who experienced widely different pressures had a high span.)

[3] The procedures for collecting, combining, and adjusting the performance measures used in this study were highly similar to those more fully described in Pelz and Andrews (1966). However, experience was not considered separately from seniority in adjusting the performance measures collected at Time 2.

[4] This scale was based on the *median* amount of pressure indicated by each scientist (e.g., one who experienced "slight" pressure during 30% of his work time, "moderate" pressure for 40%, and "great" pressure for 30% would be grouped with others whose median also fell in the "moderate" range). The resulting distribution was unimodal and reasonably symmetric.

[5] This scale was transformed to yield a reasonably symmetric unimodal distribution appropriate for analysis using statistics such as the Pearson correlation coefficient.

Other Variables

In addition to the performance and time pressure variables, the questionnaire asked about a wide range of other phenomena including motivation levels, communication with colleagues, role of the technical supervisor, and attitudes toward work. These other variables contributed to insights about the relationship of time pressure to performance and will be mentioned at appropriate places later in the article.

RESULTS

Distribution of Time Pressures

The amount of time pressure typically experienced by respondents in this site varied widely (see Table 1). At Time 1, 27% devoted more than half of their time to activities for which they experienced no sense of urgency. At the other extreme, 8% said half or more of their time was spent on activities for which there was "great" or "extreme" urgency. The distribution of time pressures was roughly similar at Time 2 to what it was at Time 1, though pressure levels tended to be somewhat lower.

With respect to optimum time pressure, there was again substantial variation between respondents. Almost all wanted at least some pressure, and those who *experienced* more pressure were generally the ones who also *wanted* more (correlations between typical and optimum time pressures were .5 at Time 1 and .6 at Time 2).

While some respondents experienced more pressures than they felt would be optimal (and some had less pressure than desired), experienced pressure levels were just slightly above optimal levels when averaged across all respondents. In short, these scientists and engineers—in the aggregate—were reasonably well satisfied with respect to the time pressure experienced on their jobs.

TABLE 1
MEDIAN LEVEL OF TIME PRESSURE EXPERIENCED
(PERCENT OF RESPONDENTS)

Median level of pressure	Time 1	Time 2
Relaxed	27%	39%
Slight pressure	33	30
Moderate pressure	32	24
Great pressure	7	7
Extreme pressure	1	0
Total	100%	100%
N	(117)	(118)

Experienced Time Pressure and Usefulness

Given the nature of the data, it is possible to examine how the several measures of time pressure—experienced, optimum, deviation of experienced from optimum, and span—related to three distinct aspects of performance—its innovativeness, productiveness, and usefulness. Furthermore, one can examine relationships at one point in time and also "lagged" relationships. For example, one can determine whether experiencing time pressure at Time 1 was *predictive* of performance levels during the following five years, and whether performance prior to Time 2 was predictive of *subsequent* feelings of time pressure.[6]

Figure 1 shows the interrelationships among experienced time pressure and judgments of usefulness (adjusted for experience and formal education, as described previously) at Times 1 and 2.

The most important result occurs in the cross-lagged correlations. Experienced time pressure, measured at Time 1, related +.49 to *subsequent* usefulness of scientists and engineers (Time 2); however, Time 1 usefulness was virtually unrelated (+.10) to subsequent time pressure.[7] Interestingly, experienced time pressure related more strongly to subsequent usefulness (r = .49) than it did to usefulness measured at the same time as time pressure (r = .32 at Time 1, .20 at Time 2).

These results suggest that not only did above-average sense of time pressure characterize the more useful members of a laboratory, but that

[6] The appropriate way to analyze this type of data has been the subject of lively methodological debate in recent years. Campbell and Stanley (1963), and Pelz and Andrews (1964) independently proposed the "cross-lagged panel correlation" technique. Yee and Gage (1968), Duncan (1969), Rozelle and Campbell (1969), Heise (1970), Kenny (1970), Rees (1971), and Sandell (1971) have proposed modifications to the analysis method or to the interpretations which are appropriate. The focus of attention has been on what conclusions about underlying causal dynamics could be drawn on the basis of an observed difference in cross-lagged relationships.

It seems clear that a statistically significant difference between two cross-lagged panel correlations provides strong evidence that the co-variation between two variables is not solely the result of their relationships to some third variable (i.e., a "common factor"). Moreover, Farris (1969c) has argued that a lagged correlation different from zero provides a basis for considering causal hypotheses in dynamic social systems, provided that certain other conditions have been met. Having rejected alternative explanations, certain causal hypotheses may be considered. The choice among these hypotheses will depend on particular assumptions or additional data.

[7] The difference between the cross-lagged relationships was statistically significant at the .005 level using the Pearson–Filon test (Peters & Van Voorhis, 1940).

Heise (1970) has proposed that rather than examining correlations for the cross-lagged relationships one should examine path coefficients. The path coefficients turn out to be +.3 and .0, showing essentially the same pattern as the correlations.

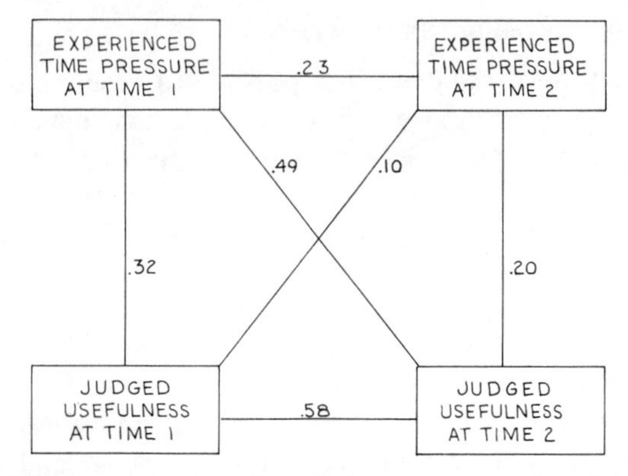

FIG. 1. Relationships among experienced time pressure and usefulness at two time periods (Pearson correlations).
Note.—See "Method" section for numbers of cases.

their sense of time pressure may well have been partially *responsible* for their higher usefulness. However, it was *not* the case that scientists and engineers who were judged more useful subsequently found themselves under markedly above-average time pressure.[8]

These findings are in sharp contrast to results reported by Farris (1969a, 1969b) for a number of other factors. In three industrial laboratories, he found that scientists' job involvement, influence, salary, and number of subordinates each tended to relate more strongly to prior performance than subsequent performance. His findings held with different time lags—when the performance measurement referred to the five years immediately prior to the measurement of the organizational factor (what we are calling "simultaneous" relationships in this panel study) or when there was a five year time lag between the measurement of performance and the measurement of the organizational factors (as in the lagged relationships in the present study).

These findings were sufficiently interesting that a number of additional analyses were run to see whether the time pressure–usefulness connection could be easily explained away. In adjusting the performance measures for differences in training and experience, had some artifact been introduced? No, a parallel analysis on the unadjusted measures showed a highly similar pattern. Could it be attributed to mixing supervisors and nonsupervisors? No, when the analysis was carried out just for people

[8] Figure 1 also shows substantial stability ($r = .58$) in judgments about a man's usefulness over the five year period, and also some tendency for stability ($r = .23$) in the amount of time pressure experienced.

who had been in nonsupervisory roles at both Times 1 and 2, the same pattern again emerged.

What about the level of time pressure? Was the relationship underlying the positive correlations linear, or did performance tend to drop at the highest levels of pressure? An extensive check showed that all of the relationships depicted in Fig. 1 were essentially linear. Figure 2 shows the two relationships involved in the cross-lagged comparison. The solid line, which corresponds to what we believe to be the underlying causal dynamics, is the most interesting. Note that scientists who indicated (at Time 1) that at least half of their time was spent under "relaxed" conditions scored, on the average, at the 34th percentile with respect to usefulness five years later when compared to others of similar training and experience. In contrast, those few who at Time 1 had said half or more of their work time was spent under "great" pressure, averaged at the 74th percentile on usefulness five years later. Similarly the dashed line in Fig. 2 shows the essentially linear relationship underlying the +.10 correlation in Fig. 1.

Why then, do these findings differ from the earlier longitudinal relationships reported by Farris (1969a, 1969b)? Perhaps "performance feedback loops"—relationships between performance and subsequent characteristics of a scientist's working environment—are in fact stronger in industrial laboratories like those studied by Farris than in government

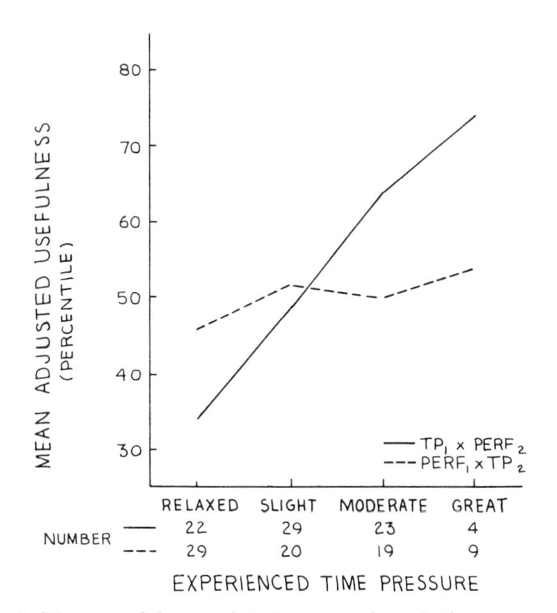

FIG. 2. Mean usefulness related to experienced time pressure.

TABLE 2
CORRELATIONS AMONG EXPERIENCED TIME PRESSURE AND TWO PERFORMANCE MEASURES AT TIMES 1 AND 2[a]

	Cross-lagged relationships		Simultaneous relationships		Stability relationship[b]
Performance measure	TP_1 & $Perf_2$	$Perf_1$ & TP_2	TP_1 & $Perf_1$	TP_2 & $Perf_2$	$Perf_1$ & $Perf_2$
Productiveness	.21	.05	.29	.09	.45
Innovation	.25	−.14	.23	.07	.60

[a] See "Method" section for numbers of cases.
[b] For stability of time pressure, see Fig. 1.

laboratories such as those in the present study. Or alternatively—unlike involvement, influence, salary, or number of subordinates—time pressure may indeed be a factor which relates more strongly to subsequent performance than to prior performance in scientific laboratories.

Time Pressure, Innovation, and Productiveness

When analyses parallel to those shown in Fig. 1 were carried out for the other performance measures—judged innovation and judged productiveness—similar patterns were obtained in the cross-lagged correlations, though trends were weaker. Table 2 provides the results.

Looking first at productiveness, one again sees a positive relationship ($r = .21$) between time pressure and subsequent productiveness, but practically no relationship ($r = .05$) between productiveness and subsequent time pressure. Again, it would appear that time pressure may have enhanced performance.[9]

Turning next to results for innovativeness, one again encounters a substantial positive difference in the cross-lagged relationships. As before, time pressure was positively related ($r = .25$) to subsequent performance (innovativeness). But note, also, the mild *negative* relationship between innovativeness and subsequent time pressure![10] This is particularly intriguing in view of the folklore, mentioned previously, that scientists, par-

[9] This statement is based on the similarity in trends to that observed previously. With the number of cases available this particular cross-lagged differential was statistically significant only at the .15 level. Heise's path coefficients (.09 and −.02) showed a pattern similar to that of the cross-lagged correlations. The underlying relationships were essentially linear.

[10] The cross-lagged difference was statistically significant at the .005 level. The Heise path coefficients were .12 and −.20, again matching the pattern of the cross-lagged correlations. Relationships were generally linear.

ticularly those doing creative work, need a relaxed environment. Among these scientists, those judged more innovative at Time 1 showed a mild tendency to experience lower-than-average time pressures at Time 2. Perhaps research management saw the reduction of time constraints as an appropriate way to encourage further creativity. Our data suggest, however, that innovation prospered under time pressure just as did other more routine aspects of scientific performance.

Characteristics of Scientists Who Felt High and Low Time Pressure

Other data provide insights into the differences between scientists who experienced high and low levels of time pressure and contribute to the validity of the time pressure measure.

Scientists who felt less than average time pressure tended to be rather isolated, free from influence and interference from their supervisor, relatively uninvolved in their work, and with below-average administrative duties. In contrast, those who were highly motivated, in vigorous contact with colleagues and supervisors, and with some administrative responsibilities in addition to their technical work were the ones on whom time pressures impinged the most. (These relationships produced correlations in the range .1 to .5—data not shown.)

TABLE 3
CORRELATIONS AMONG EXPERIENCED TIME PRESSURE AND FIVE OTHER JOB CHARACTERISTICS AT TIMES 1 AND 2[a]

	Cross-lagged relationships		Simultaneous relationships		Stability relationship[b]
Job characteristics	TP_1 & $Char_2$	$Char_1$ & TP_2	TP_1 & $Char_1$	TP_2 & $Char_2$	$Char_1$ & $Char_2$
Freedom provided by supervisor	.06	−.25	−.08	−.19	.37
Preferences for working alone	−.38	−.08	−.13	−.31	.56
Involvement in technical work	.16	.03	.25	.18	.66
Time on administrative duties	.40	.27	.34	.19	.37
Number of close colleagues	.43	.48	.39	.33	.51

[a] See "Method" section for numbers of cases.
[b] For stability of time pressure, see Fig. 1.

Table 3, which shows cross-lagged analyses for time pressure and these variables, suggests that the provision of freedom by a supervisor may be one *cause* of a scientist's later feeling under reduced time pressure; on the other hand, time pressure itself seemed to have a causal role in a person's not preferring to work alone.[11] Although the cross-lagged differentials for work involvement and time on administrative duties did not reach conventional levels of statistical significance, the trends in Table 3 suggest that feelings of work involvement and administrative duties were more likely to result from previous time pressures than was a feeling of time pressure likely to result from them. Finally, although time pressure and the number of close colleagues a person worked with were substantially related, there was no clear evidence that either had causal priority over the other.

Do these results imply that if a supervisor provides freedom for his subordinates their performance will fall? Not necessarily. [In fact, Pelz and Andrews (1966) found that among scientists within the same career level freedom was positively related to scientific performance.] However, if substantial freedom is provided, some additional actions may be required to ensure that scientists stay "hot." We would not want to imply that time pressure is the only motivator, though the results described suggest it may be one important source of motivation.

Other Time Pressure Measures and Performance

In addition to the time pressure actually experienced by a scientist, the study included three other time pressure measures—the time pressure which the scientists themselves felt would be optimal, the difference between experienced and optimum pressure (one indication of "overload") and the "span" of different pressures experienced during an average month's work. Each of these measures was analyzed in a manner similar to that just described for experienced pressure.

Optimal pressure. Optimal pressure, which itself correlated +.6 with experienced pressure, gave results generally similar to those shown in Figs. 1 and 2 and Table 2. Scientists who *wanted* above average levels of pressure at Time 1 showed a mild tendency to be the better performers during the following five years (r's averaged .22). Performance at Time 1, however, showed weak and inconsistent relationships to time pressure desires five years later (r's averaged .02).

[11] In both cases the differences in cross-lagged relationships were highly statistically significant. The items were worded as follows: "My supervisor provides considerable freedom for people under him to explore, discuss, and challenge ideas on their own." "I'm rather a lone wolf; prefer to work on my own." To answer, respondents indicated how accurate the statement was, using a 7-point scale.

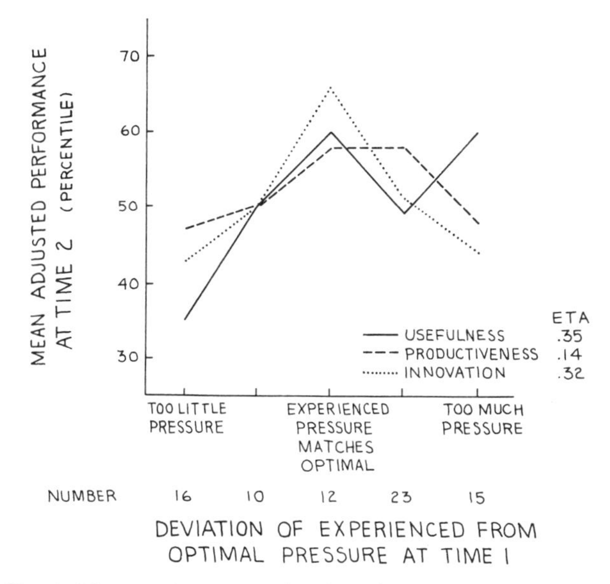

FIG. 3. Mean performance related to time pressure overload.

Overload. The overload measure showed some very interesting curvilinearities. Figure 3 presents the lagged relationships between overload at Time 1 and performance at Time 2. Scientists who had less pressure at Time 1 than they wanted tended to be low performers during the following five years. Those for whom experienced pressure closely matched what they felt would be optimal showed above-average performance in the following period. When pressures exceeded what was desired, subsequent innovation and productivity fell, though usefulness tended to be high. Also of interest was the fact that these lagged relationships (note the ETA's in Fig. 3) were consistently stronger than the comparable simultaneous relationships (not shown). In short, having more time pressure than was desired had more to do with subsequent performance than recent past performance.

These curvilinearities provide an important additional insight into the meaning of the relationships between experienced pressure and performance described previously. While it was true that the higher the pressure, the higher the performance, we now see that this could occur only because the scientists who experienced high pressures also wanted high pressures. Figure 3 shows that being subject to more pressure than was felt appropriate was *followed* by relatively low innovation and productiveness (but not usefulness). From a practical standpoint, it would appear that laboratory managers must take account of what pressures scientists feel are appropriate when setting pressure levels in their labs. Otherwise, the managers may find that they sacrifice some quality in the scientists' work in order to make it more useful to the organization. Of course, managers might also attempt to influence what are seen as appropriate levels of time pressure.

Span. The fourth time pressure measure—the "span" of pressures experienced—showed no interpretable relationships. Performance was unrelated to this aspect of time pressure.

CONCLUSIONS

Contrary to the folklore which holds that scientists perform best when in a relaxed "academic" environment, these data suggest that a sense of time pressure can enhance several qualities of scientific performance—including innovation. In addition to experiencing the most time pressure, the highest performing scientists also tended to *want* relatively large amounts of pressure. When the pressure actually experienced was markedly out of line with the pressure desired—either in being too low or too high—performance was likely to suffer. (Exception: excess pressure did not seem to hurt a scientist's judged usefulness to his lab.)

The fact that these findings are based on panel data collected over a five-year interval provides suggestions of causal dynamics not possible when relationships are among variables measured at just a single point in time. At the very least, the findings above represent *predictive* relationships (time pressure related to *subsequent* performance), and it seems most unlikely that they result from the spurious effect of some third factor.

Scientists who experienced above average time pressures tended to be those who were in active communication with colleagues, motivated by their jobs, and involved in some administrative duties as well as technical activities. In short, they were well integrated into the social processes of their laboratories. Or, in the language of Kahn *et al.* (1966), they received expectations from a number of role senders and were more susceptible to role conflict and role overload. The range of role conflict and role overload experienced by the scientists in this study appeared to enhance performance—provided that optimal and experienced time pressures were not greatly out of line.

The implication for management is that the imposition of deadlines and other forms of time pressures need not be feared—at least with respect to their effect on a man's performance—so long as the resulting pressure stays within the bounds of what is felt to be appropriate by the man involved. Some attempt to boost scientist's own *desire* for time pressures may permit the acceptance of higher pressures.

However, two cautions need also to be mentioned: (a) although we did not encounter a pressure level that was "too great" among the scientists we studied, there presumably is such a level, and pressures would need to be kept below it; (b) the present study had no information about other effects of pressure (e.g., on physical or mental health). The work of Kahn *et al.* (1964), Sloate (1969), and French and Caplan (1970) on organizational stress, and of Miller (1960) on information overload, suggests that time pressure may have negative consequences not considered in this study. Further research should clarify relationships between time pressure and other aspects of stress and overload, and consider effects of time pressure on factors other than performance. The present study, combined with the other studies just cited, indicates that different kinds of pressure may have positive as well as negative effects, depending on whether the criterion is performance, long-run organizational effectiveness, or the health of those in stressful situations.

REFERENCES

Campbell, D. T., & Stanley, J. C. Experimental and quasi-experimental designs for research on teaching. In N. L. Gage (Ed.), *Handbook of research on teaching.* Chicago: Rand McNally, 1963.

Duncan, O. D. Some linear models for two-wave two-variable panel analysis. *Psychological Bulletin,* 1969, **72**, 177–182.

Farris, G. F. Organizational factors and individual performance: A longitudinal study. *Journal of Applied Psychology,* 1969, **53**, 87–92. (a)

Farris, G. F. Some antecedents and consequences of scientific performance. *IEEE Transactions on Engineering Management,* 1969, **EM-16**, 9–16. (b)

Farris, G. F. Toward a non-experimental method for causal analyses of social phenomena. *Australian Journal of Psychology,* 1969, **21**, 259–276. (c)

Farris, G. F. Executive decision making in organizations: Identifying the key men and managing the process. M. I. T. Sloan School of Management Working Paper No. 551–71, 1971.

French, J. R. P., Jr., & Caplan, R. D. Psychosocial factors in coronary heart disease. *Industrial Medicine and Surgery,* 1970, **39**(9), 31–45.

Guilford, J. P. *Psychometric methods.* New York: McGraw–Hill, 1954.

Hall, D. T., & Lawler, E. E., III. Job pressures and research performance. *American Scientist,* 1971, **59**, 64–73.

Heise, D. R. Causal inference from panel data. In E. F. Borgatta (Ed.), *Sociological methodology (1970)*. San Francisco: Jossey–Bass, 1970.

Kahn, R. L., Wolfe, D. M., Quinn, R. P., Snoek, J. D., & Rosenthal, R. A. *Organizational stress: Studies in role conflict and ambiguity.* New York: Wiley, 1964.

Katz, D., & Kahn, R. L. *The social psychology of organizations.* New York: Wiley, 1966.

Kenny, D. A. Testing a model of dynamic causation. Paper presented to the Conference on Structural Equations, University of Wisconsin, November, 1970.

Miller, J. G. Information input, overload, and psychopathology. *American Journal of Psychiatry,* 1960, **116**, 513–531.

PELZ, D. C., & ANDREWS, F. M. Detecting causal priorities in panel study data. *American Sociological Review,* 1964, **29,** 836–848.

PELZ, D. C., & ANDREWS, F. M. *Scientists in organizations: Productive climates for research and development.* New York: Wiley, 1966.

PETERS, C. C., & VAN VOORHIS, W. R. *Statistical procedures and their mathematical bases.* New York: McGraw–Hill, 1940.

REES, M. B. A comparison of cross-lagged, path, and multivariate causal inference techniques applied to interest, information, and aspiration among high school students. Unpublished doctoral dissertation, Northwestern University, 1971.

ROZELLE, R. M., & CAMPBELL, D. T. More plausible rival hypotheses in the cross-lagged panel correlation technique. *Psychological Bulletin,* 1969, **71,** 74–80.

SANDELL, R. G. Note on choosing between competing interpretations of cross-lagged panel correlations. *Psychological Bulletin,* 1971, **75,** 367–368.

SLOATE, A. *Termination: The closing at Baker Plant.* Indianapolis, Ind.: Bobbs–Merrill, 1969.

SWAIN, R. L. Catalytic colleagues in a government R & D organization. Unpublished M. S. thesis, Massachusetts Institute of Technology, 1971.

YEE, A. H., & GAGE, N. L. Techniques for estimating the source and direction of causal influence in panel data. *Psychological Bulletin,* 1968, **70,** 115–126.

SUPERVISORY PRACTICES

Frank M. Andrews and George F. Farris

In these days of large scale research and development, most investigations are conducted by teams of scientists or engineers. It seems a reasonable assumption that the supervisors of these groups might affect their subordinates' performance.

For example, a supervisor might make a technical contribution through skillful selection of important but solvable problems, through his own ability to solve a problem, or through guiding subordinates toward a solution. In addition, a supervisor might affect performance by altering the climate within his group. He might inspire subordinates to high achievement, protect them from debilitating outside pressures, or structure the group so that subordinates stimulate one another.

Despite the reasonableness and potential usefulness of these hypotheses, they have not been well tested with respect to scientists.

This paper reports results of two analyses exploring the relationship between supervisory practices and scientific performance. First, does the team or supervisory group a scientist is in matter at all? That is, what qualities of scientific performance, if any, vary with team membership? Second, if performance does vary from team to team, is this related to practices employed by the supervisor?

[1] These data were collected and analyzed under NASA grant NsG-489/23-005-014 as part of a long range investigation of scientists and engineers. Dr. Donald C. Pelz is the general director of this research program.

This article originally appeared as "Supervisory Practices and Innovation in Scientific Teams" in *Personnel Psychology*, Volume 20, Number 4, Winter 1967 and is reprinted here with the permission of the publisher.

Source of the Data

The study was conducted in a NASA research center and focused on 94 non-supervisory scientists who comprised 21 small teams. These men were engaged in exploring the effects of extreme physical conditions on various materials. The number of scientists in each team, not counting the supervisor, ranged from 1 to 11 (median = 5). Each team had its own supervisor and was a reasonably stable entity (two-thirds of the non-supervisors had worked under their present chief for at least two years).

Performance and the Supervisory Group

Measures of Performance

Data were obtained about each non-supervisor's performance. The non-supervisors themselves provided information about their output of a) *technical reports* (over the past five years). Also, four qualities of each man's performance were judged by other professionals within his lab. The qualities were:

b) *Innovation*—the extent the man's work had "increased knowledge in his field through lines of research or development which were useful and new."

c) *Productiveness*—the extent the man's work had "increased knowledge in his field along established lines of research or development or as extensions or refinements of previous lines."

d) *Contribution*—the extent the man's work had "contributed to general technical or scientific knowledge in his field."

e) *Usefulness*—the extent the man's work had been "useful or valuable in helping his R & D organization carry out its responsibilities."

These qualities were independently assessed by an average of 4.4 judges, each of whom claimed to be familiar with the man's work. Two-thirds of the judges were supervisors (the man's own chief might be among them), one-third were senior-level non-supervisors. Since the judges showed reasonably good agreement (median gamma for 21 pairs of judges = .8 on the quality of innovation[2]), their evaluations were combined into a single percentile score (on each quality) for each respondent.

As is usually found for scientists and engineers, these performance measures varied according to the respondent's length of experience, seniority, and formal training (Pelz & Andrews, 1967). Since these effects could mask relationships between supervisory practices and performance, all performance measures were adjusted by adding or subtracting appropriate constants to remove such background effects.[3] Thus the final performance measures for individuals expressed how well or poorly each person performed *relative* to others with similar experience, seniority, and training.

The interrelationships among these various criteria of performance were about as expected. All evaluations were substantially related to one another (correlations ranged .7 to .8), with innovation being least similar to the others. The objective measure, output of reports, was positively related to each of the evaluations (correlations were all about .4), with the relationship to evaluated productiveness being slightly stronger than to the other criteria.

Was Performance Related to Team Membership?

To find out whether performance was related to team membership, the scientists were classified according to supervisory groups, and a one-way analysis of variance was carried out for each performance measure.

There was clear evidence that there *were* differences in subordinates' innovation that were related to team membership. Differences in innovation between scientists in different super-

[2] Since these data did not represent a probability sample from some defined population, and since the purposes of the study were descriptive rather than inferential, it would have been inappropriate to compute tests of "statistical significance." The criteria for reaching conclusions throughout this article were that a trend be clear and, where appropriate, reasonably consistent. Readers accustomed to looking for tests of statistical significance, however, can be assured that many of the trends would appear as "significant" if tested in conventional ways and that the general conclusions were not altered by the decision not to test "significance."

[3] The procedures for collecting, combining, and adjusting the performance measures used in this study were highly similar to those more fully described in Pelz and Andrews (1967).

visory groups were markedly greater than differences between scientists within the same group ($F = 1.89$). Surprisingly, there was *no* evidence that differences in the other performance measures were related to team membership! (The differences between groups were no greater than the differences within them—all F's < 1.00.)

The first finding, if replicated in subsequent studies, may be of considerable importance. It says that a scientist's innovation varied according to the particular supervisory group of which he was a member.

The second finding—that several measures of performance seemed not to be affected by team-related phenomena—is not subject to clear interpretation. Possibly there simply were no effects. However, a more likely alternative is that complex interactions cancelled out any general effects. For example, a supervisor who tended to act in a certain way may have enhanced the performance of some subordinates, but lowered the performance of others. More data than were available in the present study would be needed to adequately explore these possibilities.

Innovation and Supervisory Practices

The finding that innovation varied systematically between supervisory groups called for additional analysis. Could the supervisors have accounted, at least in part, for differences in subordinates' innovation? If so, what distinguished the supervisors of more innovative groups from supervisors of less innovative groups?

To answer these questions ten different measures of supervisory behavior were related to group-wide innovation. As will be evident, various combinations of supervisory behavior were examined in addition to the ten simple "zero order" relationships.

Measurement of Group Innovation

An innovation score for each group was computed by averaging the (adjusted) innovation scores of its members. (Recall that each member's score was itself based on several independent assessments of his innovation.)

Measures of Supervisory Behavior

Each non-supervisor had answered a lengthy questionnaire which included 36 items inquiring about the respondent's immediate chief. These items asked about a wide variety of supervisory practices suggested by previous research or theory. Most items consisted of a simple descriptive statement (e.g., my supervisor "tends to leave me pretty much on my own"), and the respondent indicated how closely it described his supervisor.

An examination of the interrelationships among these 36 items showed they could be reduced to a smaller number of measures. The ten measures shown in Exhibit 1 seemed the most efficient way of tapping the various aspects of the original items.

The ten measures were derived with the help of a Guttman-Lingoes Smallest Space Analysis (Guttman, 1967; Lingoes, 1965). For readers not familiar with this technique, it can be considered as a means of obtaining oblique factors from a non-metric factor analysis. The resulting measures, of course, were themselves somewhat interrelated (as discussed below), but they were thought to be superior to what would have resulted from a conventional orthogonal factor analysis since they more closely mirrored the actual nature of the data.

Measures A, B, and C all concerned task functions of the supervisor—his technical competence, effectiveness at critical evaluation, and influence in choosing goals and objectives.

Measures D, E, and F concerned relations between the supervisor and his subordinates, including several practices emphasized by writers such as Likert (1961), McGregor (1960) and Bennis (1964). Measure D was constructed from five highly related items, all of which dealt with the effectiveness of the supervisor in motivating others and getting them to work well together. Measures E and F concerned the supervisor's effectiveness in communicating with people, and his sensitivity to differences between them, respectively.

Measures G and H were both concerned with administrative functions of the supervisor—his effectiveness at planning and scheduling, and at handling inter-group relations. Administra-

EXHIBIT 1

*Ten Measures of Supervisory Behavior and Items from which They were Derived**

Task functions

A. Technical skills (an index based on the sum of:)
  "He knows a great deal about doing the jobs in my special area"
  "He has a good understanding of the body of knowledge that is relevant to my work"
  "He has a good understanding of the techniques and methods I use in my work"

B. Effectiveness at "providing critical evaluation"

C. Influence in choosing goals and objectives for subordinate's work
  ("Consider the choice of goals or objectives of the various technical activities for which you are responsible . . . Estimate the relative percent of weight exerted by each of the following." Items included "my immediate supervisor.")

Human relations functions

D. Effectiveness at motivating others (an index based on the sum of:)
  "He is effective for providing enthusiasm for the work"
  "He is effective for providing appreciation and encouragement"
  "He is effective at getting people to work well together"
  "He is effective for giving recognition for a job well done"
  "He is very concerned that I grow and get ahead professionally"

E. Effectiveness at "letting people know just where they stand"

F. Sensitivity to "differences between people"

Administrative functions

G. Effectiveness at "carrying out needed planning and scheduling"

H. Effectiveness at "handling relations between his group and other groups"

Leadership styles

I. Use of consultation in decision-making (an index based on the sum of:)
  "He makes most important decisions affecting group activity himself, after consulting others"
  "He makes most important decisions affecting group activity himself, without consulting others" (SCALE REVERSED)

J. Provision of "freedom for people under him to explore, discuss, and challenge ideas on their own"

* For all items except that in measure C, the respondent checked a seven-point scale to indicate how closely the statement described his supervisor. For Measure C, the respondent showed the percent of weight exerted by his supervisor.

tive skill was one of the key factors in Mann's conceptualization of supervisory "skill mix" (Mann, 1965). (Other factors stressed by Mann were technical skills and human relations skills. These are separately included in the data as indicated above.)

Finally, two measures of supervisory "style" emerged in the

data. Measure I concerned the extent to which the supervisor consulted others before making an important decision. (Subordinate participation in decision-making is an important feature of Likert's (1961) theorizing.[4])

A second "style" measure, J, considered the extent of freedom allowed subordinates by their supervisor. The topic of freedom in science has received considerable debate and some empirical investigation (see Pelz, 1964; Pelz & Andrews, 1966). Conceptually freedom is related to the often-discussed dimension of close-vs-general supervision.

As has been pointed out, these ten measures of supervisory behavior were similar to dimensions discussed by various previous investigators. Their work suggested many of the 36 items included in the present study, but there was no requirement that the items would cluster together in the meaningful categories which were actually obtained. The fact that they did so is itself an interesting finding and suggests that a replicable set of concepts for describing supervisors is becoming available.

The ten measures of supervisory behavior indicated how each non-supervisor perceived his particular chief. For these scores to be useful, they should meet two conditions. First, there should be some evidence of reliability—i.e., perceptions of non-supervisors who described the same chief should be more similar than those describing different chiefs. Second, there should be evidence that the chiefs themselves behaved differently. A one-way analysis of variance performed on each of the ten measures showed that these conditions were met. On each measure, descriptions of chiefs differed from team to team, and the differences within teams were less than the differences between them. (All F's > 1.00, median $F = 2.02$. Of necessity, three teams containing just one non-supervisor were omitted from this particular analysis.)

Group scores. Answers from all non-supervisors under a particular chief were averaged into a single score (on each of the

[4] Likert also considers *joint* decision-making. In addition to the two items composing Measure I, the questionnaire contained an item asking specifically about the extent to which the supervisor encouraged his group to make decisions jointly. This item proved to be highly related to both Measures D and I, and therefore has not been scored separately.

EXHIBIT 2

Correlations among Average Perceptions of Supervisor's Behavior
(N = 21 teams)

Measure	A	B	C	D	E	F	G	H	I
A. Technical skills									
B. Critical evaluation	.3								
C. Influence on goals	.6	.0							
D. Motivating others	.0	.6	.0						
E. Let know where stand	−.1	.4	−.1	.5					
F. Sensitivity	.1	.4	−.3	.6	.5				
G. Plan and schedule	−.1	.4	.1	.3	.2	.3			
H. Inter-group relations	.1	.5	−.2	.6	.4	.8	.5		
I. Use of consultation	−.3	.2	−.3	.4	.3	.3	−.1	.1	
J. Freedom	−.1	−.1	−.3	.0	−.1	.1	.1	.2	.2

ten measures).[5] Thus each supervisor was described in terms of what was common among the perceptions of his subordinates, thereby reducing the effects of idiosyncracies of a particular individual's perception, or special features of the relationship between a particular subordinate and his chief.

The interrelationships among these group scores are shown in Exhibit 2. One can see that most of the measures which seemed conceptually related—task functions, human relations functions, etc.—did tend to relate positively to each other.

Exhibit 2 contains two other interesting findings. Although Measure B, effectiveness at critical evaluation, seemed to have a task content and did relate to one of the other task items, Measure B related even more strongly to several measures in the human relations and administrative areas. Apparently among these supervisors, the successful exercise of critical evaluation was not solely a task function. Similarly, the supervisor's administrative skill at handling relations between his group and other groups was substantially related to his skills in the human relations area.

While interrelationships among skills in various areas might suggest that classification into three functional areas was not

[5] These mean scores were transformed to rectangular distributions. A seven-point scoring was used whenever correlations were computed. Two- and three-point scorings were also used, as will be evident.

useful, one should reserve judgment on this matter. Data presented in Exhibit 3 will show that skills in the same area related to performance in similar ways, but that skills in different areas related differently. On this criterion, the present classification was indeed useful. Also, it should be noted that Mann (1965) found that relationships between skills in the various areas varied markedly in different kinds of organizations and for workers at different levels in the hierarchy. This suggests that relationships shown in Exhibit 2 should be viewed with caution until further studies can assess their representativeness.

Relationships of Supervisory Behavior to Innovation

Exhibit 3 shows how each of the ten aspects of supervisory behavior, as perceived by the supervisor's own group of subordinates, related to average innovation in his team. In the light of previous research and theory, some of these relationships were surprising, but discussion is withheld for a later section.

With respect to task functions, a reasonably consistent positive trend appeared. Supervisors who were perceived as being skilled or influential in the task area tended to have subordinates whose work was judged as more innovative than supervisors who were less effective in this area. The trend for Measure A, technical skills, was sharpest. Those for critical evaluation and influence on goals were in the same direction but weaker.

In the human relations area, only one of the three measures showed any substantial relationship to innovation. There was, however, a consistent, and unexpected, tendency for supervisors scoring highest on human relations to have the least innovative subordinates! Highest innovation tended to occur under supervisors who scored moderate on human relations. This finding is different from results reported for non-professional workers (Likert, 1961, provides a summary).

The two measures concerned with the supervisor's administrative functions showed consistent, and again surprising, relationships with subordinates' innovation. The more effective the supervisor was at administration, the *lower* the judged in-

EXHIBIT 3

Relationships between Subordinates' Innovation and Various Practices of Their Immediate Supervisors

Measure	Trend	Eta	F
Task functions			
A. Technical skills	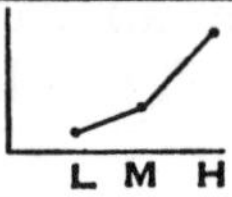 L M H	.52	3.36
B. Critical evaluation	 L M H	.33	1.13
C. Influence on goals	 L M H	.28	0.78
Human relations functions			
D. Motivating others	 L M H	.47	2.48
E. Let know where stand	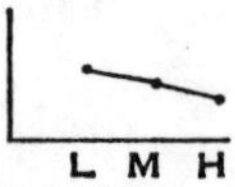 L M H	.04	0.01
F. Sensitivity	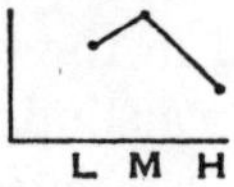 L M H	.22	0.44
Administrative functions			
G. Plan and schedule	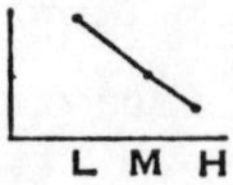 L M H	.30	0.89
H. Inter-group relations	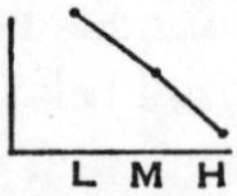 L M H	.53	3.57
Leadership styles			
I. Use of consultation	 L M H	.43	2.00
J. Freedom	 L M H	.33	1.12

Note: Vertical scales show mean innovation. Horizontal scales show three sets of supervisors: those scoring low, medium, and high on the designated measure. Each point is based on data from approximately 7 supervisory groups.

novation of his group! For Measure H, skill at handling intergroup relationships, this trend was substantial. Measure G, effectiveness at planning and scheduling, showed the same trend in weaker form. Although organizational scholars have generally assumed that the exercise of administrative functions would enhance a group's performance, this was clearly not the case for innovation by these scientists.

Finally, the two measures of leadership style, use of consultation and provision of freedom, showed moderate relationships to innovation. For both measures, innovation was higher when supervisors scored either high or low than if they scored in the middle. Further analysis, discussed in the next section, clarified the meaning of these trends.

Thus Exhibit 3 suggests that innovation flourished under supervisors who were effective at task functions. But human relations and administrative functions were not positively related to innovation—in fact, relationships tended to be curvilinear and/or negative.

Combinations of Supervisory Practices

In addition to examining the simple relationships between each supervisory function and subordinates' innovation (shown in Exhibit 3), relationships involving all possible *pairs* of supervisory practices were also examined.

Freedom. Exhibit 3 showed a curvilinear, though generally positive relationship between a supervisor's provision of freedom and subordinates' innovation. The meaning of this relationship became clearer when several other supervisory practices were considered in combination with freedom. Results appear in Exhibit 4.

Note that provision of freedom showed substantial positive relationships with innovation in teams headed by supervisors who scored *low* on task functions (Measures A and B), low on human relations functions (Measures D, E, and F), or low on administrative functions (Measure H). But in teams headed by supervisors who were effective in these areas, provision of freedom mattered less, and sometimes even related negatively.

This finding made good sense and suggests that provision of freedom was a substitute for skillful leadership! In teams

EXHIBIT 4

Correlations between Subordinates' Innovation and their Supervisor's Provision of Freedom, Separately for Designated Groups of Supervisors

Characteristics of Supervisor	Correlation
A. Technical skill	
High (N = 10 teams)	.0
Low (N = 11 teams)	.6
B. Effectiveness at critical evaluation	
High (N = 10 teams)	.0
Low (N = 11 teams)	.4
D. Skill at motivating others	
High (N = 10 teams)	−.3
Low (N = 11 teams)	.5
E. Effectiveness at letting others know where they stand	
High (N = 11 teams)	−.4
Low (N = 10 teams)	.6
F. Sensitivity to differences between people	
High (N = 11 teams)	.1
Low (N = 10 teams)	.4
H. Skill at handling inter-group relations	
High (N = 10 teams)	−.2
Low (N = 11 teams)	.6

Note: Measures C and G do not appear in this exhibit since the relationship between innovation and freedom was not markedly affected by them.

headed by less skillful leaders, innovation was high if subordinates were given freedom to explore on their own. But less skillful leadership combined with lack of freedom was associated with less innovative work.[6]

Consistency in leadership practices. Other particularly interesting sets of relationships suggested that there were certain combinations of leadership practices which should occur together.

For example, critical evaluation went with innovation *if* the supervisor was technically skilled ($r = +.5$ for supervisors

[6] These findings suggest that provision of freedom was affecting innovation rather than the reverse. If high innovation were the cause of a group of subordinates' being awarded freedom, one would expect to see positive correlations between freedom and innovation in teams headed by the more skillful supervisors, i.e., the more skillful supervisors would be more likely to match the reward of freedom with innovation. But this was not the case. Rather, Exhibit 4 shows that in groups headed by more skillful supervisors freedom and innovation were only weakly related.

with high technical skills—no table shown). But innovation was low when supervisors low in technical skills attempted to evaluate subordinates' work ($r = -.5$). Thus the relationship between critical evaluation and innovation depended on the supervisor's technical skill. Exercise of critical evaluation needed to be consistent with possession of adequate technical skill.

A second example of the need for consistency occurred when the practices of providing freedom and of consulting others were examined. Freedom was unrelated to innovation if the supervisor failed to precede his own decision-making by some consultation with others. But the relationship was substantial if freedom for subordinates was combined with a chance to influence decisions being made by the supervisor. (Among supervisors making use of consultation, $r = +.7$; for supervisors making little use of consultation, $r = -.1$—no table shown.) This also seemed intuitively reasonable.

Administrative functions. It has generally been assumed that the skillful exercise of administrative functions would result in high performance. Yet Exhibit 3 provided surprising negative relationships—exactly opposite to the usual assumption.

Exhibit 5 provides further information and shows that the negative relationships occurred mainly for supervisors who scored high in the human relations area. Among supervisors who were skilled in motivating others, effective in letting others know where they stood, and sensitive to differences between people, administrative functioning seemed incompatible with innovation. The highest innovation occurred under supervisors who were seen as relatively *poor* administrators; low innovation occurred under good administrators! (Note that five out of six correlations were strongly negative.) But among supervisors who scored low on human relations, administration was only weakly related to innovation.

Task and human relations skills. In addition to the combinations of practices already described, a careful examination was made of various skills in the task and human relations areas. Blake and Mouton (1964) have suggested that relationships between skills in one area and productivity should be espe-

EXHIBIT 5

Correlations between Subordinates' Innovation and Two Measures of their Supervisor's Administrative Skills, Separately for Supervisors High and Low in Human Relations Skills

Characteristics of Supervisor	Correlations between Innovation and:	
	Effectiveness at planning and scheduling	Effectiveness at handling inter-group relations
D. Skill at motivating others		
High (*N* = 10 teams)	−.7	−.5
Low (*N* = 11 teams)	.2	−.3
E. Skill at letting others know where they stand		
High (*N* = 11 teams)	.1	−.6
Low (*N* = 10 teams)	−.2	−.1
F. Sensitivity to differences between people		
High (*N* = 11 teams)	−.5	−.5
Low (*N* = 10 teams)	.5	−.1

cially strong when skills in the other area are also present. Oaklander and Fleishman (1964), Kahn (1956), and many others have also suggested the same idea. There was no evidence, however, that such a phenomenon occurred for these scientists. Whether the supervisor was skilled in the human relations area had little effect on the generally positive relationships between task functioning and innovation. Similarly, skills in the task area did not affect relationships between human relations functions and innovation.

Discussion

This paper set out to explore relationships between supervisory practices and scientific performance. A key preliminary question was whether the supervisor mattered at all. When performance was measured in terms of innovation, the answer was "yes": systematic differences between supervisory groups were clearly evident. Furthermore, these differences were related to supervisory practices.

Thus, while firm statements of cause and effect are not appropriate with these data, the findings do suggest that the supervisor may play an important role in enhancing or depressing innovation.[7] For several other aspects of performance, however, there was no evidence that the supervisor had a group-wide effect. Possible reasons for this have already been presented.

When specific supervisory practices were examined, the data presented two surprises and also some potentially useful findings. The surprises are discussed here; the implications in the next section.

Surprise #1. One surprise was that none of the several measures of supervisory skill in the human relations area related to innovation. Organizational scholars have suggested that human relations functions serve to enhance employee motivation and to facilitate the flow of information. Why did the present results differ?

One reason may be that scientists are different from the non-professional "rank and file workers" who were subjects in most previous studies. For example, it may be that scientists, compared to non-professionals, are stimulated more by the work itself and less by the social conditions which accompany it. Our findings on the importance of the supervisor's task functions support this speculation.

Alternatively, the differences between our results and those of many previous studies might be attributed to the nature of the task rather than to the nature of the people performing it. Fiedler (1965) found that directive, task-oriented leaders were more effective either in situations which were very favorable for the leader (where the leader had strong power, good relations with his subordinates, and a highly structured task) or in situations which were very unfavorable for the leader. In situations falling midway between these extremes, Fiedler found human-relations-oriented leaders more effective.[8] Future re-

[7] Although causal directions cannot be firmly identified, it should be noted that the measures of supervisory practices and performance derive from completely different sources. Thus the findings cannot be attributed to a simple "halo effect." (See also footnote 6.)

[8] Although Fiedler's idea seems useful, his method of typing leaders forced them into either the task or human relations orientations. It did not permit separate examination of leaders high (or low) in both orientations. Furthermore, he hypothesizes that creativity will be "forbidden" under task-oriented leaders, a result clearly not supported by the present data.

search on supervision probably should consider the nature of the task more carefully than has been done here and in most previous studies.

And finally, the difference between our findings and those of previous studies may be attributable to examination of different criteria. The present study used performance—particularly, innovation—as the criterion of good supervision. In contrast, the criteria used by previous studies have included satisfaction, grievances, turnover, absenteeism, scrap loss, morale, and stress. Human relations may be more important with respect to these criteria than for innovation.

Surprise #2. A second surprise was the markedly negative relationships between the supervisor's performance of administrative functions and his subordinates' innovation. Once again, the findings were opposite to assumptions usually made (but rarely tested) by organizational scholars. Of particular importance was the discovery that these negative relationships were most likely to occur when the supervisor scored high on human relations.

A major issue here would seem to be the direction of causality. On the one hand, the effective exercise of administrative functions, such as planning and scheduling, may have produced a rather rigid setting which left subordinates little room for innovation. This interpretation would be in close accord with a previous study which showed that creative ability paid off only in flexible situations (Andrews, 1967).

On the other hand, it is possible that causality operated in the other direction. Perhaps supervisors who headed innovative groups were less effective in the administrative area precisely because innovation made administration more difficult.

Both interpretations seem plausible, and, in fact, both may have been operating. There seems to be a fundamental dilemma here for organizations seeking innovation: How can organizations, which always require a certain degree of coordination and interdependence among people and work groups, remain sufficiently flexible to encourage innovation from within? Unfortunately, few studies have been explicitly concerned with organizational conditions enhancing innovation.

Summary and Implications

What does this study imply for the director of a research laboratory, the supervisor of an R & D team, or others attempting to encourage innovation within an organization? Let us speculate, recognizing that further research will be needed to assess the generality of our results and their causal dynamics.

Greatest innovation occurred under supervisors who knew the technical details of their subordinates' work, who could critically evaluate that work, and who could influence work goals. Thus the widespread practice of including technical competence among the criteria for choosing supervisors seems to be sound. This does not mean that a supervisor should constantly "meddle" in his subordinates' activities. But he should be available, competent in the current "state of the art," actively interested in the project, and informed about it. These, in turn, imply that the supervisor should "keep his hand in," perhaps by actually conducting research himself. He probably should not spend all his time monitoring the work of others. Furthermore, for a supervisor to be technically "close" to the activities of his subordinates, his supervisory responsibilities probably should be limited to only a few projects.

What if this kind of structure is not possible, or if a supervisor's technical competence has become obsolete? Again, the data were clear: provide substantial freedom for subordinates. Freedom acted as a partial substitute for skilled supervision. But even when subordinates have freedom, the supervisor still makes some kinds of decisions. For freedom to be effective, the data showed that the supervisor must consult with his subordinates before making these decisions.

Previous research (Pelz & Andrews, 1966) suggests that under complete freedom subordinates may engage in trivial problems, become lazy, and stagnate. To avoid this, the wise supervisor who cannot exercise task functions should attempt to combine freedom with stimulation from sources other than himself. For example, he might arrange meetings where subordinates could present, discuss, and critically evaluate one another's work. The present study showed that a technically weak supervisor should not undertake critical evaluation himself.

What about the human relations and administrative functions? Here our recommendations must differ from those often made to supervisors of non-professional workers. We found that innovation tended to be *low* when supervisors were thought to be effective at human relations or administration, and especially low when supervisors were effective at both.

If the supervisor is not particularly skilled as an administrator and is somewhat cool toward his men, this need not cause great concern. In fact, freeing supervisors from responsibilities in the human relations and administrative areas may enhance innovation. But since these areas do need attention, there may need to be some other person with responsibility for them. Some organizations assign this responsibility to an "assistant director" or "executive head."

On the other hand, what should a supervisor do if he happens to be skilled at administration and/or human relations—consciously act ineffectively? Probably not. Should he simply turn subordinates loose? The data say this would not help. One suggestion is that he might attempt to increase the size of each subordinate's own "professional arena." Each subordinate might be encouraged to take on additional tasks, evolve contacts with outside groups, or try new methodologies. These should result in an increased self-reliance and an increased capacity to move in innovative directions.

References

ANDREWS F. M. "Creative Ability, the Laboratory Environment, and Scientific Performance." *IEEE Transactions in Engineering Management,* 1967, in press.

BENNIS, W. G. "Goals and Meta-Goals of Laboratory Training," In Bennis, W. G., Schein, E. H. Berlew, D. E., and Steele, F. I. (Editors), *Interpersonal Dynamics: Essays and Readings on Human Interaction.* Homewood, Ill.: Dorsey Press, 1964.

BLAKE, R. R. AND MOUTON, JANE S. *The Managerial Grid.* Houston, Texas: Gulf Publishing Company, 1964.

FIEDLER, F. E. "Leadership: A New Model." *Discovery,* April 1965, 12–17.

GUTTMAN, L. "A General Non-Metric Technique for Finding the Smallest Euclidean Space for a Configuration of Points." *Psychometrika,* 1967, in press.

KAHN, R. L. "The Prediction of Productivity." *Journal of Social Issues,* XII (1956), 41–49.

LIKERT, R. *New Patterns of Management*. New York: McGraw-Hill, 1961.

LINGOES, J. C. "An IBM-7090 Program for Guttman-Lingoes Smallest Space Analysis: I." *Behavioral Science,* X (1965), 183–184.

MANN, F. C. "Toward an Understanding of the Leadership Role in Formal Organization," In Dubin, R., Homans, G. C., Mann, F. C. and Miller, D. C., *Leadership and Productivity*. San Francisco: Chandler, 1965.

MCGREGOR, D. *The Human Side of Enterprise*. New York: McGraw-Hill, 1960.

OAKLANDER, H. AND FLEISHMAN, E. A. "Patterns of Leadership Related to Organizational Stress in Hospital Settings." *Administrative Science Quarterly,* VIII (1964), 520–532.

PELZ, D. C. "Freedom in Research." *International Science and Technology,* February 1964, 54–66.

PELZ, D. C. AND ANDREWS, F. M. "Autonomy, Coordination, and Stimulation in Relation to Scientific Achievement." *Behavioral Science,* XI (1966), 89–97.

PELZ, D. C. AND ANDREWS, F. M. *Scientists in Organizations: Productive Climates for Research and Development*. New York: John Wiley and Sons, 1967.